Engineering Thermodynamics:
Simulation with Entropy

Engineering Thermodynamics: Simulation with Entropy

Edited by Keenan Murphy

CLANRYE
INTERNATIONAL
www.clanryeinternational.com

Clanrye International,
750 Third Avenue, 9th Floor,
New York, NY 10017, USA

ISBN: 978-1-64726-679-0

Cataloging-in-Publication Data

Engineering thermodynamics : simulation with entropy / edited by Keenan Murphy.
    p. cm.
Includes bibliographical references and index.
ISBN 978-1-64726-679-0
1. Thermodynamics. 2. Heat. 3. Heat-engines. 4. Quantum theory. I. Murphy, Keenan.
TJ265 .E54 2023
621.402 1--dc23

For information on all Clanrye International publications
visit our website at www.clanryeinternational.com

CLANRYE
INTERNATIONAL

Contents

Preface

The main aim of this book is to educate learners and enhance their research focus by presenting diverse topics covering this vast field. This is an advanced book which compiles significant studies by distinguished experts in the area of analysis. This book addresses successive solutions to the challenges arising in the area of application, along with it; the book provides scope for future developments.

Thermodynamics is a branch of physics that deals with matter and conversion of energy. Engineering thermodynamics is a subject of mechanical engineering that involves applying the principles of thermodynamics to engineering design of processes, devices, and systems, which involve effective utilization of energy and matter. Conversion between heat and work is fundamental in engineering thermodynamics. Entropy refers to a measurable physical characteristic and a scientific concept that is connected to a state of uncertainty, unpredictability or disorder. It is utilized in a variety of fields, including the principles of information theory, classical thermodynamics, and the microscopic description of nature in statistical physics. Entropy is central to the second law of thermodynamics, which asserts that the entropy of isolated systems left to unstructured expansion cannot decrease over time. This is due to the isolated systems always reaching thermodynamic equilibrium with the highest entropy. Different scenarios involving entropy production such as optimizing power and waste heat reduction can be studied by taking the help of simulation tools. This book aims to shed light on some of the unexplored aspects of engineering thermodynamics and the latest researches on simulation with entropy. It will serve as a valuable source of reference for graduate and postgraduate students.

It was a great honour to edit this book, though there were challenges, as it involved a lot of communication and networking between me and the editorial team. However, the end result was this all-inclusive book covering diverse themes in the field.

Finally, it is important to acknowledge the efforts of the contributors for their excellent chapters, through which a wide variety of issues have been addressed. I would also like to thank my colleagues for their valuable feedback during the making of this book.

Editor

Toward Improved Understanding of the Physical Meaning of Entropy in Classical Thermodynamics

Ben Akih-Kumgeh

Department of Mechanical and Aerospace Engineering, Syracuse University, 263 Link Hall, Syracuse, NY 13244, USA; bakihkum@syr.edu

Abstract: The year 2015 marked the 150th anniversary of "entropy" as a concept in classical thermodynamics. Despite its central role in the mathematical formulation of the Second Law and most of classical thermodynamics, its physical meaning continues to be elusive and confusing. This is especially true when we seek a reconstruction of the classical thermodynamics of a system from the statistical behavior of its constituent microscopic particles or vice versa. This paper sketches the classical definition by Clausius and offers a modified mathematical definition that is intended to improve its conceptual meaning. In the modified version, the differential of specific entropy appears as a non-dimensional energy term that captures the invigoration or reduction of microscopic motion upon addition or withdrawal of heat from the system. It is also argued that heat transfer is a better model process to illustrate entropy; the canonical heat engines and refrigerators often used to illustrate this concept are not very relevant to new areas of thermodynamics (e.g., thermodynamics of biological systems). It is emphasized that entropy changes, as invoked in the Second Law, are necessarily related to the non-equilibrium interactions of two or more systems that might have initially been in thermal equilibrium but at different temperatures. The overall direction of entropy increase indicates the direction of naturally occurring heat transfer processes in an isolated system that consists of internally interacting (non-isolated) sub systems. We discuss the implication of the proposed modification on statements of the Second Law, interpretation of entropy in statistical thermodynamics, and the Third Law.

Keywords: entropy; heat transfer; heat engines; thermal non-equilibrium; Second Law

---

1. Introduction

In the first half of the 19th Century, von Mayer and Joule discovered the energy conservation principle, demonstrating that heat (internal energy) and work are inter-convertible, and that the total energy of the universe is constant. The Second Law forbids the realization of perpetual motion machines and limits the effectiveness of heat engines and cold machines (heat pumps and refrigerators). Based on Carnot's theorems (in the more analytical form developed by Clapeyron), Thomson and Clausius greatly improved the theory of heat engines. Clausius eventually introduced the concept of entropy to represent the transformation or *Verwandlung* of heat into work and vice versa [1,2]. The year 2015 marked the 150th anniversary of this definition, but it continues to be a confusing concept in thermodynamics. It is one of the challenges encountered in the attempt to reconcile the microscopic view of matter offered by statistical thermodynamics with the macroscopic view of classical thermodynamics. In Clausius' work [2,3], the term emerges from his attempt to quantify the transformation of heat into mechanical work and mechanical work into heat. It should be noted that, although the equality of work and heat was established in the First Law, the physical units were often not the same, so that a conversion factor was often needed in quantifying the mechanical equivalence of heat. Work was generally given in units of kilogram-meters (kgm), from which in modern terms, the corresponding potential energy in newton-meter or joule would be obtained as the product of that quantity in kilogram-meter and the acceleration due to gravity. On the other hand, heat was often quantified in terms of the work-equivalence of heat, whereby 423.55 kgm (or about 4155 J) of work are needed to raise the temperature of 1 kg of water by 1 °C [2].

Confusion about the meaning of entropy is not a new problem [4–6]. A few years after its definition, Tait proposed a redefinition of entropy as the useful part of energy [7]. This idea was initially taken up by Maxwell [8]. Tait seemed to have pursued this line of thought in order to demonstrate that the entropy definition proposed by Clausius had initially been obtained by Thomson in the latter's earlier publications [7,9,10]. The widespread misconception about entropy continued into the 20th century with von Neumann allegedly advising Shannon to adopt the name entropy in his information theory, with the justification that "... nobody really knows what entropy is, so in a debate, you will always have the advantage" (adapted from [6]).

Clausius' analysis focused on the interaction of a heat engine cycle with heat reservoirs and he arrived at the cyclic integral, $\oint \frac{Q}{\tau}$, that equals zero for a reversible cycle. One of the critics of the foundations of Thermodynamics, Truesdell, attacks the rather circular definition of irreversible processes offered by Clausius: irreversible processes are those processes that are not reversible [11]. One needs to understand irreversible processes in order to understand reversible ones. Truesdell is looking for a mathematical expression for irreversible processes since he criticizes Thermodynamics as a subject with an unusually high ratio of words to equations. In modern thermodynamics textbooks, this difficulty is circumvented by defining reversible, internally reversible, and externally or fully reversible processes [12,13]. Uffink, in his criticism of the liberal extrapolation of the Second Law to the concept of the Arrow of Time, points out the problem with the opaque distinction between reversible and irreversible processes [6]. It should also be noted that little attention is paid to the heat reservoirs in Clausius' derivation in modern discussions of entropy. Difficulties in reconciling classical thermodynamics with statistical thermodynamics [14–16] also seem to be related to differences in the conceptual understanding of entropy in these two fields. The need to clarify classical entropy continues to attract attention with a number of new articles devoted to this topic.

Another area where problems arise owing to the current view of entropy is in metrology, where it is used in defining the thermodynamic temperature scale [17]. Recommendations have been made to the international system of units to consider changing the temperature unit from kelvin to a unit based on the Boltzmann constant, an energy parameter. The argument for this change is that precise temperature measurement based on the kelvin is still dependent on the equilibrium properties of a given material, whereas relating temperature to the more universal energy constant, k_B, will make temperature measurement independent of the material, method of realization, and temperature range. Could some of these conceptual problems be resolved by only including energy variables in entropy definition and dispensing with the central role of temperature? Furthermore, in quantum metrology, some authors argue that the conditions for thermodynamic equilibrium and thermal contact are not met; this renders energy quantification through the kelvin meaningless [18,19]. These issues are all related to entropy definition in the sense that its definition was seen as a basis for the realization of a thermodynamic temperature scale. In the process, the implicit energy unit was lost and the notion of absolute temperature was given more significance.

In its historical context, entropy is wedded to the concepts of heat engines and cold machines. It is important today to clarify the physical meaning of this important property, since thermodynamics is increasingly used in the analysis of a wide range of problems that are very remote from the concept of heat engines. Examples include biological processes and quantum mechanical systems.

In this work, we offer a modified definition of entropy aimed at clarifying its physical meaning. We suggest that heat transfer between two systems initially at different temperatures is a better physical model for entropy explanation than the usual heat engines. From this perspective, entropy is closely related to non-equilibrium thermodynamics: we start with two systems at equilibrium but such that their temperatures are different; their interaction leads to the flow of heat in the direction of the temperature gradient. This, together with the heat exchange sign convention, leads to the principle of entropy increase. Alternatively, the entropy increase principle could be associated with the evolution of an isolated system from a prepared non-equilibrium state toward a new equilibrium state. We also show that further modification of the suggested entropy definition makes it possible to arrive at the proposition of the Third Law or Nernst Theorem, without the need for a separate law. In essence, the Third Law establishes the entropy difference for systems or processes at two temperatures close to zero.

We start by restating the definition of entropy and the related heat engine analysis. We then discuss some of the conceptual difficulties arising from this definition. This is followed by a presentation of our modification. We further assess the use of this modified definition in the analysis of model processes. We then discuss how the modified version of entropy aligns with statistical thermodynamics entropy and the Third Law.

2. Clausius Approach to Entropy Definition

The kinetic theory, thermodynamic process relations, and the First Law were well established before Clausius' work on entropy.

The ideal gas provided a link between macroscopic and microscopic thermodynamics. The pressure exerted by an ideal gas is related to the average kinetic energy of a constituent particle, the system volume, and the number of gas particles:

$$p = \frac{2}{3} \frac{N}{V} \left[\frac{1}{2} m \bar{v}^2 \right]. \tag{1}$$

In terms of temperature:

$$p = \frac{NkT}{V}, \tag{2}$$

where the purely thermodynamic property, temperature, is taken to be related to the average kinetic energy of the particles as:

$$\text{with } T = \frac{2}{3} \frac{\left[\frac{1}{2} m \bar{v}^2 \right]}{k}. \tag{3}$$

One could define $\epsilon = kT = 1/3 m \bar{v}^2$ as an energy variable to obtain the gas law as $p = \frac{N}{V} \epsilon$. More will be said about $\epsilon = kT$ later, but we note here that this is equivalent to $\frac{1}{\beta}$ in statistical mechanics. For the purpose of this paper, β is not preferable; it grows unbound at low temperatures, complicating the new physical interpretation we offer. It should also be noted that, in statistical mechanics, β is motivated by a constraint on the total energy as entropy is maximized. This role could as well be played by ϵ, taken as $\frac{1}{\beta}$. In Gibbsian statistical mechanics, we shall comment that ϵ is similar to an energy parameter that Gibbs called the *modulus* θ. The equation of state above is obtained using the kinetic theory, which is part of statistical mechanics. This attests to the central role of statistical physics in the foundation of classical thermodynamics.

For an adiabatic process from state 1 to state 2, if constant specific heats and ideal gas behavior are assumed, then two of the three state variables, pressure, temperature, and specific volume may be related as:

$$pv^\gamma = const., \tag{4}$$

$$\frac{T_2}{T_1} = \left(\frac{v_1}{v_2} \right)^\gamma. \tag{5}$$

The efficiency of a heat engine is generally given as the ratio of net mechanical work to the heat supplied:

$$\eta = \frac{W_{net}}{Q_{in}} = 1 - \frac{Q_{out}}{Q_{in}}. \tag{6}$$

For a four-stroke heat engine cycle with two adiabatic processes, heat is added during one of the four processes and rejected during another. The heat exchange can be determined from the First Law for closed systems:

$$dU = \delta Q + \delta W = \delta Q - pdV, \tag{7}$$

where U is the internal energy, δQ is the heat added, δW is work added, so that the internal energy of the system can increase on account of heat and/or work addition. For an ideal gas, internal energy is a function of temperature only, and $dU = \frac{\partial U}{\partial T}dT = Nc_v(T)dT$, so that for an adiabatic process:

$$Nc_v(T)dT = -pdV \text{ or } c_v(T)dT = -pdv = -nkTdv. \tag{8}$$

Isothermal heat exchange, therefore, implies that the internal energy stays constant and

$$\int \delta Q = \int pdV. \tag{9}$$

For an isothermal process in which intermediate states can be assumed to be in quasi-equilibrium, pressure and volume are related as

$$pV = const. \tag{10}$$

One can make use of the ideal gas law to express p in terms of V and T or V and ϵ

$$pV = NkT = N\epsilon. \tag{11}$$

The efficiency of this heat engine cycle can be shown to be

$$\eta = 1 - \frac{T_L}{T_H}. \tag{12}$$

What is lost in establishing Equation (12) is that the heat added to a system translates to internal energy that is characteristic of microscopic motion and can be measured in terms of NkT. This is the motivating idea that led to the discovery of entropy, namely, the search for a property of the system that scales with the heat added to a system. If we sought to maintain the energy connection as we replace Q_{in} and Q_{out}, we should have the efficiency as

$$\eta = 1 - \frac{NkT_L}{NkT_H}. \tag{13}$$

However, with our intensive energy variable, $\epsilon = kT$, this takes the form

$$\eta = 1 - \frac{\epsilon_L}{\epsilon_H}. \tag{14}$$

Clausius' entropy definition draws from Equations (6) and (12), while ours draws from Equations (6) and (14), motivated by the emphasis on the energy variable as the best physical quantity to connect heat to a mechanical property of the system receiving or rejecting heat.

Tait criticized Clausius' entropy definition, alleging that it was a restatement of Thomson's earlier work. This view seems to stem from a 1854-paper in which Joule and Thomson sum up their quest for an absolute temperature as:

> If any substance whatever, subjected to a perfectly reversible cycle of operations, takes in heat only in a locality kept at a uniform temperature, and emits heat only in another locality kept at a uniform temperature, the temperatures of these localities are proportional to the quantities of heat taken in or emitted at them in a complete cycle of operations [20,21].

This is equivalent to saying that $\frac{Q_1}{Q_2} = \frac{T_1}{T_2}$.

As mentioned before, Clausius' path to entropy definition starts from the Carnot cycle and seems to be prompted by the fact that heat and work are interchangeable but did not have a common physical unit at the time. Transformations of work to heat and vice versa are such that the following relation holds,

$$W = Q \times f(T). \tag{15}$$

The equivalence value of the transformation of heat to work (*original German: Äquivalenzwert der Umwandlung von Wärme in die mechanische Arbeit*) can be determined as

$$N = \frac{Q}{\tau}. \tag{16}$$

For a series of reversible heat engines in contact with many reservoirs, the equivalence value of all transformations can be determined from

$$N = -\sum \frac{\delta Q}{\tau}. \tag{17}$$

For a very large number of reversible heat engines, the summation can be replaced by integration

$$N = -\int_1^n \frac{\delta Q}{\tau}. \tag{18}$$

The term *entropy* is then introduced such that its differential corresponds to the transformation

$$dS = \frac{\delta Q}{\tau}. \tag{19}$$

Clausius then seeks a suitable form of the function τ and $f(T)$, where he considers that $f(T) = 1/\tau$. Drawing from the previous derivation of the thermal efficiency of a Carnot heat engine using an ideal gas with constant specific heat, the equality $\frac{Q_1}{Q_2} = \frac{T_1}{T_2}$, is used to suggest that $\tau = T$. However, Clausius recognizes that this choice is not unique, given that the ideal gas uses many assumptions and a number of temperature functions could equally be acceptable [3].

One of the weaknesses of Clausius' approach to entropy definition is the sole focus on the reversible heat engine (Carnot) cycle, for which the net entropy in a cycle is zero. The fact that the efficiency is independent of material properties, is often presented as a special feature of the Carnot cycle. However, it is known from the analysis of four-process heat engine cycles with two adiabats and heat exchange at constant volume, temperature, or pressure, that the thermal efficiency is $\eta = 1 - \frac{T_1}{T_2}$. Here, T_1 is the temperature at which the working medium loses contact with the heat sink and T_2 is the temperature at which the working medium establishes contact with the heat source [12,13]. Such cycles include the Otto cycle with heat exchange processes at constant volume, the Brayton or Joule cycle of a gas turbine with heat exchange processes at constant pressure, and the Carnot cycle with heat exchange at constant temperature. It happens that in the case of the Carnot heat engine, T_1 and T_2 are also the minimum and maximum temperatures of the cycle, respectively, while the maximum temperatures in Otto and Brayton cycles are higher than T_2. Clearly, their efficiencies can be represented without invoking the materials of their working media. Analyzed without explicit focus on entropy changes of the heat reservoirs, all of these heat engine cycles have no net entropy generation. Therefore, focusing on the cycles, and not including the heat reservoirs or the walls enclosing the working medium, obscures the origin of entropy generation. Including the analysis of the heat source and heat sinks in entropy analysis is more complicated than would be achieved with a simple explanatory model process.

Furthermore, in theoretical analysis of heat engines, no attention is paid to the walls of the device, but they tend to be sources of entropy generation through friction, heat storage, and heat transfer.

It is therefore difficult to transfer analysis of heat engines to thermodynamic systems that are not concerned with work and heat interconversion. Furthermore, heat engines and refrigerators rely on thermal non-equilibrium and the associated tendency to effect heat transfer in quest for thermal equilibrium. Heat transfer is a more appropriate thermodynamic model process to explain entropy generation in heat engine systems as well as in processes that are not related to heat engines but potentially subjected to temperature gradients.

1   We start with a simple heat transfer problem. Two systems, 1 and 2, are separately in thermal equilibrium but at different temperatures, that is $T_1 \neq T_2$ (see Figure 1).
2   We allow the systems to interact, and we seek a parameter whose sign will unambiguously indicate the natural tendency of heat to flow from hot to cold.

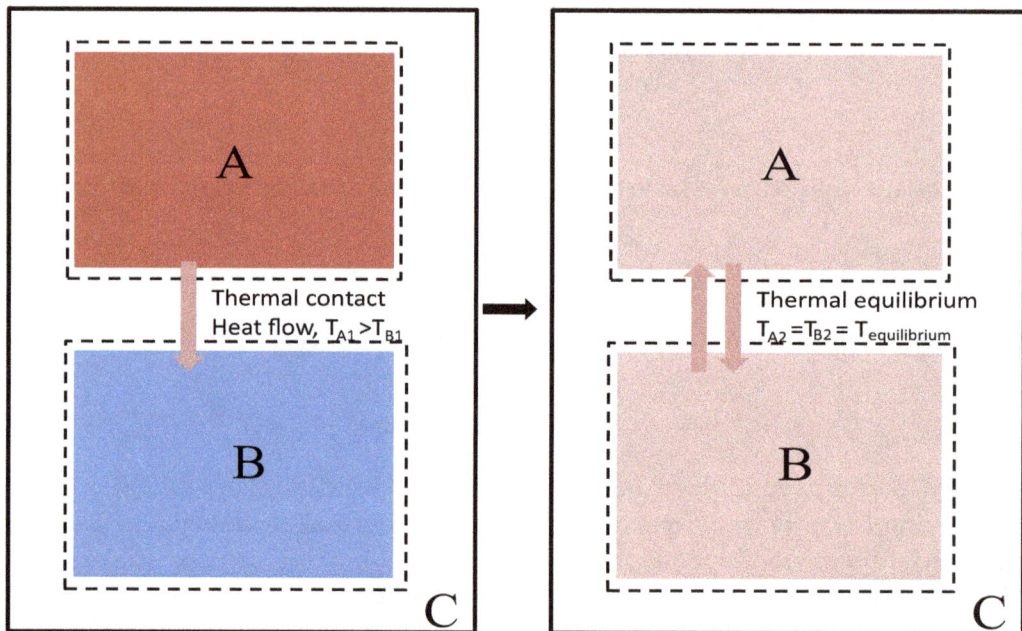

Figure 1. Heat transfer model: non-equilibrium interaction of system A and system B, both previously in thermal equilibrium. Heat transfer occurs until both systems attain a new thermal equilibrium state. During the heat transfer process, body A has a temperature $T < T_{A1}$, and B has a different (lower) temperature, $T > T_{B1}$. Both attain a final equilibrium temperature with $T_{A2} = T_{B2} = T_{equilibrium}$. The entropy increase principle can only be invoked from the perspective of the *universe*, C, and not in isolated discussion of one of the subsystems, although it is clear that the entropy of system B always increases as it receives heat from B.

Since Clausius' initial motivation is to seek an equivalence between heat and its transformation to mechanical work, he presents an interesting conceptual scheme according to which heat transfer proceeds through initial conversion of heat from the source to mechanical work, followed by conversion of this mechanical work back to heat at the lower temperature region. That is, heat going to work corresponds to reducing the intensity of random microscopic motion (corresponding to very high degrees of freedom) by transferring the related energy to macroscopically ordered motion that is observable to us as work (bulk motion with up to three degrees of freedom); conversely, work going to heat is equivalent to transforming the energy of macroscopically ordered motion to more intensified random microscopic motion that becomes unavailable to directional bulk motion. If we leave aside Clausius' mechanistic picture of heat transfer, we simply view the hotter region as one with average particles that are faster and the lower region as one with average particles that are slower. Allowed to interact by collisions, the slower particles gain energy while the faster ones become less energetic.

3   A fundamental indicator of the natural tendency of this transformation is the specific entropy change. It is preferable that it takes the form of a non-dimensional energy parameter to correspond to our notion of invigorating or reducing microscopic motion (a relative assessment):

$$ds = \frac{\delta q}{kT}, \tag{20}$$

where s is the specific entropy, q is the amount of heat added per unit particle of the system, k is the Boltzmann constant and T is the temperature. The convention is such that for heat added δq is positive; hence, s increases.

We use the microscopic energy variable, $\epsilon = kT$, playing the role of an intensive energy of the system, with units of joule per particle. A derived energy unit for ϵ could be named after Clausius or to avoid confusion between specific internal energy and our new intensive energy, ϵ, the latter could be called Clausius energy. This microscopic energy is basically the average kinetic energy of a representative particle without the 1/3 factor that accounts for the three translational degrees of freedom:

$$ds = \frac{\delta q}{\epsilon}. \tag{21}$$

This modification does not change existing thermodynamic relations, but the units are now aligned to give better insight to the physical meaning of specific entropy. The fundamental thermodynamic potential becomes $du = \epsilon ds + \Sigma F_i dx_i$ in specific terms and $dU = \epsilon dS + \Sigma F_i dX_i$ in units of energy, where $dS = Nds$. Here, $\epsilon = \frac{\partial U}{\partial S}|x_i$ is a driving force with $S = Ns$ as the associated coordinate.

4   In order to determine the sign of entropy change associated with natural heat transfer from a hot to a colder region, we consider heat transfer from the hot body, 1, to the colder body, 2. For simplicity, let both have the same mass, m, and the same temperature-independent specific heat capacity at constant volume, c_v. Let us further assume that no volume-change work occurs, so that all heat transfer only results in changes in the internal energy of each system and $\delta q = du = c_v dT = c_v/kd\epsilon$. The common final temperature, hence final intensive energy, can be determined from:

$$\int_{T_1}^{T_f} dU_1 + \int_{T_2}^{T_f} dU_2 = 0, \tag{22}$$

$$\int_{T_1}^{T_f} mc_v dT + \int_{T_2}^{T_f} mc_v dT = 0, \tag{23}$$

$$\implies T_f = \frac{1}{2}(T_1 + T_2), \tag{24}$$

$$\text{so that } \epsilon_f = \frac{1}{2}(\epsilon_1 + \epsilon_2), \text{ where } \epsilon = kT. \tag{25}$$

We note here that $\epsilon_2 < \epsilon_f < \epsilon_1$ or $T_2 < T_f < T_1$, correspondingly.
The specific entropy change, Δs, associated with the process can be determined from:

$$\int ds = \int_{\epsilon_1}^{\epsilon_f} ds_1 + \int_{\epsilon_2}^{\epsilon_f} ds_2, \tag{26}$$

$$\Delta s = \int_{\epsilon_1}^{\epsilon_f} \frac{\delta q}{\epsilon} + \int_{\epsilon_2}^{\epsilon_f} \frac{\delta q}{\epsilon}, \tag{27}$$

$$\Delta s = \frac{c_v}{k}\int_{\epsilon_1}^{\epsilon_f} \frac{d\epsilon}{\epsilon} + \frac{c_v}{k}\int_{\epsilon_2}^{\epsilon_f} \frac{d\epsilon}{\epsilon}, \tag{28}$$

$$\Delta s = \frac{c_v}{k}\ln\left(\frac{\epsilon_f}{\epsilon_1}\right) + \frac{c_v}{k}\ln\left(\frac{\epsilon_f}{\epsilon_2}\right) = \frac{c_v}{k}\ln\left(\frac{\epsilon_f^2}{\epsilon_1\epsilon_2}\right) = \frac{c_v}{k}\ln\left(\frac{(\epsilon_1 + \epsilon_2)^2}{4\epsilon_1\epsilon_2}\right). \tag{29}$$

The last expression in Equation (30) can be rewritten as

$$\Delta s = \frac{c_v}{k}\ln\left(\frac{(\epsilon_1 - \epsilon_2)^2 + 4\epsilon_1\epsilon_2}{4\epsilon_1\epsilon_2}\right) \geq 0. \tag{30}$$

That is, two systems initially at different temperatures, such that 1 is hotter than 2, if allowed to interact by heat exchange will reach a new equilibrium state with a temperature intermediate between the initial body temperatures. The associated entropy change is always positive. Thus, the process is that of the non-equilibrium interaction of systems seeking a new thermal equilibrium state.

The problem considered above imposed a direction by requiring that both systems achieve an equilibrium temperature; heat must flow from the hotter to the colder until thermal equilibrium is attained. We found that this natural process is associated with an increase in entropy. Now, suppose we have two bodies, 1 and 2, of equal mass and heat capacity, but we are not told which of them is hotter, and we seek to establish the condition that has to be fulfilled for the associated entropy change to be positive as they go from states 1 to 1' and from 2 to 2'. Assuming that T_1 decreases by ΔT, then by virtue of energy conservation, T_2 increases by ΔT. Similarly, $\epsilon_1' = \epsilon_1 - \Delta\epsilon$ and $\epsilon_2' = \epsilon_2 + \Delta\epsilon$:

$$\Delta s = \frac{c_v}{k}\ln\left(\frac{\epsilon_1'}{\epsilon_1}\right) + \frac{c_v}{k}\ln\left(\frac{\epsilon_2'}{\epsilon_2}\right), \tag{31}$$

$$\Delta s = \frac{c_v}{k}\ln\left(\frac{\epsilon_1 - \Delta\epsilon}{\epsilon_1}\right) + \frac{c_v}{k}\ln\left(\frac{\epsilon_2 + \Delta\epsilon}{\epsilon_2}\right), \tag{32}$$

$$\Delta s = \frac{c_v}{k}\left[\ln\left(\frac{\epsilon_1 - \Delta\epsilon}{\epsilon_1}\right) + \ln\left(\frac{\epsilon_2 + \Delta\epsilon}{\epsilon_2}\right)\right] = \frac{c_v}{k}\left[\ln\left(\frac{\epsilon_1 - \Delta\epsilon}{\epsilon_1}\right)\left(\frac{\epsilon_2 + \Delta\epsilon}{\epsilon_2}\right)\right], \tag{33}$$

$$\Delta s = \frac{c_v}{k}\left[\ln\left(1 + \frac{\Delta\epsilon}{\epsilon_2} - \frac{\Delta\epsilon}{\epsilon_1} - \frac{(\Delta\epsilon)^2}{\epsilon_1\epsilon_2}\right)\right] = \frac{c_v}{k}\left[\ln\left(1 + \frac{\epsilon_1 - \epsilon_2 - \Delta\epsilon}{\epsilon_1\epsilon_2}\Delta\epsilon\right)\right]. \tag{34}$$

From Equation (34), two observations can be made. Firstly, for natural heat transfer, that is, $\Delta s \geq 0$, it is necessary that $\epsilon_1 > \epsilon_2$, bearing in mind that it was assumed that body 1 transfers heat to body 2. Secondly, if $\epsilon_1 - \epsilon_2 = \Delta\epsilon$, we are dealing with fluctuations around thermal equilibrium, so that $\Delta s = 0$. The result, $\Delta s \geq 0$, can be directly connected to the statistical mechanics result, $\Delta s = k\ln(\frac{W_f}{W_i}) \geq 0$, where, for a naturally occurring process, the number of ways of realizing the microscopic energy distribution or multiplicity of the final state, W_f, must be greater than that of the initial state, W_i.

5   Having freed the entropy increase principle from the unnecessary analysis of heat engines and refrigerators by favoring the heat transfer model, we now focus on how the heat transfer model can be used to explain the performance of heat engines and refrigerators as well as two common formulations of the Second Law.

Figure 2 is a schematic of a heat engine embedded within a natural heat transfer process, with a hot reservoir maintained at T_H and a heat sink maintained at T_L. By means of a piston-cylinder device working in a cycle, mechanical work can be obtained from the configuration, with the effect that less heat is transferred to the heat sink. The working medium leaves the heat sink at state 1 and is raised adiabatically to a higher pressure and temperature until it reaches state 2, where it establishes contact with the heat source. It loses contact with the heat source at state 3 and adiabatically expands to yield useful work before re-establishing contact with the heat sink at state 4. If the heat transfer process is infinitely fast, the heat exchange processes occur at constant volume, and the device establishes contact with the source at $T_2 < (T_H = T_3)$ and with the heat sink at $T_4 > (T_L = T_1)$. Viewed from the "universes" C_H and C_L, all heat transfer processes are natural, and thus, are accompanied by entropy increase so that from the perspective of the "overall universe" C, the entropy increase principle is fulfilled. In the most efficient case, heat transfer occurs such that $T_2 = T_H = T_3$ and $T_4 = T_L = T_1$, which requires infinite time for the isothermal heat exchange process (will have to rely on equilibrium fluctuations). In this limiting case, no entropy is generated in the universes C_H, C_L, and C.

The operation of cold machines (heat pumps and refrigerators) is illustrated in Figure 3. The processes take place in the reverse order of those of heat engines. However, the externally perceived transfer of heat from a region of low temperature to one of high temperature is assured by embedded heat transfer processes that obey the natural law. Net entropy is generated in the universe, C, in all realizable devices, with the exception of the ideal case where heat exchange between the working medium and the reservoirs occur as fluctuations around thermal equilibrium at all times.

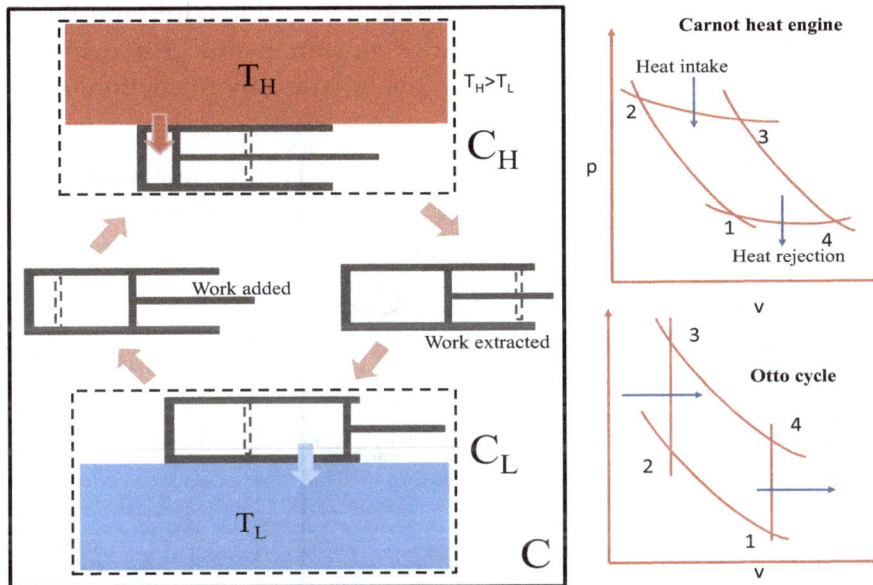

Figure 2. Heat engine illustrating thermal contact and heat transfer between the device and reservoirs. The best engine obtains when heat transfer processes are isothermal, but these require infinite time. For infinitely fast heat transfer, the exchange is isochoric.

The apparent counter-intuitive flow of heat from a region of lower temperature to one of higher temperature relies on the conversion of work to internal energy so that the working medium attains a temperature that is higher than that of the hotter region to which heat is rejected. In other words, a thermal non-equilibrium condition is brought about by converting mechanical energy into energy of random microscopic motion.

From the foregoing discussion, we can combine the two commonly stated formulations of the Second Law into a single heat transfer principle and discuss the various formulations of the Second Law as implications of this main principle. We recall that the heat engine or Kelvin–Planck formulation [22] is stated as: *"It is impossible to devise an engine which, working in a cycle, shall produce no effect other than the extraction of heat from a reservoir and the performance of an equal amount of mechanical work".*

Similarly, the refrigerator or Clausius statement is often stated as: *"It is impossible to devise a machine which, working in a cycle, shall produce no effect other than the transfer of heat from a colder to a hotter body".*

These two formulations can be taken to arise from a general heat transfer principle, that is, the Main Principle expressed by the Second Law is: *"In the absence of other driving forces, heat flows from a region of higher temperature to one of lower temperature, as a consequence of which the net entropy of the two regions increases".*

Implication 1: Heat engines (cyclic devices that convert heat to mechanical work) operate by intercepting some of the heat as it flows from a hot to a colder body. The inner workings of such devices are such that the heat transfer principle holds at all stages of heat exchange. The most effective heat engine is one in which heat intake and heat rejection occur while the working medium is in thermal equilibrium with constant temperature heat source and sink, respectively. This limiting case corresponds to net zero entropy generation in the universe by the action of the device and its heat reservoirs.

Implication 2: Cold machines or refrigerators/heat pumps (cyclic devices that use mechanical work to transfer heat against a temperature gradient) operate by incorporating heat exchange processes between the heat reservoirs so that the principle of heat transfer holds true. The heat exchange is enabled by converting added mechanical work to internal energy. The most effective refrigerator is one in which heat intake and heat rejection occur while the working medium is in thermal equilibrium with the constant temperature heat source and sink, respectively. This limiting case corresponds to net zero entropy generation in the universe by the action of the device.

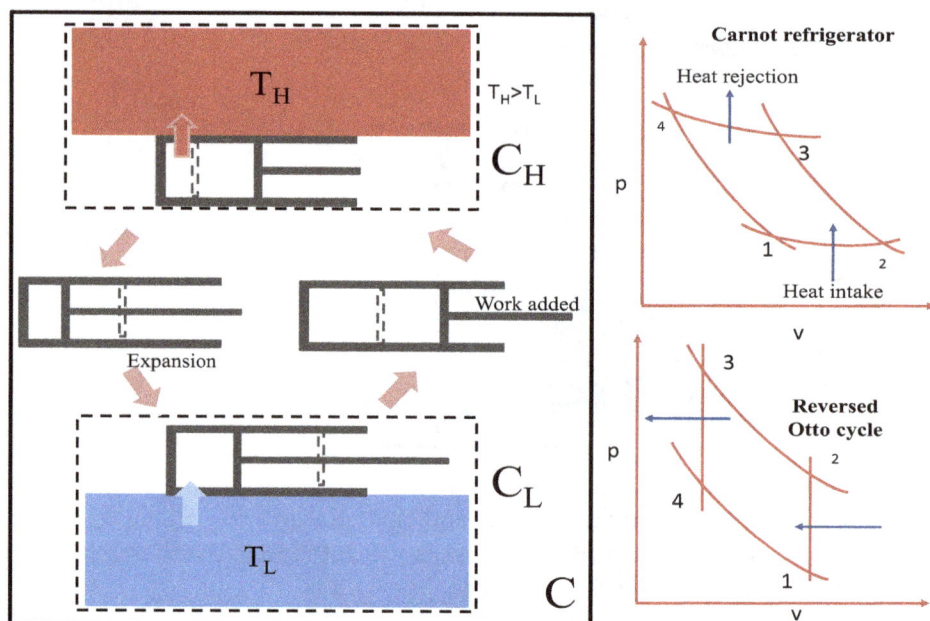

Figure 3. Refrigerator or heat pump cycle illustrating heat transfer processes internal to the device.

Thus, it is clear that the Second Law of Thermodynamics should properly be called the law of natural heat transfer processes; it is clearly about non-equilibrium thermodynamics, and not thermostatics. Specific entropy changes capture the existing non-equilibrium potential of heat flow.

We notice that the law focuses on heat transfer but is silent concerning the direction of adiabatic processes induced by purely mechanical non-equilibrium interactions.

6   Ideal adiabatic or isentropic processes are induced by non-thermal forces or gradients. Mechanical non-equilibrium results from an imbalance of forces or pressures acting on a system. The direction of a natural response of such non-equilibrium system or interacting systems can be determined using Newton's First and Second Axioms. In the case of concentration gradients between two systems in thermal equilibrium, diffusion proceeds in accordance with Fick's theory [23], seeking a new equilibrium concentration state.

The mixing problem and Gibbs' mixing paradox are related to diffusion due to concentration non-equilibrium. They are usually analyzed using statistical mechanics, but, in my view, entropy has been tacitly redefined to measure the diffusion interaction. Let us consider the mixing of two ideal gases. Since $\delta q = 0$ for the process, the differential of the fundamental potential becomes $du = \Sigma F_i dx_i$, that is, the classical entropy term does not even feature in the equation. If the only driving force arises from concentration differences, then this force is the chemical potential, μ_i, and the associated coordinate is the number density of one of the gases, n_i. The statistical mechanical problem thus seems to evaluate an implicitly modified entropy definition, $ds_\mu = \frac{\mu_i dn_i}{T}$. Viewed from this perspective, classical thermodynamics would suggest that the Gibbs mixing paradox is a non-problem because if the gases separated by the membrane are dissimilar, there is a non-zero driving force, namely, the chemical potential, which leads to non-equilibrium interaction. For similar gases, however, the chemical potential is zero and no measurable interaction/mixing occurs. This shows that modified definitions of entropy as measures of various non-equilibrium interactions of systems can be defined, but they may not be measuring thermal non-equilibrium interactions as initially intended by Clausius. Such problems would not be correctly handled by the heat transfer model of the Second Law as proposed here, but the delineation it offers is of epistemic value.

There is a tendency among some scientists and philosophers to try to bring the concept of entropy to bear on cosmological debates such as the nature of time and the dynamics of the universe. Entropy as a measure of thermal non-equilibrium interactions will undoubtedly face challenges related to system definition and the other non-thermal non-equilibrium processes mentioned. We need to know how non-thermal driving forces give rise to thermal non-equilibrium of a given system, from which deductions about entropy changes can be made. A rigorous reconstruction of processes from the present to the remote past would also inevitably bring us to the problem of infinite regress.

How can non-thermal non-equilibrium lead to conditions of thermal non-equilibrium, and hence induce entropy generation through heat transfer? Consider the sequence of processes shown in Figure 4. The piston-cylinder device consists of two cylinders separated by a diathermal wall, while the walls of the enclosure are perfectly insulated. If we focus on the friction-free adiabatic compression from state 1 to 2, we see that the system can be returned to state 1 by friction-free expansion, giving off exactly the same amount of mechanical work as was applied during compression. This speaks to the fact that process 1 to 2 is not accompanied by entropy generation.

If we now focus on the compression from state 1 to 2, unbeknownst to us that the system comprises two cylinders separated by a diathermal wall, we may continue to hold the view that the compression process has not increased the entropy of the universe. We do see that after the compression, gas A is at a higher temperature and therefore not in thermal equilibrium with gas B. If, after compression, both gases are held at constant volume, and have the same mass and heat capacity, then the resulting heat transfer is analogous to the heat transfer model process previously discussed. Figure 4 thus summarizes a scenario in which, if it were not known to us that the device comprises two cylinders separated by a diathermal wall, we would conclude that no entropy is generated, whereas the partition leads to the creation of thermal non-equilibrium and subsequent relaxation to a new thermal equilibrium, with an overall increase in the entropy of the universe.

The example discussed here is connected to Implication 2 above, namely, the operation of cold machines by means of mechanical work addition. The equivalence of heat and work means that mechanical energy can be transformed to microscopic thermal motion, which is manifested as internal energy and higher temperature or intensive energy, ϵ. Since the Second Law is mostly concerned with thermal non-equilibrium and the direction of heat transfer, one can add as an Auxiliary Statement to the Main Principle expressed by the Second Law:

> Although entropy is operationally defined through heat transfer, and would therefore be absent in interactions of two subsystems initially in thermal equilibrium, thermal non-equilibrium can be brought about in such a system of subsystems by mechanical work addition, whereby the internal energy of one subsystem is raised. This accords with the principle of equivalence of work and heat as they relate to changes in the energy state of a given system.

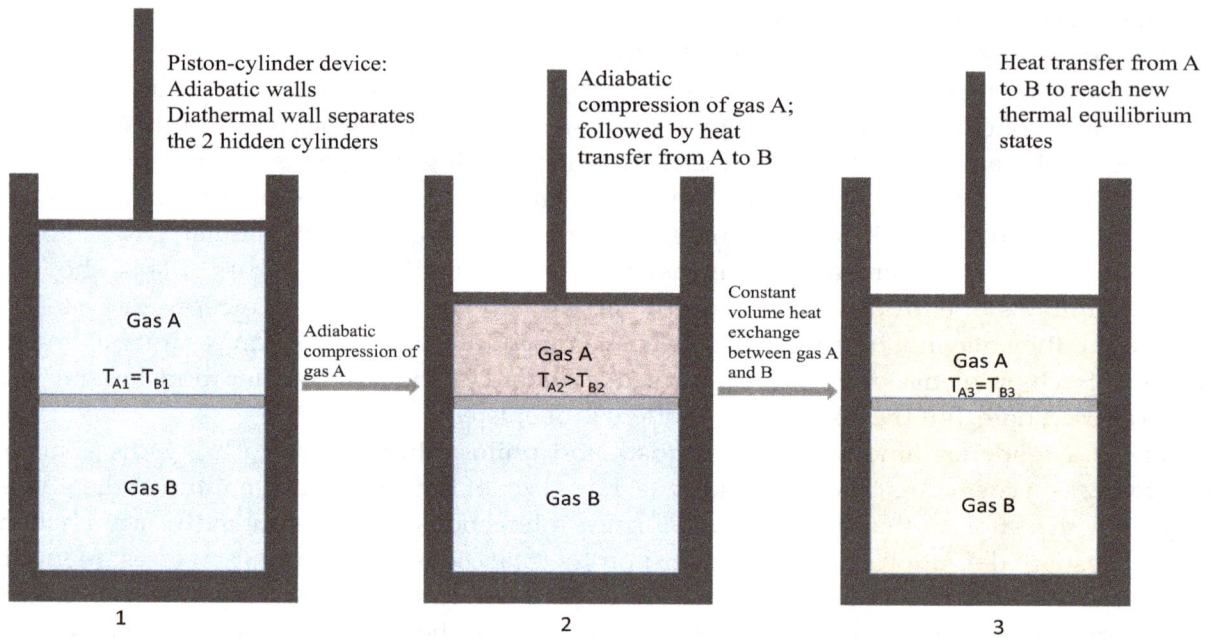

Figure 4. Invisible entropy generation problem. A completed adiabatic compression process of a subsystem is followed by thermal non-equilibrium interaction with another subsystem, generating entropy.

One more thing to consider is whether the emphasized role of heat transfer in the formulation of the Second Law and the entropy increase principle can be used to establish the impossibility of perpetual motion machines. Two classical thought experiments bring out the essence of most machines alleged to be capable of violating the Second Law. One is the Maxwell's Demon [24] and the other is the Feynman–Smoluchowski (Brownian) ratchet [25]. These thought experiments shift their focus from macroscopic thermal non-equilibrium to the mechanical non-equilibrium of the microscopic particle motion. This microscopic mechanical non-equilibrium differs from that of macroscopic systems in the randomness of particle velocity magnitudes and directions (equal probability distribution).

The Maxwell's Demon thought experiment concerns the transfer of heat from a colder to a hotter region by taking advantage of the Maxwellian particle velocity distribution. This system is supposed to succeed thanks to a special mechanical agent capable of detecting approaching high-velocity particles from the colder region and preferentially letting these pass through a gate to the hotter system, without suffering an equal or greater flux of particles from the hotter to the colder system. Were such an agent practically realizable, work would have to be expended to detect and regulate the opening and closing of the gate. The non-equilibrium process would have to be decided by classical or quantum mechanics; the randomness, however, precludes the possibility of selective directional motion without enormous energy expenditure on particle sorting. Macroscopically considered, according to the heat transfer model, it is impossible to effect transfer of mass and heat from a cold to a hotter region without expenditure of mechanical work, be it by a demon or a mechanical contraption.

The Feynman–Smoluchowski engine is supposed to be able to extract work from a single heat reservoir, contradicting the Kelvin–Planck statement and the view that heat engines merely intercept and convert to work part of the heat as it undergoes natural transfer down a temperature gradient. The device is supposed to succeed by tapping into mechanical non-equilibrium of constituent particles (velocity distribution); it restricts motion in one direction and turns a shaft in the other, when hit by more energetic particles on one side than on the other. Such a device will necessarily establish thermal equilibrium with the system of particles and the restricting ratchet and pawl system will fail to secure directional preference as pointed out by Feynman. These two thought experiments fail to tap into mechanical non-equilibrium of the microscopic particles on account of their randomness and uniform mix with slower particles. Thus, the focus on interaction of non-equilibrium systems, such as heat transfer and mechanical non-equilibrium of microscopic particles, serves as a simple and clear explanatory framework for all issues related to the Second Law.

2.1. The Third Law of Thermodynamics

The Third Law of thermodynamics states that the change in entropy between two states approaches zero as both states approach the absolute zero temperature. In statistical thermodynamics; however, entropy is evaluated at a given state. The question arises of whether, in this instance, statistical thermodynamics recovers the result of classical thermodynamics.

The Third Law, otherwise known as the Nernst theorem, grew out of investigations of systems at low temperatures, and the recognition that specific heats tend to zero as the absolute zero temperature is approached. The fact that the entropy change between two states close to absolute zero is zero does not arise directly from the current entropy definition. However, we can explore whether it is possible to arrive at this result if we adopt the temperature function in Clausius' equation, $f(\tau)$, to be $f(\tau) = \epsilon + \epsilon_0$, where ϵ_0 is the microscopic or intensive energy at absolute zero for a given substance. If ϵ_0 is small but non-zero, this definition can render the Third Law of Thermodynamics (Nernst Theorem) unnecessary. That is, it follows that specific entropy differences tend to zero as $T \to 0K$. Assuming that we are cooling from ϵ_1 to ϵ_2, so that $(\epsilon_2 + \epsilon_0) < (\epsilon_1 + \epsilon_0)$ and $\epsilon_1 << \epsilon_0$:

$$\Delta s = \int_{\epsilon_1}^{\epsilon_2} \frac{\delta q}{\epsilon + \epsilon_0}, \tag{35}$$

$$\Delta s = \int_{\epsilon_1}^{\epsilon_2} \frac{c_v/kd\epsilon}{\epsilon + \epsilon_0}, \tag{36}$$

$$\Delta s = \frac{c_v}{k}\ln\left(\frac{\epsilon_2 + \epsilon_0}{\epsilon_1 + \epsilon_0}\right) \approx \frac{c_v}{k}\ln\left(\frac{\epsilon_0}{\epsilon_1 + \epsilon_0}\right) = 0, \tag{37}$$

$$\lim_{\epsilon_1,\epsilon_2 \to 0}(\Delta s) = \frac{c_v}{k}\ln\left(\frac{\epsilon_0}{\epsilon_1 + \epsilon_0}\right) = 0, \tag{38}$$

where we have used $\epsilon_1 << \epsilon_0$ in the last equation. This does not eliminate the need to define an entropy value at absolute zero. The question is whether a law is needed for this, seeing that similar definitions are used for enthalpy, internal energy, Gibbs function, and other energies for pragmatic reasons. The validity of this analysis relies on the assumption that $f(\tau) = \epsilon + \epsilon_0$, which may initially be considered inapplicable to the ideal gas. However, near absolute zero temperature, it is more realistic to assume that $\epsilon_0 \neq 0$ for all substances. The assumptions on which the ideal gas model is based fail at very low temperatures, so that it is reasonable to assume that $f(\tau) = \epsilon + \epsilon_0$ is more general. At temperatures where ideal gas assumptions hold, $\epsilon >> \epsilon_0$, so that ϵ_0 can be neglected.

It has been pointed out by one of the reviewers that the modified definition explains the entropy differential or entropy change, but it does not say what entropy is. One way to work around this is to explain entropy at a given state with reference to a hypothetical heat addition process that brings the system from absolute zero to the state in question. The entropy at the state is, therefore, a measure of the invigoration of microscopic motion associated with heat addition, bringing the system from absolute zero kelvin to the given state. This follows from $s(\epsilon) = s_0 + \int_0^\epsilon \frac{\delta q}{\epsilon + \epsilon_0}$, where s_0 can be set to zero by definition (neither true nor false and not empirically demonstrable).

2.2. Connection to Entropy in Statistical Mechanics

In his approach as presented in [26] (pp. 33, 44), Gibbs seeks a canonical distribution in phase space, with the requirement that its probability be single-valued, and, for each phase space, it should neither be negative nor imaginary:

$$\int_{phases}^{all} \cdots \int P dp_1 ... dq_n = 1. \tag{39}$$

The options, $P = const. \times \epsilon$ and $P = const.$, are considered impossible while the Boltzmann factor is taken as the simplest conceivable case that meets all requirements:

$$P = e^{\eta}; \text{ with } \eta = \frac{\psi - \epsilon}{\theta} \text{ or} \eta = -\frac{\epsilon - \psi}{\theta}, \tag{40}$$

where η is the average probability index, ψ is a constant corresponding to the energy for which the probability is unity, ϵ is the energy of the phase space differential volume, and θ is called the modulus. The modulus is considered to play the same role as temperature in classical thermodynamics. The subject is further developed to arrive at the differential of the average energy:

$$d\bar{e} = -\theta d\eta - \Sigma(A_i d\alpha_i), \tag{41}$$

which is then compared to the differential form of the Second Law, after ignoring the negative signs arising from the definition of η:

$$d\epsilon = T d\eta + \Sigma(A_i d\alpha_i). \tag{42}$$

That is, it is compared to the fundamental potential, du:

$$du = T ds + \Sigma(F_i dx_i), \tag{43}$$

where u is the specific internal energy, s—the specific entropy, F_i—the generalized force, and x_i, the generalized coordinate. In Gibbs' presentation, the average probability index, η, taken with its negative sign, is thought to correspond to entropy in classical thermodynamics. There are clearly conceptual issues with this approach that need further clarifications, but a striking point is the difference in the dimensions of T and θ—the former being temperature and the latter, energy per unit of amount of matter. One of the challenges in statistical mechanics is the axiomatic acceptance of previous results established in classical thermodynamics. The proposed modification here identifies the Gibbs modulus θ as analogous to $\epsilon = kT$, instead of T. The connection is natural because in classical thermodynamics, $\epsilon = kT$ is the statistical average energy of the microscopic constituents and in Gibbs' approach to statistical mechanics, the modulus θ is a normalizing energy quantity for an elemental volume of phase space.

The case is slightly different for the Boltzmann entropy. Entropy enters in his statistical mechanics through the H-theorem. According to Boltzmann, H is the sum of all values of the logarithm of the distribution function f corresponding to designated molecules in a volume element $d\omega$:

$$H = \int f \ln f d\omega, \tag{44}$$

$$s = -H = -\int f \ln f d\omega, \tag{45}$$

$$\text{alternatively } s = const. \ln W, \tag{46}$$

$$\text{with } W \propto 1/f. \tag{47}$$

Mathematically, H is the first moment of $\ln f$. However, specific entropy is $-H$, corresponding to the first moment of $\ln(1/f)$. Increase in H denotes an increase in the weighted mean of $\ln f$. It is well known that the transition from Equation (45) to Equation (46) involves very subtle arguments, some of which are still debatable. The constant in Equation (46) was introduced by Planck and named after Boltzmann (k or k_B). Its units are chosen such that they are consistent with classical thermodynamics.

Since our modified entropy definition views specific entropy as a non-dimensional energy variable, the constant in Equation (46) becomes unity and one simply interprets the increase in the number of ways upon addition of heat to the canonical ensemble to be similar to the present notion of invigorating and increasing the average microscopic energy variable, ϵ. Implicit in Equation (45) is the assumption that $s_0(0K) = 0$; otherwise, it should be $s - s_0$ in order to highlight the fact that the probability is assessed with reference to absolute rest. If we carry out the integration in Equation (36) in the limits from $\epsilon = 0$ to an arbitrary ϵ, we obtain $s - s_0 = (c_v/k)[\ln(\epsilon) - \ln(\epsilon_0)]$. We see that $s = (c_v/k)\ln(\epsilon)$, if it is assumed or set by convention that $s_0 = (c_v/k)\ln(\epsilon_0) \approx 0$. In essence, Boltzmann's entropy captures the invigoration or reduction of microscopic motion through counting the number of ways energy distribution can be realized in a given macro state with implicit reference to a single possible distribution at absolute zero temperature; classical entropy on the contrary, focuses on the average intensive energy of the particles, which can increase or reduce based on heat transfer. However, since the average energy of a system at a given state can be obtained from the number of ways the microscopic energy is distributed, there is a connection between statistical and classical entropy centered on the idea that $\ln\epsilon \propto \ln W$. In an isothermal heat exchange process (constant ϵ or T), the notion of invigoration or reduction of microscopic motion following heat exchange is retained by properly interpreting the associated volume and pressure change. In the case of an ideal gas, the entropy change is simply $s_2 - s_1 = \ln(v_2/v_1) = -\ln(p_2/p_1)$. Here, entropy as a non-dimensional energy measure indicates that isothermal heat addition amounts to invigoration of microscopic motion in order to maintain the same intensive energy (temperature) but in a new state characterized by a larger volume or a lower pressure. In statistical mechanics, this isothermal entropy change would be reflected in the multiplicity, W, as well. Much of statistical mechanics is developed with the assumption of a correct classical thermodynamics framework. It is not surprising, then, that concepts of entropy go back and forth between these two branches. We have shown that, this notwithstanding, a clear conceptual connection can be found once we focus on differential entropy as a non-dimensional measure of thermal non-equilibrium interactions. In classical thermodynamics, the differential is non-dimensional energy and in statistical thermodynamics, it can be energy or number of ways of realizing energy distribution among microstates.

3. Conclusions

This work contributes to improved understanding of entropy in classical thermodynamics by suggesting a modification to the central argument from which the entropy definition by Clausius arises. The proposed modification points to the fact that heat transferred to or from a system affects the internal energy of the system, which is a macroscopic representation of the microscopic motion. It is also recognized that some residual energy is possible at absolute zero, which is not properly represented by temperature. Admitting that substances possess non-zero but very small energy at absolute zero temperature, it follows naturally that as the temperature approaches absolute zero, the entropy change between two states at low temperatures also approaches zero. This is very similar to the Third Law of thermodynamics, albeit obtained not as an axiom. It is shown that the modified definition of entropy preserves the entropy increase principle and direction of entropy change for heat transfer problems. We also discuss how adiabatic mechanical non-equilibrium processes can give rise to thermal non-equilibrium, initiating heat transfer and entropy generation. It is also shown that the modified definition of entropy offers a clearer connection between entropy in classical thermodynamics defined on the basis of average microscopic energy and entropy in statistical thermodynamics defined in terms of probabilities or ways of realizing microscopic energy distributions.

Acknowledgments: Support is acknowledged from the Syracuse University College of Engineering and Computer Science.

References

1.   Clausius, R.   Über verschiedene für die Anwendung bequeme Formen der Hauptgleichungen der mechanischen Wärmetheorie. *Annalen der Physik* **1865**, *201*, 353–400. (In German)
2.   Clausius, R. *The Mechanical Theory of Heat*; Macmillan and Company: London, UK, 1879.
3.   Clausius, R.   Über eine veränderte Form des zweiten Hauptsatzes der mechanischen Wärmetheorie. *Annalen der Physik* **1854**, *169*, 481–506. (In German)
4.   Wehrl, A. General properties of entropy. *Rev. Modern Phys.* **1978**, *50*, 221.
5.   Wehrl, A. The many facets of entropy. *Rep. Math. Phys.* **1991**, *30*, 119–129.
6.   Uffink, J. Bluff your way in the second law of thermodynamics. *Stud. Hist. Philos. Sci. Part B* **2001**, *32*, 305–394.
7.   Tait, P.G. *Sketch of Thermodynamics*; David Douglas: Edinburgh, UK, 1877.
8.   Maxwell, J.C. *Theory of Heat*; Longmans, Green and Company: London, UK, 1888.
9.   Thomson, W. IX.—*On the Dynamical Theory of Heat*. Part V. Thermo-electric Currents. *Trans. R. Soc. Edinb.* **1857**, *21*, 123–171.
10.   Thomson, W. 2. On a Universal Tendency in Nature to the Dissipation of Mechanical Energy. *Proc. R. Soc. Edinb.* **1857**, *3*, 139–142.
11.   Truesdell, C.A. *The Tragicomic History of Thermodynamics, 1822–1854*; Springer: New York, NY, USA, 1980.
12.   YA, C.; Boles, M. *Thermodynamics: An Engineering Approach*; McGraw-Hill: New York, NY, USA, 2008.
13.   Moran, M.J.; Shapiro, H.N.; Boettner, D.D.; Bailey, M.B. *Fundamentals of Engineering Thermodynamics*; John Wiley & Sons: Hoboken, NJ, USA, 2010.
14.   Sklar, L. The reduction (?) of thermodynamics to statistical mechanics. *Philos. Stud.* **1999**, *95*, 187–202.
15.   Callender, C. Reducing thermodynamics to statistical mechanics: The case of entropy. *J. Philos.* **1999**, *96*, 348–373.
16.   Callender, C. Taking Thermodynamics Too Seriously. *Stud. Hist. Philos. Sci. Part B* **2001**, *32*, 539–553.
17.   Fischer, J.; Gerasimov, S.; Hill, K.D.; Machin, G.; Moldover, M.; Pitre, L.; Steur, P.; Stock, M.; Tamura, O.; Ugur, H.; et al. Report to the CIPM on the Implications of Changing the Definition Of the Base Unit Kelvin. Available online: http://temperatures.ru/pdf/Kelvin_CIPM.pdf (accessed on 21 July 2016).
18.   Nawrocki, W. Revising the SI: The joule to replace the kelvin as a base unit. *Metrol. Meas. Syst.* **2006**, *13*, 171–181.
19.   Nawrocki, W. The Quantum SI-Towards the New Systems of Units. *Metrol. Meas. Syst.* **2010**, *17*, 139–150.
20.   Joule, J.P.; Thomson, W. On the Thermal Effects of Fluids in Motion. Part II. *Philos. Trans. R. Soc. Lond.* **1854**, *144*, 321–364.
21.   Chang, H. *Inventing Temperature: Measurement and Scientific Progress*; Oxford University Press: Oxford, UK, 2004.
22.   Pippard, A. *Elements of Classical Thermodynamics: For Advanced Students Of Physics*; Cambridge University Press: Cambridge, UK, 1964.
23.   Fick, A. Ueber diffusion. *Annalen der Physik* **1855**, *170*, 59–86. (In German)
24.   Thomson, W. The sorting demon of Maxwell. *R. Soc. Proc.* **1879**, *9*, 113–114.
25.   Feynman, R.P.; Leighton, R.B.; Sands, M. *Feynman Lectures on Physics, Vol. 1: Mainly Mechanics, Radiation and Heat*; Addison-Wesley: Boston, MA, USA, 1963.
26.   Gibbs, J.W. *Elementary Principles in Statistical Mechanics*; Dover Publications: Mineola, NY, USA, 2014.

2

Entropy and the Second Law of Thermodynamics—The Nonequilibrium Perspective

Henning Struchtrup (ID)

Mechanical Engineering, University of Victoria, Victoria, BC V8W 2Y2, Canada; struchtr@uvic.ca

Abstract: An alternative to the Carnot-Clausius approach for introducing entropy and the second law of thermodynamics is outlined that establishes entropy as a nonequilibrium property from the onset. Five simple observations lead to entropy for nonequilibrium and equilibrium states, and its balance. Thermodynamic temperature is identified, its positivity follows from the stability of the rest state. It is shown that the equations of engineering thermodynamics are valid for the case of local thermodynamic equilibrium, with inhomogeneous states. The main findings are accompanied by examples and additional discussion to firmly imbed classical and engineering thermodynamics into nonequilibrium thermodynamics.

Keywords: 2nd law of thermodynamics; entropy; nonequilibrium; entropy generation; teaching thermodynamics

Highlights

- 2nd law and entropy based on 5 simple observations aligning with daily experience.
- Entropy defined for nonequilibrium and equilibrium states.
- Positivity of thermodynamic temperature ensures dissipation of kinetic energy, and stability of the rest state.
- Global balance laws based on assumption of local thermodynamic equilibrium.
- Agreement with Classical Equilibrium Thermodynamics and Linear Irreversible Thermodynamics.
- Entropy definition in agreement with Boltzmann's H-function.
- Discussion of thermodynamic engines and devices, incl. Carnot engines and cycles, only *after* the laws of thermodynamics are established.

Preamble

This text centers on the introduction of the 2nd Law of Thermodynamics from a small number of everyday observations. The emphasis is on a straight path from the observations to identifying the 2nd law, and thermodynamic temperature. There are only few examples or applications, since these can be found in textbooks. The concise presentation aims at the more experienced reader, in particular those that are interested to see how the 2nd law can be introduced without Carnot engines and cycles.

Nonequilibrium states and nonequilibrium processes are at the core of thermodynamics, and the present treatment puts these at center stage—where they belong. Throughout, all thermodynamic systems considered are allowed to be in nonequilibrium states, which typically are inhomogeneous in certain thermodynamic properties, with homogeneous states (for the proper variables) assumed only in equilibrium.

The content ranges from simple observation in daily life over the equations for systems used in engineering thermodynamics to the partial differential equations of thermo-fluid-dynamics; short discussions of kinetic theory of gases and the microscopic interpretation of entropy are included.

The presentation is split into many short segments to give the reader sufficient pause for consideration and digestion. For better flow, some material is moved to the Appendix which provides the more difficult material, in particular excursions into Linear Irreversible Thermodynamics and Kinetic Theory.

While it would be interesting to compare the approach to entropy presented below to what was or is done by others, this will not be done. My interest here is to give a rather concise idea of the introduction of entropy on the grounds of nonequilibrium states and irreversible processes, together with a smaller set of significant problems that can be tackled best from this viewpoint. The reader interested in the history of thermodynamics, and other approaches is referred to the large body of literature, of which only a small portion is referenced below.

1. Intro: What's the Problem with the 2nd Law?

After teaching thermodynamics to engineering students for more than two decades (from teaching assistant to professor) I believe that a lot of the trouble, or the conceived trouble, with understanding of the 2nd law is related to how it is taught to the students. Most engineering textbooks introduce the 2nd law, and its main quantity, *Entropy*, following the classical Carnot-Clausius arguments [1] using reversible and irreversible engines [2–5]. To follow this approach properly, one needs a lot of background on processes, systems, property relations, engines and cycles. For instance in a widely adopted undergraduate textbook [2], the mathematical formulation of the 2nd law—the balance law for entropy—appears finally on page 380, more than 300 pages after the conservation law for energy, which is the mathematical expression of the *1st Law of Thermodynamics.*

Considering that the full discussion of processes, systems, property relations, engines and cycles requires *both*, the 1st *and* the 2nd law, it should be clear that a far more streamlined access to the body of thermodynamics can be achieved when both laws—in their full mathematical formulation—are available as early as possible. The doubtful reader will find some examples in the course of this treatise. Some authors are well aware of this advantage, which is common in Germany [6], but seems to come only slowly to the North-American market [7,8].

Early introduction of the 2nd law cannot be based on the Carnot-Clausius argument using engines and cycles, but must find other ways. In References [6,7] this problem is solved by simply postulating the 2nd law, and then showing first that is makes perfect sense, by analyzing simple systems, before using it to find less intuitive results, such as the limited efficiency of heat engines, and any other results of (applied) thermodynamics.

The problem with the postulative approach is that assumptions and restrictions of the equations are not always clearly stated. While the 1st law of thermodynamics is universally valid, this is not the case for most formulations of the 2nd law of thermodynamics. The presentation below aims at introducing the 2nd law step by step, with clear discussion of all assumptions used.

2. Outline: Developing the 2nd Law

Below, I will develop the 2nd law from a small number of observations. This approach is also outlined in my undergraduate textbook [8], but with some shortcuts and omissions that I felt are required for the target audience, that is, students who are exposed to the topic for the first time. The presentation below is aimed more at the experienced reader, therefore when it comes to the 2nd law, I aim to avoid these shortcuts, and discuss several aspects in more depth, in particular the all-important question of nonequilibrium states, while reducing the discussion of basic elements like properties and their relations as much as possible.

Many treatments of thermodynamics claim that it can only make statements on equilibrium states, and cannot deal with nonequilibrium states and irreversible processes. Would this be true, simple thermodynamic elements such as heat conductors or irreversible turbines could not be treated in the context of thermodynamics.

Indeed, classical [9–11] or more recent [12–14] treatments focus on equilibrium states, and avoid the definition of entropy in nonequilibrium.

However, in his famous treatment of kinetic theory of gases, Boltzmann gave proof of the H-theorem [15–17], which can be interpreted as the 2nd law of thermodynamics for ideal gases in *any* state. Entropy and its balance follow directly by integration from the particle distribution function and the Boltzmann equation, and there is no limitation at all to equilibrium states. Of course, for general nonequilibrium states one cannot expect to find a simple property relation for entropy, but this does not imply that entropy as an extensive system property does not exist. All frameworks of nonequilibrium thermodynamics naturally include a nonequilibrium entropy [18–22].

The arguments below focus on the approach to equilibrium from nonequilibrium states. Entropy is introduced as the property to describe equilibration processes, hence is defined for general nonequilibrium states.

Luckily, most engineering problems, and certainly those that are the typical topics in engineering thermodynamics, can be described in the framework of *local thermodynamic equilibrium* [18], where the well-known property relations for equilibrium are valid locally, but states are inhomogeneous (global equilibrium states normally are homogeneous in certain properties, see Section 40). Textbook discussions of open systems like nozzles, turbines, heat exchangers, and so forth, rely on this assumption, typically without explicitly stating this fact [2–8] . It is also worthwhile to note that local thermodynamic equilibrium arises naturally in kinetic theory, when one considers proper limitations on processes, viz. sufficiently small Knudsen number, see Appendix D.5 and References [16,17].

Obviously, the 2nd law does not exist on its own, that is for its formulation we require some background—and language—on systems, processes, and the 1st law, which will be presented first, as concisely as possible, before we will embark on the construction of the balance law for entropy, the property relations, and some analysis of its consequences. For completeness, short sections on *Linear Irreversible Thermodynamics* and *Kinetic Theory of Gases* are included in the Appendix.

For better accessibility, in most of the ensuing discussion we consider closed systems, the extension to open systems will be outlined only briefly. Moreover, mixtures, reacting or not, are not included, mainly for space reasons. Larger parts of the discussion, and the figures, are adapted from my textbook on technical thermodynamics [8], albeit suitably re-ordered, tailored, and re-formulated. The interested reader is referred to the book for many application to technical processes in open and closed systems, as well as the discussion of inert and reacting mixtures.

In an overview paper [13] of their detailed account of entropy as an equilibrium property [12], Lieb and Yngvason state in the subtitle that

"The existence of entropy, and its increase, can be understood without reference to either statistical mechanics or heat engines."

This is the viewpoint of this contribution as well, hence I could not agree more. However, these authors' restriction of entropy to equilibrium states is an unnecessary limitation. The analysis of irreversible processes is central in modern engineering thermodynamics, where systems in nonequilibrium states must be evaluated, and the entropy generation rate is routinely studied to analyze irreversible losses, and to redesign processes with the goal of entropy generation minimization [8,23]. Hence, while the goal appears to be the same, the present philosophy differs substantially from authors who restrict entropy to equilibrium states.

3. The 2nd Law in Words

In the present approach, the 2nd law summarizes everyday experiences in a few generalized observations which are then used to conclude on its mathematical formulation, that is, the balance law for entropy. Only the mathematical formulation of thermodynamics, which also includes the 1st law and property relations, allows to generalize from simple experience to the unexpected.

Below, the observations used, and the mathematical formulation of the 2nd law, are summarized as an introduction of what will come for the thermodynamically experienced reader, and for future reference for the newcomer. The definitions of terms used, explanations of the observations, and their evaluation will be presented in subsequent sections

The observation based form of the 2nd law reads:

Basic Observations

Observation 1. *A closed system can only be manipulated by heat and work transfer.*

Observation 2. *Over time, an isolated thermodynamic system approaches a unique and stable equilibrium state.*

Observation 3. *In a stable equilibrium state, the temperature of a thermally unrestricted system is uniform.*

Observation 4. *Work transfer is unrestricted in direction, but some work might be lost to friction.*

Observation 5. *Heat will always go from hot to cold by itself, but not vice versa.*

The experienced reader should not be surprised by any of these statements. The second, third and fifth probably are most familiar, since they appear as input, or output, in any treatment of the 2nd law. The first and the fourth are typically less emphasized, if stated at all, but are required to properly formulate the transfer of entropy, and to give positivity of thermodynamic temperature.

Careful elaboration will show that these observations, with the assumption of local thermodynamic equilibrium, will lead to the 2nd law for closed systems in the form

$$\frac{dS}{dt} - \sum_k \frac{\dot{Q}_k}{T_k} = \dot{S}_{gen} \geq 0 \quad \text{and} \quad T > 0. \tag{1}$$

Here, S is the concave entropy of the system, $\dot{Q}_k$ is energy transfer by heat over the system boundary at positive thermodynamic temperature T_k, and $\dot{S}_{gen}$ is the non-negative generation rate of entropy within the system, which vanishes in equilibrium.

4. Closed System

The first step in any thermodynamic consideration is to identify the system that one wishes to describe. Any complex system, for example, a power plant, can be seen as a compound of some—or many—smaller and simpler systems that interact with each other. For the basic understanding of the thermodynamic laws it is best to begin with the simplest system, and study more complex systems later as assemblies of these simple systems.

The simplest system of interest, and the one we will consider for most of the discussion, is the *closed system* where a simple substance (i.e., no chemical changes) is enclosed by walls, and no mass flows over the system boundaries, for example, the piston-cylinder device depicted in Figure 1.

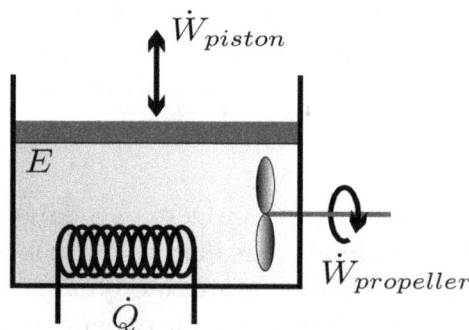

Figure 1. Closed system with energy E exchanging work $\dot{W}$ and heat $\dot{Q}$ with its surroundings.

There is only a small number of manipulations possible to change the state of a closed system, which are indicated in the figure—the volume of the system can be changed by moving the piston, the system can be stirred with a propeller, and the system can be heated or cooled by changing the temperature of the system boundary, as indicated by the heating (or cooling) coil. Another possibility to heat or cool the system is through absorption and emission of radiation, and transfer of radiation across the system boundary (as in a microwave oven)–this is just another way of heating. One could also shake the system, which is equivalent to stirring.

The statement that there is no other possible manipulation of the system than these is formulated in **Observation 1**.

These manipulative actions lead to exchange of energy between the system and its surroundings, either by work in case of piston movement and stirring, or by the exchange of heat. The transfer of energy (E) by work ($\dot{W}$) and heat ($\dot{Q}$) will be formulated in the *1st Law of Thermodynamics* (Section 12). The fundamental difference between piston and propeller work, as (possibly) reversible and irreversible processes will become clear later.

5. Properties

To get a good grip on properties that describe the system, we consider a system of volume V which is filled by a mass m of substance. To describe variation of properties in space, it is useful to divide the system into infinitesimal elements of size dV which contain the mass dm, as sketched in Figure 2.

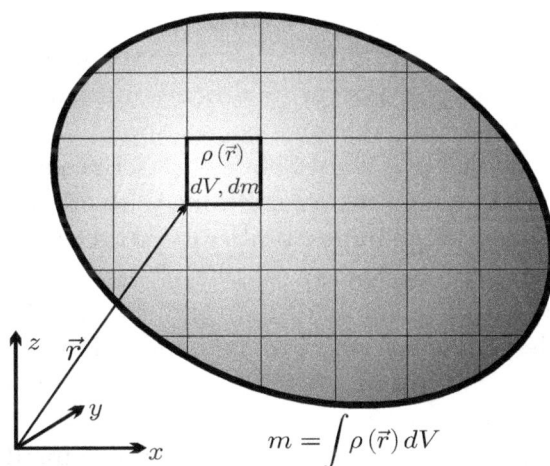

Figure 2. Inhomogeneous distribution of mass density $\rho\left(\vec{r}\right)$ in a system.

The volume $V = \int dV$ filled by the substance can, in principle, be measured by means of a ruler. The mass $m = \int dm$ of the substance can be measured using a scale. The pressure p of the substance can be measured as the force required to keep a piston in place, divided by the surface area of the piston.

One distinguishes between extensive properties, which are related to the size of the system, and intensive properties, which are independent of the overall size of the system. Mass m and volume V are extensive quantities, for example, they double when the system is doubled; pressure p and temperature T (yet to be defined) are intensive properties, they remain unchanged when the system is doubled.

A particular class of intensive properties are the specific properties, which are defined as the ratio between an extensive property and the corresponding mass. In inhomogeneous states intensive and specific properties vary locally, that is they have different values in different volume elements dV.

The *local* specific properties are defined through the values of the extensive property $d\Phi$ and the mass dm in the volume element,

$$\phi = \frac{d\Phi}{dm} . \tag{2}$$

For example, the local specific volume v, and the local mass density ρ, are defined as

$$v = \frac{1}{\rho} = \frac{dV}{dm} . \tag{3}$$

The values of the extensive properties for the full system are determined by integration of the specific properties over the mass or volume elements,

$$\Phi = \int d\Phi = \int \phi dm = \int \rho \phi dV . \tag{4}$$

As an example, Figure 2 shows the inhomogeneous distribution of mass density ρ in a system (i.e., $\phi = 1$). Note that due to inhomogeneity, the density is a function of location $\vec{r} = \{x, y, z\}$ of the element dV, hence $\rho = \rho(\vec{r})$.

For homogeneous states, the integrands can be taken out of the integrals, and we find simple relations such as

$$v = \frac{1}{\rho} = \frac{V}{m} \quad , \quad \phi = \frac{\Phi}{m} . \tag{5}$$

6. Micro and Macro

A macroscopic amount of matter filling the volume V, say a steel rod or a gas in a box, consists of an extremely large number—to the order of 10^{23}—of atoms or molecules. These are in constant interaction which each other and exchange energy and momentum, for example, a gas particle in air at standard conditions undergoes about 10^9 collisions per second.

From the viewpoint of mechanics, one would have to describe each particle by its own (quantum mechanical) equation of motion, in which the interactions with all other particles would have to be taken into account. Obviously, due to the huge number of particles, this is not feasible. Fortunately, the constant interaction between particles leads to a collective behavior of the matter already in very small volume elements dV, in which the state of the matter can be described by few macroscopic properties like pressure, mass density, temperature and others. This allows us to describe the matter not as an assembly of atoms, but as a continuum where the state in each volume element dV is described by these few macroscopic properties.

Note that the underlying assumption is that the volume element contains a sufficiently large number of particles, which interact with high frequency. Indeed, the continuum hypothesis breaks down under certain circumstances, in particular for highly rarefied gases [17]. In all of what follows, however, we shall only consider systems in which the assumption is well justified. Appendix D provides a short discussion of kinetic gas theory, where macroscopic thermodynamics arises in the limit of high collision frequency between particles (equivalent to small mean free path).

7. Processes and Equilibrium States

A *process* is any change in one or more properties occurring within a system. The system depicted in Figure 1 can be manipulated by moving the piston or propeller, and by changing the temperature of the system boundary (heating/cooling coil). Any manipulation changes the state of the system locally and globally—a process occurs.

After all manipulation stops, the states, that is, the values of the local intensive properties in the volume elements, will keep changing for a while—that is the process continues—until a stable final state is assumed. This stable final state is called the *equilibrium state*. The system will remain in the equilibrium state until a new manipulation commences.

Simple examples from daily life are:

(a)   A cup of coffee is stirred with a spoon. After the spoon is removed, the coffee will keep moving for a while until it comes to rest. It will stay at rest indefinitely, unless stirring is recommenced or the cup is moved.

(b)   Milk is poured into coffee. Initially, there are light-brown regions of large milk content and dark-brown regions of low milk content. After a while, however, coffee and milk are well-mixed, at mid-brown color, and remain in that state. Stirring speeds the process up, but the mixing occurs also when no stirring takes place. Personally, I drink standard dip-coffee into which I pour milk: I have not used a spoon for mixing both in years.

(c)   A spoon used to stir hot coffee becomes hot at the end immersed in the coffee. A while after it is removed from the cup, it will have assumed a homogeneous temperature.

(d)   Oil mixed with vinegar by stirring will separate after a while, with oil on top of the vinegar.

In short, observation of daily processes, and experiments in the laboratory, show that a system that is left to itself for a sufficiently long time will approach a stable equilibrium state, and will remain in this state as long as the system is not subjected to further manipulation. This experience is the content of **Observation 2**. Example (d) shows that not all equilibrium states are homogeneous; however, temperature will always be homogeneous in equilibrium, which is laid down as **Observation 3**.

The details of the equilibrium state depend on the constraints on the system, in particular material, size, mass, and energy; this will become clear further below (Section 39).

The time required for reaching the equilibrium state, and other details of the process taking place, depend on the initial deviation from the equilibrium state, the material, and the geometry. Some systems may remain for rather long times in metastable states—these will not be further discussed.

Physical constraints between different parts of a system can lead to different equilibrium states within the parts. For instance, a container can be divided by a rigid wall, with different materials at both sides. Due to the physical division, the materials in the compartments might well be at different pressures, and different temperatures, and they will not mix. However, if the wall is diathermal, that is, it allows heat transfer, then the temperature will equilibrate between the compartments. If the wall is allowed to move, it will do so, until the pressures in both parts are equal. If the wall is removed, depending on their miscibility the materials might mix, see examples (b) and (d).

Unless otherwise stated, the systems discussed in the following are free from internal constraints.

8. Reversible and Irreversible Processes

When one starts to manipulate a system that is initially in equilibrium, the equilibrium state is disturbed, and a new process occurs.

All real-life applications of thermodynamics involve some degree of nonequilibrium. For the discussion of thermodynamics it is customary, and useful, to consider idealized processes, for which the manipulation happens sufficiently slow. In this case, the system has sufficient time to adapt so that it is in an equilibrium state at *any* time. Slow processes that lead the system through a series of equilibrium states are called *quasi-static*, or *quasi-equilibrium*, or *reversible*, processes.

If the manipulation that causes a quasi-static process stops, the system is already in an equilibrium state, and no further change will be observed.

Equilibrium states are simple, quite often they are homogenous states, or can be approximated as homogeneous states (see Section 40). The state of the system is fully described by few extensive properties, such as mass, volume, energy, and the corresponding pressure and temperature.

When the manipulation is fast, so that the system has no time to reach a new equilibrium state, it will be in nonequilibrium states. If the manipulation that causes a nonequilibrium process stops, the system will undergo further changes until it has reached its equilibrium state.

The equilibration process takes place while no manipulation occurs, that is, the system is left to itself. Thus, the equilibration is an uncontrolled process.

Nonequilibrium processes typically involve inhomogeneous states, hence their proper description requires values of the properties at all locations $\vec{r}$ (i.e., in all volume elements dV) of the system. Accordingly, the detailed description of nonequilibrium processes is more complex than the description of quasi-static processes. This is the topic of theories of nonequilibrium thermodynamics, where the processes are described through partial differential equations, see Appendix C. For instance, the approach of Linear Irreversible Thermodynamics yields the Navier-Stokes and Fourier laws that are routinely used in fluid dynamics and heat transfer. Apart from giving the desired spatially resolved description of the process, these equations are also useful in examining under which circumstances a process can be approximated as quasi-static. For the moment, we state that a process must be sufficiently slow for this to be the case.

The approach to equilibrium introduces a timeline for processes—As time progresses, an isolated system, that is, a system that is not further manipulated in any way, so that heat and work vanish, will talways approach, and finally reach, its unique equilibrium state. The opposite will not be observed, that is an isolated system will never be seen spontaneously leaving its equilibrium state when no manipulation occurs.

Indeed, we immediately detect whether a movie of a nonequilibrium process is played forward or backwards: well mixed milk coffee will not separate suddenly into milk and coffee; a spoon of constant temperature will not suddenly become hot at one end, and cold at the other; a propeller immersed in a fluid at rest will not suddenly start to move and lift a weight (Figure 3); oil on top of water will not suddenly mix with the water; and so forth. We shall call processes with a time-line *irreversible*.

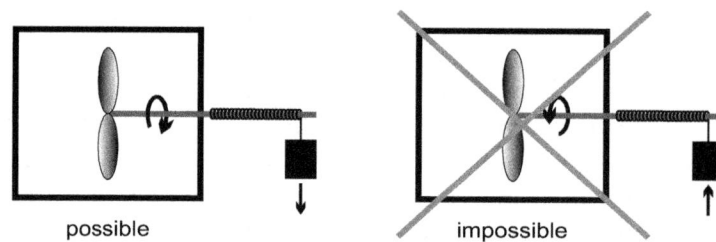

Figure 3. A possible and an impossible process.

Only for quasi-static processes, where the system is always in equilibrium states, we cannot distinguish whether a movie is played forwards or backwards. This is why these processes are also called *reversible*. Since equilibration requires time, quasi-static, or reversible, processes typically are slow processes, so that the system always has sufficient time to adapt to an imposed change.

To be clear, we define quasi-static processes as reversible. One could consider irreversible slow processes, such as the compression of a gas with a piston subject to friction. For the gas itself, the process would be reversible, but for the system of gas *and* piston, the process would be irreversible.

9. Temperature and the 0th Law

By touching objects we can distinguish between hot and cold, and we say that hotter states have a higher temperature. Objective measurement of temperature requires (a) a proper definition, and (b) a proper device for measurement—a thermometer.

Experience shows that physical states of systems change with temperature. For instance, the gas thermometer in Figure 4 contains a certain amount of gas enclosed in a container at fixed volume V. Increase of its temperature T by heating leads to a measurable change in the gas pressure p. Note that pressure is a mechanical property, which is measured as force per area. An arbitrary temperature scale can be defined, for example, as $T = a + bp$ with arbitrary constants a and b.

To study temperature, we consider two systems, initially in their respective equilibrium, both not subject to any work interaction, that is, no piston or propeller motion in Figure 1, which are manipulated by bringing them into physical contact, such that energy can pass between the systems (thermal contact), see Figure 5. Then, the new system that is comprised of the two initial systems will exhibit a process towards its equilibrium state. Consider first equilibration of a body **A** with the gas thermometer, so that the compound system of body and thermometer has the initial temperature $\bar{T}_A$, which can be read of the thermometer. Next, consider the equilibration of a body **B** with the gas thermometer, so that the compound system of body and thermometer has the initial temperature $\bar{T}_B$, as shown on the thermometer.

$$T = a + bp$$

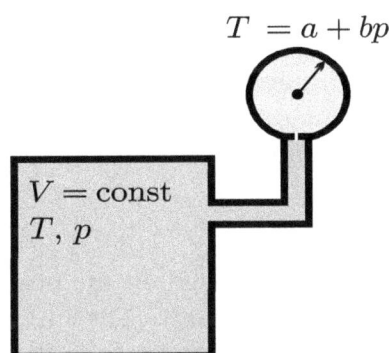

Figure 4. In a gas thermometer, temperature T is determined through measurement of pressure p.

Now, we bring the two bodies and the thermometer into thermal contact, and let them equilibrate. It is observed that both systems change their temperature such the hotter system becomes colder, and vice versa. Independent of whether the thermometer is in thermal contact only with system **A** or in thermal contact with system **B**, it shows the same temperature. Hence, the equilibrium state is characterized by a common temperature of both systems. Since no work interaction took place, one speaks of the *thermal equilibrium* state.

Expressed more formally, we conclude that if body **C** (the thermometer in the above) is in thermal equilibrium with body **A** and in thermal equilibrium with body **B**, than also bodies **A** and **B** will be in thermal equilibrium. All three bodies will have the same temperature. The extension to an arbitrary number of bodies is straightforward, and since any system under consideration can be thought of as a compund of smaller subsystems, we can conclude that a system in thermal equilibrium has a homogeneous temperature.

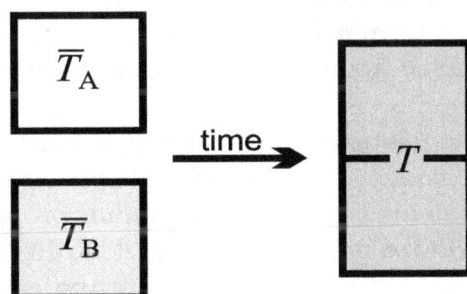

Figure 5. Two bodies of different temperatures $\bar{T}_A$, $\bar{T}_B$ assume a common temperature T a while after they are brought into thermal contact.

The observation outlined above defines temperature, hence its important enough to be laid out as a law (**Observation 3**)

The 0th Law of Thermodynamics

In a stable equilibrium state, the temperature of a thermally unrestricted system is uniform. Or, two bodies in thermal equilibrium have the same temperature.

The 0th law introduces temperature as a measurable quantity. Indeed, to measure the temperature of a body, all we have to do is to bring a calibrated thermometer into contact with the body and wait until the equilibrium state of the system (body and thermometer) is reached. When the size of the thermometer is sufficiently small compared to the size of the body, the final temperature of body *and* thermometer will be (almost) equal to the initial temperature of the body.

10. Ideal Gas Temperature Scale

For proper agreement and reproducibility of temperature measurement, it is helpful to agree on a temperature scale.

Any gas at sufficiently low pressures and large enough temperatures, behaves as an ideal gas. From experiments one observes that for an ideal gas confined to a fixed volume the pressure increases with temperature. The temperature scale is *defined* such that the relation between pressure and temperature is linear, that is

$$T = a + bp \tag{6}$$

The Celsius scale was originally defined based on the boiling and freezing points of water at $p = 1$ atm to define the temperatures of $100\,°C$ and $0\,°C$. For the Celsius scale one finds $a = -273.15\,°C$ independent of the ideal gas used. The constant b depends on the volume, mass and type of the gas in the thermometer.

By shifting the temperature scale by a, one can define an alternative scale, the *ideal gas temperature scale*, as

$$T\,(\mathrm{K}) = bp\,. \tag{7}$$

The ideal gas scale has the unit Kelvin [K] and is related to the Celsius scale as

$$T\,(\mathrm{K}) = T\,(°\mathrm{C})\,\frac{\mathrm{K}}{°\mathrm{C}} + 273.15\mathrm{K}\,. \tag{8}$$

It will be shown later that this scale coincides with the thermodynamic temperature scale that follows from the 2nd law (Section 27).

11. Thermal Equation of State

Careful measurements on simple substances show that specific volume v (or density $\rho = 1/v$), pressure p and temperature T cannot be controlled independently. Indeed, in equilibrium states they are linked through a relation of the form $p = p\,(v, T)$, or $p = p\,(\rho, T)$, known as the *thermal equation of state*. For most substances, this relation cannot be easily expressed as an actual equation, but is laid down in property tables.

The thermal equation of state relates measurable properties. It suffices to know the values of two properties to determine the values of others. This will still be the case when we add energy and entropy in equilibrium states to the list of thermodynamic properties, which can be determined through measurement of any two of the measurable properties, that is, (p, T) or (v, T) or (p, v).

To summarize: If we assume local thermal equilibrium, the complete knowledge of the macroscopic state of a system requires the values of two intensive properties in each location (i.e., in each infinitesimal volume element), and the local velocity. The state of a system in global equilibrium, where properties are homogeneous, is described by just two intensive properties (plus the size of the system, that is either total volume, or total mass). In comparison, full knowledge of the microscopic state would require the knowledge of location and velocity of each particle.

The ideal gas is one of the simplest substances to study, since it has simple property relations. Careful measurements have shown that for an ideal gas pressure p, total volume V, temperature T (in K), and mass m are related by an explicit thermal equation of state, the *ideal gas law*

$$pV = mRT\,. \tag{9}$$

Here, R is the *gas constant* that depends on the type of the gas. With this, the constant in (7) is $b = V / (mR) = v/R$.

Alternative forms of the ideal gas equation result from introducing the specific volume $v = V/m$ or the mass density $\rho = 1/v$ so that

$$pv = RT \quad , \quad p = \rho RT. \tag{10}$$

12. The 1st Law of Thermodynamics

It is our daily experience that heat can be converted to work, and that work can be converted to heat. A propeller mounted over a burning candle will spin when the heated air rises due to buoyancy: heat is converted to work. Rubbing our hands makes them warmer: work is converted to heat. Humankind has a long and rich history of making use of both conversions.

While the heat-to-work and work-to-heat conversions are readily observable in simple and more complex processes, the governing law is not at all obvious from simple observation. It required groundbreaking thinking and careful experiments to unveil the *Law of Conservation of Energy*. Due to its importance in thermodynamics, it is also known as the *1st Law of Thermodynamics*, which expressed in words, reads:

1st Law of Thermodynamics

Energy cannot be produced nor destroyed, it can only be transferred, or converted from one form to another. In short, energy is conserved.

It took quite some time to formulate the 1st law in this simple form, the credit for finding and formulating it goes to Robert Meyer (1814–1878), James Prescott Joule (1818–1889), and Hermann Helmholtz (1821–1894). Through careful measurements and analysis, they recognized that thermal energy, mechanical energy, and electrical energy can be transformed into each other, which implies that energy can be transferred by doing work, as in mechanics, and by heat transfer.

The 1st law is generally valid, no violation was ever observed. As knowledge of physics has developed, other forms of energy had to be included, such as radiative energy, nuclear energy, or the mass-energy equivalence of the theory of relativity, but there is no doubt today that energy is conserved under all circumstances.

We formulate the 1st law for the simple closed system of Figure 1, where all three possibilities to manipulate the system from the outside are indicated. For this system, the conservation law for energy reads

$$\frac{dE}{dt} = \dot{Q} - \dot{W}, \tag{11}$$

where E is the total energy of the system, $\dot{Q}$ is the total heat transfer rate in or out of the system, and $\dot{W} = \dot{W}_{piston} + \dot{W}_{propeller}$ is the total power—the work per unit time—exchanged with the surroundings. Energy is an extensive property, hence also heat and work scale with the size of the system. For instance, doubling the system size, doubles the energy, and requires twice the work and heat to observe the same changes of the system.

This equation states that the change of the system's energy in time (dE/dt) is equal to the energy transferred by heat and work per unit time $(\dot{Q} - \dot{W})$. The sign convention used is such that heat transferred *into* the system is positive, and work done *by* the system is positive.

13. Energy

There are many forms of energy that must be accounted for. For the context of the present discussion, the total energy E of the system is the sum of its kinetic energy E_{kin}, potential energy E_{pot}, and internal—or thermal—energy U,

$$E = U + E_{kin} + E_{pot}. \tag{12}$$

The kinetic energy is well-known from mechanics. For a homogeneous system of mass m and barycentric velocity $\mathcal{V}$, kinetic energy is given by

$$E_{kin} = \frac{m}{2}\mathcal{V}^2 . \tag{13}$$

For inhomogeneous states, where each mass element has its own velocity, the total kinetic energy of the system is obtained by integration of the specific kinetic energy e_{kin} over all mass elements $dm = \rho dV$;

$$e_{kin} = \frac{1}{2}\mathcal{V}^2 \quad \text{and} \quad E_{kin} = \int \rho e_{kin} dV = \int \frac{\rho}{2}\mathcal{V}^2 dV . \tag{14}$$

Also the potential energy in the gravitational field is well-known from mechanics. For a homogeneous system of mass m, potential energy is given by

$$E_{pot} = m g_n \bar{z} , \tag{15}$$

where $\bar{z}$ is the elevation of the system's center of mass over a reference height, and $g_n = 9.81\frac{m}{s^2}$ is the gravitational acceleration on Earth. For inhomogeneous states the total potential energy of the system is obtained by integration of the specific potential energy e_{pot} over all mass elements $dm = \rho dV$; we have

$$e_{pot} = g_n z \quad \text{and} \quad E_{pot} = \int \rho e_{pot} dV = \int \rho g_n z dV . \tag{16}$$

Even if a macroscopic element of matter is at rest, its atoms move about (in a gas or liquid) or vibrate (in a solid) fast, so that each atom has microscopic kinetic energy. The atoms are subject to interatomic forces, which contribute microscopic potential energies. Moreover, energy is associated with the atoms' internal quantum states. Since the microscopic energies cannot be observed macroscopically, one speaks of the *internal energy*, or *thermal energy*, of the material, denoted as U.

For inhomogeneous states the total internal energy of the system is obtained by integration of the specific internal energy u over all mass elements $dm = \rho dV$. For homogeneous and inhomogeneous systems we have

$$U = mu \quad \text{and} \quad U = \int \rho u dV . \tag{17}$$

14. Caloric Equation of State

Internal energy cannot be measured directly. The *caloric equation of state* relates the specific internal energy u to measurable quantities in equilibrium states, it is of the form $u = u(T, v)$, or $u = u(T, p)$. Recall that pressure, volume and temperature are related by the thermal equation of state, $p(v, T)$; therefore it suffices to know two properties in order to determine the others.

We note that internal energy summarizes all microscopic contributions to energy. Hence, a system, or a volume element within a system, will always have internal energy u, independent of whether the system is in (local) equilibrium states or in arbitrarily strong nonequilibrium states. Only in the former, however, does the caloric equation of state provide a link between energy and measurable properties.

The caloric equation of state must be determined by careful measurements, where the response of the system to heat or work supply is evaluated by means of the first law. For most materials the results cannot be easily expressed as equations, and are tabulated in property tables.

We consider a closed system heated slowly at constant volume (*isochoric* process), with homogeneous temperature T at all times. Then, the first law (27) reduces to (recall that $U = mu(T, v)$ and $m = const.$)

$$m\left(\frac{\partial u}{\partial T}\right)_v \frac{dT}{dt} = \dot{Q} . \tag{18}$$

Here, we use the standard notation of thermodynamics, where $\left(\frac{\partial u}{\partial T}\right)_v = \frac{\partial u(T,v)}{\partial T}$ denotes the partial derivative of internal energy with temperature at constant specific volume $v = V/m$. This derivative is known as the *specific heat* (or *specific heat capacity*) *at constant volume*,

$$c_v = \left(\frac{\partial u}{\partial T}\right)_v . \tag{19}$$

As defined here, based on SI units, the specific heat c_v is the amount of heat required to increase the temperature of 1kg of substance by 1K at constant volume. It can be measured by controlled heating of a fixed amount of substance in a fixed volume system, and measurement of the ensuing temperature difference; its SI unit is $\left[\frac{kJ}{kgK}\right]$.

In general, internal energy $u(T,v)$ is a function of a function of temperature and specific volume. For incompressible liquids and solids the specific volume is constant, $v = const$, and the internal energy is a function of temperature alone, $u(T)$. Interestingly, also for ideal gases the internal energy turns out to be a function of temperature alone, both experimentally and from theoretical considerations. For these materials the specific heat at constat volume depends only on temperature, $c_v(T) = \left(\frac{du}{dt}\right)$ and its integration gives the caloric equation of state as

$$u(T) = \int_{T_0}^{T} c_v(T') \, dT' + u_0 . \tag{20}$$

Only energy differences can be measured, where the first law is used to evaluate careful experiments. The choice of the energy constant $u_0 = u(T_0)$ fixes the energy scale. The actual value of this constant is relevant for the discussion of chemical reactions [8]. Note that proper mathematical notation requires to distinguish between the actual temperature T of the system, and the integration variable T'.

For materials in which the specific heat varies only slightly with temperature in the interval of interest, the specific heat can be approximated by a suitable constant average c_v, so that the caloric equation of state assumes the particularly simple linear form

$$u(T) = c_v(T - T_0) + u_0 . \tag{21}$$

15. Work and Power

Work, denoted by W, is the product of a force and the displacement of its point of application. Power, denoted by $\dot{W}$, is work done per unit time, that is the force times the velocity of its point of application. The total work for a process is the time integral of power over the duration $\Delta t = t_2 - t_1$ of the process,

$$W = \int_{t_1}^{t_2} \dot{W} dt . \tag{22}$$

For the closed system depicted in Figure 1 there are two contributions to work: *moving boundary work*, due to the motion of the piston, and *rotating shaft work*, which moves the propeller. Other forms of work, for example, spring work or electrical work could be added as well.

Work and power can be positive or negative. We follow the sign convention that work done *by* the system is positive and work done *to* the system is negative.

For systems with homogeneous pressure p, which might change with time as a process occurs (e.g., the piston moves), one finds the following expressions for moving boundary work with finite and infinitesimal displacement, and for power,

$$W_{12} = \int_{1}^{2} p dV \quad , \quad \delta W = p dV \quad , \quad \dot{W} = p \frac{dV}{dt} . \tag{23}$$

Moving boundary work depends on the process path, so that the work exchanged for an infinitesimal process step, $\delta W = pdV = \dot{W}dt$, is not an exact differential (see next section). Closed equilibrium systems are characterized by a single homogeneous pressure p, a single homogeneous temperature T, and the volume V. In quasi-static (or reversible) processes, the system passes through a series of equilibrium states which can be indicated in suitable diagrams, for example, the p-V-diagram.

In a closed system the propeller stirs the working fluid and creates inhomogeneous states. The power is related to the torque $\mathbf{T}$ and the revolutionary speed $\dot{n}$ (revolutions per unit time) as $\dot{W} = 2\pi\dot{n}\mathbf{T}$. Fluid friction transmits fluid motion (i.e., momentum and kinetic energy) from the fluid close to the propeller to the fluid further away. Due to the inherent inhomogeneity, stirring of a fluid in a closed system cannot be a quasi-static process, and is *always* irreversible.

In general, there might be several work interactions $\dot{W}_j$ of the system, then the total work for the system is the sum over all contributions; for example, for power

$$\dot{W} = \sum_j \dot{W}_j \, . \tag{24}$$

For reversible processes with additional work contributions, one has $\dot{W} = \sum_j x_j \frac{dY_j}{dt}$, where $\{x_j, Y_j\}$ are pairs of conjugate work variables, such as $\{p, V\}$.

Finally, we know from the science of mechanics that by using gears and levers, one can transfer energy as work from slow moving to fast moving systems and vice versa, and one can transmit work from high pressure to low pressure systems and vice versa. However, due to friction within the mechanical system used for transmission of work, some of the work may be lost. This experience is formulated in **Observation 4**.

16. Exact and Inexact Differentials

Above we have seen that work depends on the process path. In the language of mathematics this implies that the work for an infinitesimal step is not an exact differential, and that is why a Greek delta (δ) is used to denote the work for an infinitesimal change as δW. As will be seen in the next section, heat is path dependent as well.

State properties like pressure, temperature, volume and energy describe the momentary state of the system, or, for inhomogeneous states, the momentary state in the local volume element. State properties have exact differentials for which we write, for example, dE and dV. The energy change $E_2 - E_1 = \int_1^2 dE$ and the volume change $V_2 - V_1 = \int_1^2 dV$ are independent of the path connecting the states.

It is important to remember that work and heat, as path functions, do not describe states, but the processes that leads to changes of the state. Hence, for a process connecting two states $1, 2$ we write $W_{12} = \int_1^2 \delta W$, $Q_{12} = \int_1^2 \delta Q$, where W_{12} and Q_{12} are the energy transferred across the system boundaries by heat or work.

A state is characterized by state properties (pressure, temperature, etc.), it does not possess work or heat.

Quasi-static (reversible) processes go through well defined equilibrium states, so that the whole process path can be indicated in diagrams, for example, the p-V-diagram.

Nonequilibrium (irreversible) processes, for which typically the states are different in all volume elements, cannot be drawn into diagrams. Often irreversible processes connect homogeneous equilibrium states which can be indicated in the diagram. It is recommended to use dashed lines to indicate nonequilibrium processes that connect equilibrium states. As an example, Figure 6 shows a p-V-diagram of two processes, one reversible, one irreversible, between the same equilibrium states 1 and 2. We emphasize that the dashed line does not refer to actual states of the system. The corresponding work for the nonequilibrium process cannot be indicated as the area below the curve, since its computation requires the knowledge of the—inhomogeneous!—pressures at the piston surface at all times during the process.

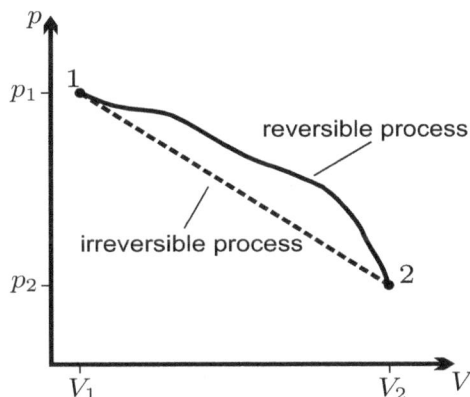

Figure 6. A reversible (quasi-static) and an irreversible (nonequilibrium) process between the equilibrium states 1 and 2.

17. Heat Transfer

Heat is the transfer of energy due to differences in temperature. Experience shows that for systems in thermal contact the direction of heat transfer is restricted, such that heat will always go from hot to cold by itself, but not vice versa. This experience is formulated in **Observation 5**.

This restriction of direction is an important difference to energy transfer by work between systems in mechanical contact, which is not restricted.

Since heat flows only in response to a temperature difference, a quasi-static (reversible) heat transfer process can only be realized in the limit of infinitesimal temperature differences between the system and the system boundary, and for infinitesimal temperature gradients within the system.

We use the following notation: $\dot{Q}$ denotes the heat transfer rate, that is the amount of energy transferred as heat per unit time. Heat depends on the process path, so that the heat exchanged for an infinitesimal process step, $\delta Q = \dot{Q}dt$, is not an exact differential. The total heat transfer for a process between states 1 and 2 is

$$Q_{12} = \int_1^2 \delta Q = \int_{t_1}^{t_2} \dot{Q}dt \ . \tag{25}$$

By the convention used, heat transferred into the system is positive, heat transferred out of the system is negative.

A process in which no heat transfer takes place, $\dot{Q} = 0$, is called *adiabatic process*.

In general, there might be several heat interactions $\dot{Q}_k$ of the system, then the total heat for the system is the sum over all contributions; for example, for the heating rate

$$\dot{Q} = \sum_k \dot{Q}_k \ . \tag{26}$$

For the discussion of the 2nd law we will consider the $\dot{Q}_k$ as heat crossing the system boundary at locations where the boundary has the temperature T_k.

18. 1st Law for Reversible Processes

The form (11) of the first law is valid for *all* closed systems. When only reversible processes occur within the system, so that the system is in equilibrium states at any time, the equation can be simplified as follows: From our discussion of equilibrium states we know that for reversible processes the system will be homogeneous and that all changes must be very slow, which implies very small velocities relative to the center of mass of the system. Therefore, kinetic energy, which is velocity squared, can be ignored, $E_{kin} = 0$. Stirring, which transfers energy by moving the fluid and friction, is irreversible, hence in a reversible process only moving boundary work can be transferred, where piston friction is absent. As long as the system location does not change, the potential energy does not change, and we can set $E_{pot} = 0$.

With all this, for reversible (quasi-static) processes the 1st law of thermodynamics reduces to

$$\frac{dU}{dt} = \dot{Q} - p\frac{dV}{dt} \quad \text{or} \quad U_2 - U_1 = Q_{12} - \int_1^2 p\,dV \,, \tag{27}$$

where the second form results from integration over the process duration.

19. Entropy and the Trend to Equilibrium

The original derivation of the 2nd law is due to Sadi Carnot (1796–1832) and Rudolf Clausius (1822–1888), where discussions of thermodynamic engines combined with **Observation 5** were used to deduce the 2nd law [1]. Even today, many textbooks present variants of their work [2–5]. As discussed in the introduction, we aim at introducing entropy without the use of heat engines, only using the 5 observations.

We briefly summarize our earlier statements on processes in closed systems: a closed system can be manipulated by exchange of work and heat with its surroundings only. In nonequilibrium—that is, irreversible—processes, when all manipulation stops, the system will undergo further changes until it reaches a final equilibrium state. This equilibrium state is stable, that is the system will not leave the equilibrium state spontaneously. It requires new action—exchange of work or heat with the surroundings—to change the state of the system. This paragraph is summarized in **Observations 1–3**.

The following nonequilibrium processes are well-known from experience, and will be used in the considerations below:

(a) Work can be transferred without restriction, by means of gears and levers. However, in transfer some work might be lost to friction (**Observation 4**).

(b) Heat goes from hot to cold. When two bodies at different temperatures are brought into thermal contact, heat will flow from the hotter to the colder body until both reach their common equilibrium temperature (**Observation 5**).

The process from an initial nonequilibrium state to the final equilibrium state requires some time. However, if the actions on the system (only work and heat!) are sufficiently slow, the system has enough time to adapt and will be in equilibrium states at all times. We speak of quasi-static—or reversible—processes. When the slow manipulation is stopped at any time, no further changes occur.

If a system is not manipulated, that is there is neither heat or work exchange between the systems and its surroundings, we speak of an *isolated system*. The behavior of isolated systems described above—a change occurs until a stable state is reached—can be described mathematically by an inequality. The final stable state must be a maximum (alternatively, a minimum) of a suitable property describing the system. For a meaningful description of systems of arbitrary size, the new property should scale with system size, that is it must be extensive.

We call this new extensive property *entropy*, denoted S, and write an inequality for the isolated system,

$$\frac{dS}{dt} = \dot{S}_{gen} \geq 0 \,. \tag{28}$$

$\dot{S}_{gen}$ is called the *entropy generation rate*. The entropy generation rate is positive in nonequilibrium ($\dot{S}_{gen} > 0$), and vanishes in equilibrium ($\dot{S}_{gen} = 0$). The new Equation (28) states that in an isolated system the entropy will grow in time ($\frac{dS}{dt} > 0$) until the stable equilibrium state is reached ($\frac{dS}{dt} = 0$). Non-zero entropy generation, $\dot{S}_{gen} > 0$, describes the irreversible process towards equilibrium, for example, through internal heat transfer and friction. There is no entropy generation in equilibrium, where entropy is constant. Since entropy only grows before the isolated system reaches its equilibrium state, the latter is a maximum of entropy.

While this equation describes the observed behavior in principle, it does not give a hint at what the newly introduced quantities S and $\dot{S}_{gen}$—entropy and entropy generation rate—are, or how they can be determined. Hence, an important part of the following discussion concerns the relation of entropy to measurable quantities, such as temperature, pressure, and specific volume. Moreover, it will be seen that entropy generation rate describes the irreversibility in, for example, heat transfer across finite temperature difference, or frictional flow.

The above postulation of an inequality is based on phenomenological arguments. The discussion of irreversible processes has shown that over time all isolated systems will evolve to a unique equilibrium state. The first law alone does not suffice to describe this behavior. Nonequilibrium processes aim to reach equilibrium, and the inequality is required to describe the clear direction in time.

As introduced here, entropy and the above rate equation describe irreversible processes, where initial nonequilibrium states evolve towards equilibrium. Not only is there no reason to restrict entropy to equilibrium, but rather, in this philosophy, it is essential to define entropy as a nonequilibrium property.

In the next sections we will extend the second law to non-isolated systems, identify entropy as a measurable property—at least in equilibrium states—and discuss entropy generation in irreversible processes.

20. Entropy Transfer

In non-isolated systems, which are manipulated by exchange of heat and work with their surroundings, we expect an exchange of entropy with the surroundings which must be added to the entropy inequality. We write

$$\frac{dS}{dt} = \dot{\Gamma} + \dot{S}_{gen}, \quad \text{with } \dot{S}_{gen} \geq 0, \tag{29}$$

where $\dot{\Gamma}$ is the *entropy transfer rate*. This equation states that the change of entropy in time (dS/dt) is due to transport of entropy over the system boundary ($\dot{\Gamma}$) and generation of entropy within the system boundaries ($\dot{S}_{gen}$). This form of the second law is valid for all processes in closed systems. The entropy generation rate is positive ($\dot{S}_{gen} > 0$) for irreversible processes, and it vanishes ($\dot{S}_{gen} = 0$) in equilibrium and for reversible processes, where the system is in equilibrium states at all times.

All real technical processes are somewhat irreversible, since friction and heat transfer cannot be avoided. Reversible processes are idealizations that can be used to study the principal behavior of processes, and best performance limits.

We apply **Observation 1**: Since a closed system can only be manipulated through the exchange of heat and work with the surroundings, the transfer of any other property, including the transfer of entropy, must be related to heat and work, and must vanish when heat and work vanish. Therefore the entropy transfer $\dot{\Gamma}$ can only be of the form

$$\dot{\Gamma} = \sum_k \beta_k \dot{Q}_k - \sum_j \gamma_j \dot{W}_j, \tag{30}$$

Recall that total heat and work transfer are the sum of many different contributions, $\dot{Q} = \sum_k \dot{Q}_k$ and $\dot{W} = \sum_j \dot{W}_j$. In the above formulation, the coefficients β_k and γ_j are used to distinguish heat and work transfer at different conditions at that part of the system boundary where the transfer ($\dot{Q}_k$ or $\dot{W}_j$) takes place. Since work and heat scale with the size of the system, and entropy is extensive, the coefficients β_k and γ_j must be intensive, that is, independent of system size.

At this point, the coefficients β_k, γ_j depend in an unknown manner on properties describing the state of the system and its interaction with the surroundings. While the relation between the entropy transfer rate $\dot{\Gamma}$ and the energy transfer rates $\dot{Q}_k, \dot{W}_j$ is not necessarily linear, the form (30) is chosen to clearly indicate that entropy transfer is zero when no energy is transferred, $\dot{\Gamma} = 0$ if $\dot{Q}_k = \dot{W}_j = 0$ (isolated system).

With this expression for entropy transfer, the 2nd law assumes the form

$$\frac{dS}{dt} + \sum_j \gamma_j \dot{W}_j - \sum_k \beta_k \dot{Q}_k = \dot{S}_{gen} \geq 0 , \tag{31}$$

This equation gives the mathematical formulation of the trend to equilibrium for a non-isolated closed system (exchange of heat and work, but not of mass). The next step is to identify entropy S and the coefficients β_k, γ_j in the entropy transfer rate $\dot{\Gamma}$ in terms of quantities we can measure or control.

21. Direction of Heat Transfer

A temperature reservoir is defined as a large body in equilibrium whose temperature does not change when heat is removed or added (this requires that the reservoir's thermal mass, $m_R c_R$, approaches infinity).

We consider heat transfer between two reservoirs of temperatures T_H and T_L, where T_H is the temperature of the hotter reservoir. The heat is transferred through a heat conductor (**HC**), which is the thermodynamic system to be evaluated. A pure steady state heat transfer problem is studied, where the conductor receives the heat flows $\dot{Q}_H$ and $\dot{Q}_L$, and exchanges no work with the surroundings, $\dot{W} = 0$.

The left part of Figure 7 shows a schematic of the heat transfer process. For steady state conditions no change over time is observed in the conductor, so that $\frac{dE}{dt} = \frac{dS}{dt} = 0$. We emphasize that for this process the heat conductor will be in a nonequilibrium state, for example, it could be a solid heat conductor with an imposed temperature gradient, or, possibly, a gas in a state of natural convection in the gravitational field. To proceed with the argument, it is not necessary to quantify energy and entropy of the conductor, since both do not change in steady state processes.

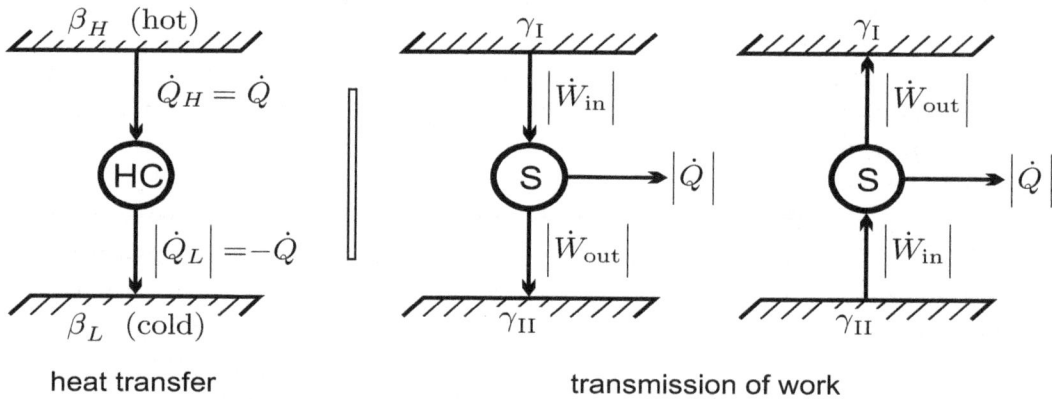

Figure 7. Heat transfer through a heat conductor HC (left) and transmission of work through a steady state system S (right).

For steady state, the first and second law (11, 31) applied to the heat conductor **HC** reduce to

$$\dot{Q}_H = -\dot{Q}_L = \dot{Q} \tag{32}$$
$$-\beta_H \dot{Q}_H - \beta_L \dot{Q}_L = \dot{S}_{gen} \geq 0 . \tag{33}$$

Here, β_H and β_L are the values of β at the hot and cold sides of the conductor, respectively. Combining both we have

$$(\beta_L - \beta_H) \dot{Q} = \dot{S}_{gen} \geq 0 . \tag{34}$$

We apply **Observation 5**: Since heat must go from hot to cold (from reservoir T_H to reservoir T_L), the heat must be positive, $\dot{Q} = \dot{Q}_H > 0$, which requires $(\beta_L - \beta_H) > 0$. Thus, the coefficient β must be smaller for the part of the system which is in contact with the hotter reservoir, $\beta_H < \beta_L$. This must be so irrespective of the values of *any* other properties at the system boundaries (L, H), that is, independent of the conductor material or its mass density, or any other material properties, and also for all possible values $\dot{Q}$ of the heat transferred. It follows, that β_L, β_H must depend on temperature of the respective reservoir *only*.

Moreover, β must be a decreasing function of reservoir temperature alone, if temperature of the hotter reservoir is defined to be higher.

22. Work Transfer and Friction Loss

For the discussion of the coefficient γ we turn our attention to the transmission of work. The right part of Figure 7 shows two "work reservoirs" characterized by different values γ_I, γ_{II} between which work is transmitted by a steady state system S.

We apply **Observation 4**. The direction of work transfer is not restricted: by means of gears and levers work can be transmitted from low to high force and vice versa, and from low to high velocity and vice versa. Therefore, transmission might occur from I to II, and as well from II to I. Accordingly, there is no obvious interpretation of the coefficient γ. Indeed, we will soon be able to remove the coefficient γ from the discussion.

According to the second part of **Observation 4**, friction might occur in the transmission. Thus, in the transmission process we expect some work being lost to frictional heating, therefore $|\dot{W}_{out}| \leq |\dot{W}_{in}|$. In order to keep the transmission system at constant temperature, some heat must be removed to a reservoir (typically the outside environment). Work and heat for both cases are indicated in the figure, the arrows indicate the direction of transfer.

The first law for both transmission processes reads (steady state, $\frac{dE}{dt} = 0$)

$$0 = -|\dot{Q}| - |\dot{W}_{out}| + |\dot{W}_{in}| \,, \tag{35}$$

where the signs account for the direction of the flows. Since work loss in transmission means $|\dot{W}_{out}| \leq |\dot{W}_{in}|$, this implies that heat must leave the system, $\dot{Q} = -|\dot{Q}| \leq 0$, as indicated in the figure.

Due to the different direction of work in the two processes considered, the second law (31) gives different conditions for both situations (steady state, $\frac{dS}{dt} = 0$),

$$-\gamma_I |\dot{W}_{in}| + \gamma_{II} |\dot{W}_{out}| + \beta |\dot{Q}| \geq 0 \,, \quad \gamma_I |\dot{W}_{out}| - \gamma_{II} |\dot{W}_{in}| + \beta |\dot{Q}| \geq 0 \,, \tag{36}$$

where, as we have seen in the previous section, β is a measure for the temperature of the reservoir that accepts the heat. Elimination of the heat $|\dot{Q}|$ between first and second laws gives two inequalities,

$$(\gamma_{II} - \beta) |\dot{W}_{out}| - (\gamma_I - \beta) |\dot{W}_{in}| \geq 0 \,, \quad (\gamma_I - \beta) |\dot{W}_{out}| - (\gamma_{II} - \beta) |\dot{W}_{in}| \geq 0, \tag{37}$$

or, after some reshuffling,

$$(\beta - \gamma_{II}) \frac{|\dot{W}_{out}|}{|\dot{W}_{in}|} \leq (\beta - \gamma_I) \quad , \quad (\beta - \gamma_I) \frac{|\dot{W}_{out}|}{|\dot{W}_{in}|} \leq (\beta - \gamma_{II}) \,. \tag{38}$$

Combining the two equations (38) gives the two inequalities

$$(\beta - \gamma_I) \left(\frac{|\dot{W}_{out}|}{|\dot{W}_{in}|} \right)^2 \leq (\beta - \gamma_I) \quad , \quad (\beta - \gamma_{II}) \left(\frac{|\dot{W}_{out}|}{|\dot{W}_{in}|} \right)^2 \leq (\beta - \gamma_{II}) \,. \tag{39}$$

From the these follows, since $0 \leq \frac{|\dot{W}_{out}|}{|\dot{W}_{in}|} \leq 1$, that $(\beta - \gamma)$ must be non-negative, $\beta - \gamma \geq 0$.

Both inequalities (38) must hold for arbitrary transmission systems, that is for all $0 \leq \frac{|\dot{W}_{out}|}{|\dot{W}_{in}|} \leq 1$, and all temperatures of the heat receiving reservoir, that is for all β. For a reversible transmission, where $\frac{|W_{out}|}{|W_{in}|} = 1$, both inequalities (38) can only hold simultaneously if $\gamma_I = \gamma_{II}$. Accordingly, $\gamma_I = \gamma_{II} = \gamma$ must be a constant, and $(\beta - \gamma) \geq 0$ for all β.

With γ as a constant, the entropy balance (31) becomes

$$\frac{dS}{dt} + \gamma \dot{W} - \sum \beta_k \dot{Q}_k = \dot{S}_{gen} \geq 0 \,, \tag{40}$$

where $\dot{W} = \sum_j \dot{W}_j$ is the net power for the system. The energy balance solved for power, $\dot{W} = \sum \dot{Q}_k - \frac{dE}{dt}$, allows us to eliminate work, so that the 2nd law becomes

$$\frac{d(S - \gamma E)}{dt} - \sum (\beta_k - \gamma) \dot{Q}_k = \dot{S}_{gen} \geq 0 \,. \tag{41}$$

23. Entropy and Thermodynamic Temperature

Without loss of generality, we can absorb the energy term γE into entropy, that is, we set

$$(S - \gamma E) \rightarrow S \,; \tag{42}$$

this is equivalent to setting $\gamma = 0$. Note that, since energy is conserved, any multiple of energy can be added to entropy without changing the principal features of the 2nd law; obviously, the most elegant formulation is the one where work does not appear.

Moreover, we have found that $(\beta - \gamma)$ is a non-negative monotonously decreasing function of temperature, and we *define* thermodynamic temperature as

$$T = \frac{1}{\beta - \gamma} > 0 \,. \tag{43}$$

Note that non-negativity of inverse temperature implies that temperature itself is strictly positive.

With this, we have the 2nd law in the form

$$\frac{dS}{dt} - \sum_k \frac{\dot{Q}_k}{T_{R,k}} = \dot{S}_{gen} \geq 0 \,. \tag{44}$$

The above line of arguments relied solely on the temperatures of the reservoirs with which the system exchanges heat; in order to emphasize this, we write the reservoir temperatures as $T_{R,k}$.

The form (44) is valid for *any* system **S**, in *any* state, that exchanges heat with reservoirs which have thermodynamic temperatures $T_{R,k}$. The entropy of the system is S, and it should be clear from the derivation that it is defined for any state, equilibrium, or nonequilibrium! Thermodynamic temperature must be positive to ensure dissipation of work due to friction. The discussion below will show that for systems in local thermal equilibrium, the reservoir temperature can be replaced by the system boundary temperature.

24. Entropy in Equilibrium: Gibbs Equation

Equilibrium entropy can be related to measurable quantities in a straightforward manner, so that it is measurable as well, albeit indirectly. We consider an equilibrium system undergoing a quasi-static processes, in contact with a heater at temperature T; for instance we might think of a carefully controlled resistance heater. Due to the equilibrium condition, the temperature of the system must be T as well (0th law!), and the entropy generation vanishes, $\dot{S}_{gen} = 0$. Then, Equation (44) for entropy becomes

$$\frac{dS_E}{dt} = \frac{\dot{Q}}{T} \,; \tag{45}$$

while for this case the the 1st law (45) reads

$$\frac{dU_E}{dt} = \dot{Q} - p\frac{dV}{dt} .$$

(46)

In both equations we added the index E to highlight the equilibrium state; p is the homogeneous pressure of the equilibrium state.

We are only interested in an infinitesimal step of the process, of duration dt. Eliminating the heat between the two laws, we find

$$TdS_E = dU_E + pdV .$$

(47)

This relation is known as the Gibbs equation, named after Josiah Willard Gibbs (1839–1903). The Gibbs equation is a differential relation between properties of the system and valid for *all* simple substances—in equilibrium states.

We note that T and p are intensive, and U, V and S are extensive properties. The specific entropy $s_E = S_E/m$ can be computed from the Gibbs equation for specific properties, which is obtained by division of (47) with the constant mass m. We ignore the subscript E for streamlined notation, so that the Gibbs equation for specific properties reads

$$Tds = du + pdv .$$

(48)

Solving the first law for reversible processes (27) for heat and comparing the result with the Gibbs equation we find, with $\dot{Q}dt = \delta Q$,

$$dS = \frac{1}{T}(dU + pdV) = \frac{1}{T}\delta Q .$$

(49)

We recall that heat is a path function, that is, δQ is an inexact differential, but entropy is a state property, that is, dS is an exact differential. In the language of mathematics, the inverse thermodynamic temperature $\frac{1}{T}$ serves as an integrating factor for δQ, such that $dS = \frac{1}{T}\delta Q$ becomes an exact differential.

It must be noted that one can always find an integrating factor for a differential form of two variables. Hence, it must be emphasized that thermodynamic temperature T remains an integrating factor if additional contributions to reversible work (conjugate work variables) are considered in the first law, which leads to the Gibbs equation in the form $TdS = dU - \sum_j x_j dY_j$, where $\{x_j, Y_j\}$ are pairs of conjugate work variables, such as $\{p, V\}$. For instance, this becomes clear in Caratheodory's axiomatic treatment of thermodynamics (for adiabatic processes) [9], which is briefly discussed in Appendix A.

From the above, we see that for reversible processes $\delta Q = TdS$. Accordingly, the total heat exchanged in a reversible process can be computed from temperature and entropy as the area below the process curve in the temperature-entropy diagram (T-S-diagram),

$$Q_{12} = \int_1^2 TdS .$$

(50)

This is analogue to the computation of the work in a reversible process as $W_{12} = \int_1^2 pdV$.

25. Measurability of Properties

Some properties are easy to measure, and thus quite intuitive, for example, pressure p, temperature T and specific volume v. Accordingly, the thermal equation of state, $p(T, v)$ can be measured with relative ease, for systems in equilibrium. Other properties cannot be measured directly, for instance internal energy u, which must be determined by means of applying the first law to a calorimeter, or equilibrium entropy s, which must be determined from other properties by integration of the Gibbs Equation (48).

The Gibbs equation gives a differential relation between properties for any simple substance. Its analysis with the tools of multivariable calculus shows that specific internal energy u, specific enthalpy $h = u + pv$, specific Helmholtz free energy $f = u - Ts$, and specific Gibbs free energy $g = h - Ts$ are potentials when considered as functions of particular variables. The evaluation of the potentials leads to a rich variety of relations between thermodynamic properties. In particular, these relate properties that are more difficult, or even impossible, to measure to those that are more easy to measure, and thus reduce the necessary measurements to determine data for all properties. The discussion of the thermodynamic potentials energy u, enthalpy h, Helmholtz free energy f and Gibbs free energy g, based on the Gibbs equation is one of the highlights of equilibrium thermodynamics [8,10]. Here, we refrain from a full discussion and only consider one important result in the next section.

To avoid misunderstanding, we point out that the following Sections 26–29 concern thermodynamic properties of systems in equilibrium states. We also stress that entropy and internal energy are system properties also in nonequilibrium states.

26. A Useful Relation

The Gibbs equation formulated for the Helmholtz free energy $f = u - Ts$ arises from a Legendre transform $T ds = d(Ts) - s dT$ in the Gibbs equation as

$$df = -s dT - p dv . \tag{51}$$

Hence, $f(T,v)$ is a thermodynamic potential [8,10], with

$$-s = \left(\frac{\partial f}{\partial T}\right)_v \quad , \quad -p = \left(\frac{\partial f}{\partial v}\right)_T \quad , \quad \left(\frac{\partial s}{\partial v}\right)_T = \left(\frac{\partial p}{\partial T}\right)_v . \tag{52}$$

The last equation is the Maxwell relation for this potential, it results from exchanging the order of derivatives, $\frac{\partial^2 f}{\partial v \partial T} = \frac{\partial^2 f}{\partial T \partial v}$. Remarkably, the Maxwell relation (52)$_3$ contains the expression $\left(\frac{\partial p}{\partial T}\right)_v$, which can be interpreted as the change of pressure p with temperature T in a process at constant volume v. Since p, T and v can be measured, this expression can be found experimentally. In fact, measurement of $\{p, T, v\}$ gives the thermal equation of state $p(T,v)$, and we can say that $\left(\frac{\partial p}{\partial T}\right)_v$ can be determined from the thermal equation of state. The other expression, $\left(\frac{\partial s}{\partial v}\right)_T$, cannot be measured by itself, since it contains entropy s, which cannot be measured directly. Hence, with the Maxwell relation the expression $\left(\frac{\partial s}{\partial v}\right)_T$ can be measured indirectly, through measurement of the thermal equation of state.

To proceed, we consider energy and entropy in the Gibbs Equation (48) as functions of temperature and volume, $u(T,v)$, $s(T,v)$. We take the partial derivative of the Gibbs equation with respect to v while keeping T constant, to find

$$\left(\frac{\partial u}{\partial v}\right)_T = T \left(\frac{\partial s}{\partial v}\right)_T - p . \tag{53}$$

With the Maxwell relation (52)$_3$ to replace the entropy derivative $\left(\frac{\partial s}{\partial v}\right)_T$ in (53)$_1$, we find an equation for the volume dependence of internal energy that is entirely determined by the thermal equation of state $p(T,v)$,

$$\left(\frac{\partial u}{\partial v}\right)_T = T \left(\frac{\partial p}{\partial T}\right)_v - p . \tag{54}$$

Since internal energy cannot be measured directly, the left hand side cannot be determined experimentally. The equation states that the volume dependence of the internal energy is known from measurement of the thermal equation of state.

27. Thermodynamic and Ideal Gas Temperatures

In the derivation of the 2nd law, thermodynamic temperature T appears as the factor of proportionality between the heat transfer rate $\dot{Q}$ and the entropy transfer rate $\dot{\Gamma}$. In previous sections we have seen that this definition of thermodynamic temperature stands in agreement with the direction of heat transfer: heat flows from hot (high T) to cold (low T) by itself. The heat flow aims at equilibrating the temperature within any isolated system that is left to itself, so that two systems in thermal equilibrium have the same thermodynamic temperature. Moreover, the discussion of internal friction showed that thermodynamic temperature must be positive.

While we have claimed agreement of thermodynamic temperature with the ideal gas temperature scale in Section 9, we have yet to give proof of this. To do so, we use (54) together with the experimental result stated in Section 14, that for an ideal gas the internal energy does *not* depend on volume, but only on temperature (see also Section 35). This implies, for the ideal gas,

$$0 = \left(\frac{\partial u}{\partial v}\right)_T = T\left(\frac{\partial p}{\partial T}\right)_v - p \quad \Rightarrow \quad p = T\left(\frac{\partial p}{\partial T}\right)_v . \tag{55}$$

Accordingly, ideal gas pressure must be a linear function of the thermodynamic temperature T,

$$p = \pi(v)\, T . \tag{56}$$

The volume dependency $\pi(v)$ must be measured, for example, in a piston cylinder system in contact with a temperature reservoir, so that the temperature is constant. Measurements show that pressure is inversely proportional to volume, so that

$$p = \pi_0 \frac{T}{v} , \tag{57}$$

with a constant π_0 that fixes the thermodynamic temperature scale.

The Kelvin temperature scale, named after William Thomson, Lord Kelvin (1824–1907), historically used the triple point of water (611 kPa, 0.01 °C) as reference. The triple point is the unique equilibrium state at which a substance can coexist in all three phases, solid, liquid and vapor. The Kelvin scales assigns the value of $T_{Tr} = 273.16$ K to this unique point, which can be reproduced with relative ease in laboratories, so that calibration of thermometers is consistent. With this choice, the constant π_0 is the specific gas constant $R = \bar{R}/M$, where $\bar{R} = 8.314$ kJ/(kmolK) is the universal gas constant, and M is the molecular mass with unit [kg/kmol] (e.g., $M_{He} = 4\frac{kg}{kmol}$ for helium, $M_{H_2O} = 18\frac{kg}{kmol}$ for water, $M_{air} = 29\frac{kg}{kmol}$ for air), so that, as already stated in Section 11,

$$pv = RT . \tag{58}$$

In 2018, the temperature scale became independent of the triple point of water. Instead, it is now set by fixing the Boltzmann constant k_B, which is the gas constant per particle, that is, $\bar{R} = k_B A_v$ where A_v is the Avogadro constant [24]. At the same time, other SI units were fredefined by assigning fixed values to physical constants, including the Avogadro constant, which defines the number of particles in one mole [25].

The historic development of the 2nd law relied on the use of Carnot engines, that is, a fully reversible engine between two reservoirs, and the Carnot process—which is a particular realization of a Carnot engine. Evaluation of the Carnot cycle for an ideal gas then shows the equivalence of ideal gas temperature and thermodynamic temperature. In the present treatment, all statements about engines are derived from the laws of thermodynamics, after they are found, based on simple experience.

The positivity of thermodynamic temperature implies positive ideal gas temperature and hence positive gas pressures. In Section 41, positive thermodynamic temperature is linked to mechanical stability. The ideal gas equation provides an intuitive example for this: A gas under negative pressure would collapse, hence be in an unstable state.

28. Measurement of Properties

Only few thermodynamic properties can be measured easily, namely temperature T, pressure p, and volume v. These are related by the thermal equation of state $p(T, v)$ which is therefore relatively easy to measure.

The specific heat $c_v = \left(\frac{\partial u}{\partial T} \right)_v$ can be determined from careful measurements. These calorimetric measurements employ the first law, where the change in temperature in response to the heat (or work) added to the system is measured.

Other important quantities, however, for example, u, h, f, g, s, cannot be measured directly. We briefly study how they can be related to measurable quantities, that is, T, p, v, and c_v by means of the Gibbs equation and the differential relations derived above.

We consider the measurement of internal energy. The differential of $u(T, v)$ is

$$du = c_v dT + \left(\frac{\partial u}{\partial v} \right)_T dv . \tag{59}$$

Therefore, the internal energy $u(T, v)$ can be determined by integration when c_v and $\left(\frac{\partial u}{\partial v} \right)_T$ are known from measurements. By (54) the term $\left(\frac{\partial u}{\partial v} \right)_T$ is known through measurement of the thermal equation of state, and we can write

$$du = c_v dT + \left[T \left(\frac{\partial p}{\partial T} \right)_v - p \right] dv . \tag{60}$$

Integration is performed from a reference state (T_0, v_0) to the actual state (T, v). Since internal energy is a point function, its differential is exact, and the integration is independent of the path chosen. The easiest integration is in two steps, first at constant volume v_0 from (T_0, v_0) to (T, v_0), then at constant temperature T from (T, v_0) to (T, v),

$$u(T, v) - u(T_0, v_0) = \int_{T_0}^{T} c_v(T', v_0) \, dT' + \int_{v_0}^{v} \left[T \left(\frac{\partial p}{\partial T} \right)_{v'} - p(T, v') \right] dv' . \tag{61}$$

Accordingly, in order to determine the internal energy $u(T, v)$ for all T and v it is sufficient to measure the thermal equation of state $p(T, v)$ for all (T, v) and the specific heat $c_v(T, v_0)$ for all temperatures T but *only one* volume v_0. For the ideal gas, the volume contribution vanishes, and the above reduces to (20).

The internal energy can only be determined apart from a reference value $u(T_0, v_0)$. As long as no chemical reactions occur, the energy constant $u(T_0, v_0)$ can be arbitrarily chosen.

Entropy $s(T, v)$ follows by integration of the Gibbs equation, for example, in the form, again with (54),

$$ds = \frac{1}{T} du + \frac{p}{T} dv = \frac{c_v}{T} dT + \frac{1}{T} \left(\left(\frac{\partial u}{\partial v} \right)_T + p \right) dv = \frac{c_v}{T} dT + \left(\frac{\partial p}{\partial T} \right)_v dv , \tag{62}$$

as

$$s(T, v) - s(T_0, v_0) = \int_{T_0}^{T} \frac{c_v(T', v_0)}{T'} dT' + \int_{v_0}^{v} \left(\frac{\partial p}{\partial T} \right)_{v'} dv' ; \tag{63}$$

Also entropy can be determined only apart from a reference value $s(T_0, v_0)$ which only plays a role when chemical reactions occur; the third law of thermodynamics fixes the scale properly.

After u and s are determined, enthalpy h, Helmholtz free energy f, and Gibbs free energy g simply follow by means of their definitions. Thus the measurement of *all* thermodynamic quantities requires only the measurement of the thermal equation of state $p(T, v)$ for all (T, v) and the measurement of the specific heat at constant volume $c_v(T, v_0)$ for all temperatures, but only one volume, for example, in a constant volume calorimeter. All other quantities follow from differential relations that are based on the Gibbs equation, and integration [8,10].

Above we have outlined the necessary measurements to fully determine all relevant thermodynamic properties for systems in equilibrium. We close this section by pointing out that all properties can be determined if just one of the thermodynamic potentials u, h, f, g is known [8,10]. Since all properties can be derived from the potential, the expression for the potential is sometimes called the *fundamental relation*.

29. Property Relations for Entropy

For incompressible liquids and solids, the specific volume is constant, hence $dv = 0$. The caloric equation of state (59) implies $du = c_v dT$ and the Gibbs equation reduces to $Tds = c_v dT$. For constant specific heat, $c_v = const.$, integration gives entropy as explicit function of temperature,

$$s(T) = c_v \ln \frac{T}{T_0} + s_0 , \tag{64}$$

where s_0 is the entropy at the reference temperature T_0.

For the ideal gas, where $\left(\frac{\partial p}{\partial T}\right)_p = \frac{R}{v}$ and the specific heat depends on T only, entropy assumes the familiar form

$$s(T, v) = \int_{T_0}^{T} \frac{c_v(T')}{T'} dT' + R \ln \frac{v}{v_0} + s_0 , \tag{65}$$

For a gas with constant specific heat, the integration can be performed to give

$$s(T, v) = c_v \ln \frac{T}{T_0} + R \ln \frac{v}{v_0} + s_0 . \tag{66}$$

Of course, a substance behaves as an ideal gas only for sufficiently low pressures or sufficiently hight temperatures, so that these relations have a limited range of applicability. In particular for low temperatures, the ideal gas law and the equations above are not valid.

30. Local Thermodynamic Equilibrium

In the previous sections, we considered homogeneous systems that undergo equilibrium processes, and discussed how to determine thermodynamic properties of systems in equilibrium states. To generalize for processes in inhomogeneous systems, we now consider the system as a compound of sufficiently small subsystems. The key assumption is that each of the subsystems is in *local thermodynamic equilibrium*, so that it can be characterized by the same state properties as a macroscopic equilibrium system. To simplify the proceedings somewhat, we consider numbered subsystems of finite size, and summation.

The exact argument for evaluation of local thermodynamic equilibrium considers infinitesimal cells dV, partial differential equations, and, to arrive at the equations for systems, integration. This detailed approach, known as *Linear Irreversible Thermodynamics* (LIT), is presented in Appendix C. The simplified argument below avoids the use of partial differential equations, and aims only on the equations for systems, hence this might be the preferred approach for use in an early undergraduate course [8].

Figure 8 indicates the splitting into subsystems, and highlights a subsystem i inside the system and a subsystem k at the system boundary. Temperature and pressure in the subsystems are given by T_i, p_i and T_k, p_k, respectively. Generally, temperature and pressure are inhomogeneous, that is adjacent subsystems have different temperatures and pressures. Accordingly, each subsystem interacts with its neighborhood through heat and work transfer as indicated by the arrows. Heat and work exchanged with the surroundings of the system are indicated as $\dot{Q}_k$ and $\dot{W}_k$.

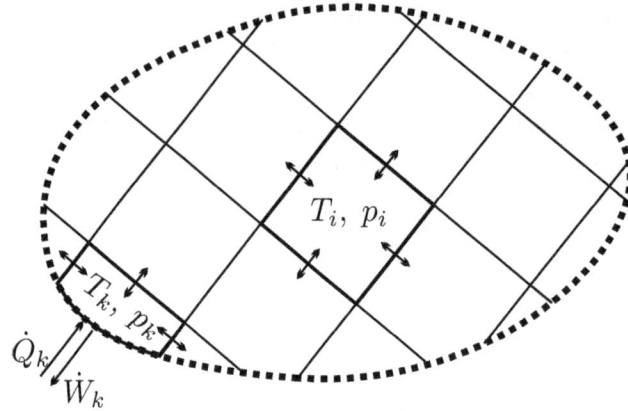

Figure 8. Non-equilibrium system split into small equilibrium subsystems. Arrows indicate work and heat exchange between neighboring elements, and the surroundings.

Internal energy and entropy in a subsystem i are denoted as E_i and S_i, and, since both are extensive, the corresponding quantities for the complete system are obtained by summation over all subsystems, $E = \sum_i E_i$, $S = \sum_i S_i$. Note that in the limit of infinitesimal subsystems the sums become integrals, as in Section 5. The balances of energy and entropy for a subsystem i read

$$\frac{dE_i}{dt} = \dot{Q}_i - \dot{W}_i \quad , \quad \frac{dS_i}{dt} = \frac{\dot{Q}_i}{T_i} + \dot{S}_{gen,i} \ , \tag{67}$$

where $\dot{Q}_i = \sum_j \dot{Q}_{i,j}$ is the net heat exchange, and $\dot{W}_i = \sum_j \dot{W}_{i,j}$ is the net work exchange for the subsystem. Here, the summation over j indicates the exchange of heat and work with the neighboring cells, such that, for example, $\dot{Q}_{i,j}$ is the heat that i receives from the neighboring cell j.

The boundary cells of temperatures T_k are either adiabatically isolated to the outside, or they exchange heat with external systems (reservoirs) of temperature $T_{R,k}$, which, in fact, are the temperatures that appear in the 2nd law in the form of Equation (44). For systems in local thermodynamic equilibrium, temperature differences at boundaries, such at those between a gas and a container wall, are typically extremely small. Hence, temperature jumps at boundaries are usually ignored, so that $T_{R,k} = T_k$, and we will proceed with this assumption. Appendix C.4 provides a more detailed discussion of temperature jumps and velocity slip within the context of Linear Irreversible Thermodynamics.

To obtain first and second law for the compound system, we have to sum the corresponding laws for the subsystems, which gives

$$\frac{dE}{dt} = \dot{Q} - \dot{W} \quad \text{with} \quad \dot{Q} = \sum_k \dot{Q}_k \ , \ \dot{W} = \sum_k \dot{W}_k \tag{68}$$

and

$$\frac{dS}{dt} = \sum_k \frac{\dot{Q}_k}{T_k} + \dot{S}_{gen} \quad \text{with} \quad \dot{S}_{gen} \geq 0 \ . \tag{69}$$

In the above, $\dot{Q}_k$ is the heat transferred over a system boundary which has temperature T_k. This subtle change from Equation (44), which has the reservoir temperatures, results from ignoring temperature jumps at boundaries. As will be explained next, the summation over k concerns only heat and work exchange with the surroundings.

Since energy is conserved, the internal exchange of heat and work between subsystems cancels in the conservation law for energy (68). For instance, in the exchange between neighboring subsystems i and j, $Q_{i,j}$ is the heat that i receives from j and $W_{i,j}$ is the work that i does on j. Moreover, $Q_{j,i}$ is the heat that j receives from i and $W_{j,i}$ is the work that j does on i. Since energy is conserved, no energy is added or lost in transfer between i and j, that is $Q_{i,j} = -Q_{j,i}$ and $W_{i,j} = -W_{j,i}$. Accordingly, the sums vanish, $Q_{i,j} + Q_{j,i} = 0$ and $W_{i,j} + W_{j,i} = 0$. Extension of the argument shows that the internal exchange of heat and work between subsystems adds up to zero, so that only exchange with the surroundings, indicated by subscript k, appears in (68).

Entropy, however, is not conserved, but may be produced. Exchange of heat and work between subsystems, if irreversible, will contribute to the entropy generation rate $\dot{S}_{gen}$. Thus, the total entropy generation rate $\dot{S}_{gen}$ of the compound system is the sum of the entropy generation rates in the subsystems $\dot{S}_{gen,i}$ plus additional terms related to the energy transfer between subsystems, $\dot{S}_{gen} = \sum_i \dot{S}_{gen,i} + \sum_{i,j} \dot{S}_{gen,i,j} > 0$. In simple substances, this internal entropy generation occurs due to internal heat flow and internal friction.

Strictly speaking, the small temperature differences for heat transfer between system and boundary, $T_k - T_{R,k}$ contribute to entropy generation as well. In typical applications, the temperature differences and the associated entropy generation are so small that both can be ignored.

We repeat that entropy generation is strictly positive, $\dot{S}_{gen} > 0$, in irreversible processes, and is zero, $\dot{S}_{gen} = 0$, in reversible processes.

To fully quantify entropy generation, that is to compute its actual value, requires the detailed local computation of all processes inside the system from the conservation laws and the second law as partial differential equations—this is outlined in Appendix C.

The above derivation of the second law Equation (69) relies on the assumption that the equilibrium property relations for entropy are valid locally also for nonequilibrium systems. This *local equilibrium hypothesis*—equilibrium in a subsystem, but not in the compound system—works well for most systems in technical thermodynamics. It should be noted that the assumption breaks down for extremely strong nonequilibrium.

31. Heat Transfer between Reservoirs

In this and the following sections we proceed by considering simple processes with the 1st and 2nd law in the form (68) and (69) found for systems in local thermodynamic equilibrium. These examples are shown to highlight the contents of the 2nd law. For instructors who prefer to *postulate* the 2nd law, these would be the examples used to show the agreement with daily experience.

We begin with the basic heat transfer process between two reservoirs of thermodynamic temperatures T_H and T_L, where $T_H > T_L$ is the temperature of the hotter system, see Figure 9. The heat is transferred through a heat conductor, which is the thermodynamic system to be evaluated. One will expect a temperature gradient in the conductor, that is the conductor is not in a homogeneous equilibrium state, but in a nonequilibrium state. A pure heat transfer problem is studied, where the conductor receives the heat flows $\dot{Q}_H$ and $\dot{Q}_L$, and exchanges no work with the surroundings, $\dot{W} = 0$. The first and second law (68) and (69) applied to the heat conductor read

$$\frac{dU}{dt} = \dot{Q}_L + \dot{Q}_H \quad , \quad \frac{dS}{dt} - \frac{\dot{Q}_L}{T_L} - \frac{\dot{Q}_H}{T_H} = \dot{S}_{gen} \geq 0 \,. \tag{70}$$

For steady state conditions no changes over time are observed in the conductor, so that $\frac{dU}{dt} = \frac{dS}{dt} = 0$. The first law shows that the heat flows must be equal in absolute value, but opposite in sign,

$$\dot{Q}_H = -\dot{Q}_L = \dot{Q} \,. \tag{71}$$

With this, the second law reduces to the inequality

$$\dot{Q}\left(\frac{1}{T_L} - \frac{1}{T_H}\right) = \dot{S}_{gen} \geq 0 \,. \tag{72}$$

With the thermodynamic temperature $T_H > T_L > 0$, the bracket is positive. According to Figure 9 the proper direction of heat transfer in accordance to Clausius' statement that *heat will go from hot to cold by itself, but not vice versa* (**Observation 5**) is for $\dot{Q}_H = -\dot{Q}_L = \dot{Q} > 0$.

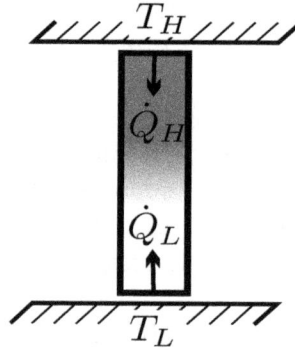

Figure 9. Heat transfer between two reservoirs at T_1 and T_2. In steady state the heat conductor does not accumulate energy, therefore $\dot{Q}_L = -\dot{Q}_H$.

Equation (72) shows that heat transfer over finite temperature differences creates entropy inside the heat conductor. In the steady state case considered here, the entropy created is leaving the system with the outgoing entropy flow $\frac{\dot{Q}_L}{T_L}$ which is larger than the incoming entropy flow $\frac{\dot{Q}_H}{T_H}$. Figure 10 gives an illustration of the allowed process, where heat goes from hot to cold, and the forbidden process, where heat would go from cold to hot by itself.

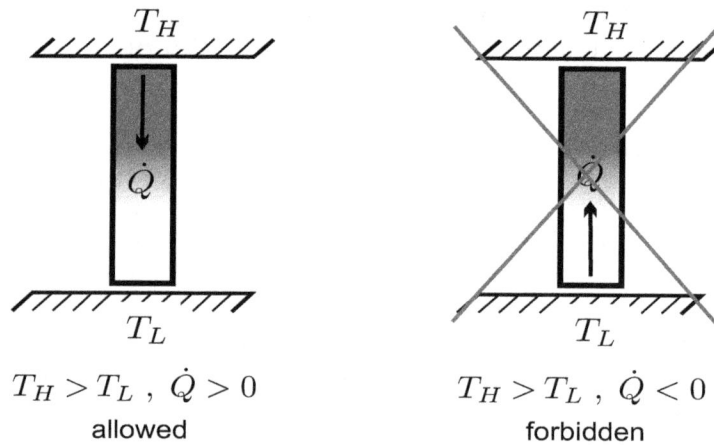

Figure 10. Heat transfer between two reservoirs with $T_H > T_L$. Heat must go from warm to cold.

32. Newton's Law of Cooling

The inequality (72) requires that $\dot{Q}$ has the same sign as $\left(\frac{1}{T_L} - \frac{1}{T_H} \right)$, a requirement that is fulfilled for a heat transfer rate

$$\dot{Q} = \alpha A \left(T_H - T_L \right) \tag{73}$$

with a positive heat transfer coefficient $\alpha > 0$, and the heat exchange surface area A. This relation is known as Newton's law of cooling, and is often used in heat transfer problems. The values of the positive coefficient α must be found from the detailed configuration and conditions in the heat transfer system. The surface area A appears due to the intuitive expectation that enlarging the transfer area leads to a proportional increase in the amount of heat transferred.

Heat transfer was introduced as energy transfer due to temperature difference with heat going from hot to cold. Newton laws of cooling states that as a result of the temperature difference one will observe a response, namely the heat flow.

The procedure to deduce Newton's law of cooling can be described as follows: The entropy generation rate (72) is interpreted as the product of a thermodynamic force—here, the difference of inverse temperatures $\left(\frac{1}{T_L} - \frac{1}{T_H} \right)$—and a corresponding flux—here, the heat flow $\dot{Q}$. To ensure positivity of the entropy generation rate, the flux must be proportional to the force, with a positive factor

αA that must be measured. This is the strategy of Linear Irreversible Thermodynamics, which can be used for all force-flux pairs, see Appendix C. A thermodynamic force is any deviation from the equilibrium state, here the temperature difference, which will vanish in equilibrium. A thermodynamic flux is a response to the force that drives a process towards equilibrium, here the heat flux.

With Newton's law of cooling it is easy to see that heat transfer over finite temperature differences is an irreversible process. Indeed, the second law (72) gives with (73)

$$\dot{S}_{gen} = \dot{Q}\left(\frac{1}{T_L} - \frac{1}{T_H}\right) = \alpha A \frac{(T_H - T_L)^2}{T_L T_H} > 0 \,. \tag{74}$$

Equation (74) quantifies the entropy generation rate in steady state heat transfer, which, for fixed heat transfer rate $\dot{Q}$, grows with the difference of inverse temperatures. Only when the temperature difference is infinitesimal, that is, $T_H = T_L + dT$, entropy generation can be ignored, and heat transfer can be considered as a reversible process. This can be seen as follows: For infinitesimal dT the entropy generation rate becomes $\dot{S}_{gen} = \alpha A \left(\frac{dT}{T_L}\right)^2$ and heat becomes $\dot{Q} = \alpha A dT$. This implies that entropy generation vanishes with the temperature difference, $\dot{S}_{gen} = 0$ $(dT \to 0)$. In this case, to have a finite amount of heat $\dot{Q}$ transferred, the heat exchange area A must go to infinity.

33. 0th Law and 2nd Law

Above we considered heat transfer between reservoirs, but the conclusion is valid for heat conduction between arbitrary systems: As long as the systems are in thermal contact through heat conductors, and their temperatures are different, there will be heat transfer between the systems. Only when the temperatures of the systems are equal, heat transfer will cease. This is the case of thermal equilibrium, where no change in time occurs anymore. This includes that the temperature of an isolated body in thermal equilibrium will be homogeneous, where equilibration occurs through heat transfer within the system; for the formal argument see Section 40 below.

The 0th law states that in equilibrium systems in thermal contact assume the same temperature. Thus, the 0th law of thermodynamics might appear as a special case of the 2nd law. It stands in its own right, however: Not only does it define temperature as a measurable quantity, but it also states the homogeneity of temperature in equilibrium, which is required to identify the Gibbs equation in Section 24.

34. Internal Friction

When coffee, or any other liquid, is stirred, it will spin a while after the spoon is removed. The motion will slow down because of internal friction, and finally the coffee will be at rest in the cup. We show that the 2nd law describes this well-known behavior, which is observed in all viscous fluids.

With the fluid in motion, all fluid elements have different velocity vectors, that is, the system is not in a homogeneous equilibrium state. We have to account for the kinetic energy of the swirling, which must be computed by summation, that is, integration, of the local kinetic energies $\frac{\rho(\vec{r})}{2} v(\vec{r})^2$ in all volume elements; see Figure 11. The 1st and 2nd law (68) and (69) now read

$$\frac{d(U + E_{kin})}{dt} = \dot{Q} - \dot{W} \,, \quad \frac{dS}{dt} - \sum \frac{\dot{Q}_k}{T_k} = \dot{S}_{gen} \geq 0 \,. \tag{75}$$

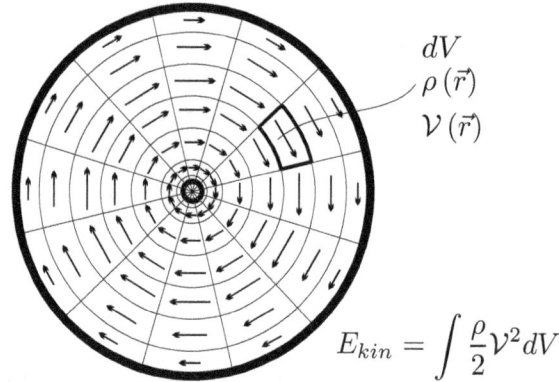

Figure 11. The kinetic energy E_{kin} of a stirred fluid is the sum of the kinetic energies in all volume elements. Friction with the container wall, and within the fluid, will slow down the fluid until it comes to rest in the final equilibrium state.

We assume adiabatic systems ($\dot{Q} = 0$) without any work exchange ($\dot{W} = 0$, this implies constant volume), so that

$$\frac{d\left(U + E_{kin}\right)}{dt} = 0 \quad , \quad \frac{dS}{dt} = \dot{S}_{gen} \geq 0 . \tag{76}$$

For simplicity we ignore local temperature differences within the stirred substance, and use the Gibbs Equation (47) so that

$$\frac{dS}{dt} = \left(\frac{\partial S}{\partial U}\right)_V \frac{dU}{dt} = \frac{1}{T}\frac{dU}{dt} = -\frac{1}{T}\frac{dE_{kin}}{dt} = \dot{S}_{gen} \geq 0 . \tag{77}$$

Since entropy generation and inverse thermodynamic temperature are non-negative, this implies

$$\frac{dE_{kin}}{dt} \leqslant 0 . \tag{78}$$

Hence, the kinetic energy $E_{kin} = \int \frac{\rho}{2}\mathcal{V}^2 dV$ decreases over time, and will be zero in equilibrium, where the stirred substance comes to rest, $\mathcal{V} = 0$.

Here we notice, again, that the sign of thermodynamic temperature is intimately linked to friction: $T > 0$ ensures that friction dissipates kinetic energy. The total entropy generation in this process is

$$S_{gen} = \int \dot{S}_{gen} dt = -\int \frac{1}{T}\frac{dE_{kin}}{dt} dt . \tag{79}$$

35. Uncontrolled Expansion of a Gas

Our next example concerns the uncontrolled expansion of an ideal gas. We consider an ideal gas in a container which is divided by a membrane, see Figure 12.

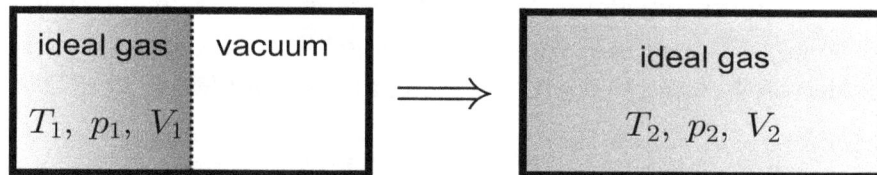

Figure 12. Irreversible adiabatic expansion of an ideal gas.

Initially the gas is contained in one part of the container at $\{T_1, p_1, V_1\}$, while the other part is evacuated. The membrane is destroyed, and the gas expands to fill the the container. The fast motion of the gas is slowed down by internal friction, and in the final homogeneous equilibrium state $\{T_2, p_2, V_2\}$

the gas is at rest and distributed over the total volume of the container. We have no control over the flow after the membrane is destroyed: this is an irreversible process.

The container is adiabatically enclosed to the exterior, and, since its walls are rigid, no work is transmitted to the exterior. Thus, the first law for closed systems (11) reduces to

$$\frac{d\left(U + E_{kin} + E_{pot}\right)}{dt} = 0 , \tag{80}$$

or, after integration,

$$U_2 + E_{kin,2} + E_{pot,2} = U_1 + E_{kin,1} + E_{pot,1} . \tag{81}$$

Since the gas it at rest initially and in the end, $E_{kin,1} = E_{kin,2} = 0$, and since potential energy has not changed $E_{pot,1} = E_{pot,2}$, the above reduces to $U_2 = U_1$. Note, however, that *during* the process $E_{kin} > 0$ and $U < U_1$.

With $U = mu$, and $m = const.$, the specific internal energy remains unchanged,

$$u\left(T_1, v_1\right) = u\left(T_2, v_2\right) . \tag{82}$$

Measurements for ideal gases show that $T_1 = T_2$, that is the initial and final temperatures of the gas are the same. With this, the previous condition becomes

$$u\left(T_1, v_1\right) = u\left(T_1, v_2\right) , \tag{83}$$

which can only hold if the internal energy of the ideal gas does not depend on volume. This experiment verifies that the internal energy of the ideal gas is independent of volume, and depends only on temperature, $u = u\left(T\right)$. In Section 27 we already used this result to show the equivalence of thermodynamic and ideal gas temperature scales.

The second law for this adiabatic process simply reads

$$\frac{dS}{dt} = \dot{S}_{gen} \geq 0 . \tag{84}$$

Integration over the process duration yields

$$S_2 - S_1 = \int_{t_1}^{t_2} \dot{S}_{gen} dt = S_{gen} \geq 0 . \tag{85}$$

The total change of entropy follows from the ideal gas entropy (66), with $T_1 = T_2$, as

$$S_2 - S_1 = m\left(s_2 - s_1\right) = mR\ln\frac{V_2}{V_1} = mR\ln\frac{v_2}{v_1} > 0 . \tag{86}$$

Since in this process the temperature of the ideal gas remains unchanged, the growth of entropy is only attributed to the growth in volume: by filling the larger volume V_2, the gas assumes a state of larger entropy. Since the container is adiabatic, there is no transfer of entropy over the boundary (i.e., $\sum \frac{\dot{Q}_k}{T_k} = 0$), and all entropy generated stays within the system, $S_{gen} = S_2 - S_1$.

In this computation, energy and entropy change, and the entropy generated can be determined from the initial and final equilibrium states. However, the process is irreversible, with states of strong nonequilibrium along the way. The rate equations for 1st and 2nd law are valid throughout the process, but do not suffice to determine values for energy and entropy at all moments in time, since they do not allow to resolve the inhomogeneity of the intermediate states. A detailed prediction of the process requires a local theory, such as the Navier-Stokes-Fourier equations of Linear Irreversible Thermodynamics (see Appendix C), or the Boltzmann equation of Kinetic Gas Theory (see Appendix D).

Values for system energy and entropy can be obtained from the local description through integration over the system.

36. Irreversibility and Work Loss

The thermodynamic laws for closed systems that exchange heat with an arbitrary number of reservoirs read

$$\frac{d\left(U + E_{kin}\right)}{dt} = \dot{Q}_0 + \sum \dot{Q}_k - \dot{W} \quad , \quad \frac{dS}{dt} - \frac{\dot{Q}_0}{T_0} - \sum \frac{\dot{Q}_k}{T_k} = \dot{S}_{gen} \geq 0 \ , \tag{87}$$

where the heat exchange $\dot{Q}_0$ with a reservoir at T_0 is highlighted. Most thermodynamic engines utilize the environment as heat source or sink, and in this case $\dot{Q}_0$ should be considered as the heat exchanged with the environment. Note that the environment is freely available, and no cost is associated with removing heat from, or rejecting heat into, the environment. Moreover the environment is large compared to any system interacting with it, hence its temperature T_0 remains constant.

Elimination of $\dot{Q}_0$ between the two laws and solving for work gives

$$\dot{W} = \sum \left(1 - \frac{T_0}{T_k}\right) \dot{Q}_k - \frac{d\left(U + E_{kin} - T_0 S\right)}{dt} - T_0 \dot{S}_{gen} \ . \tag{88}$$

This equation applies to arbitrary processes in closed systems. The generation of entropy in irreversible processes reduces the work output of work producing devices (where $\dot{W} > 0$, for example, heat engines) and increases the work requirement of work consuming devices (where $\dot{W} < 0$, for example, heat pumps and refrigerators). We note the appearance of the Carnot factor $\left(1 - \frac{T_0}{T_k}\right)$ multiplying the heating rates $\dot{Q}_k$.

The amount of work lost to irreversible processes is

$$\dot{W}_{\text{loss}} = T_0 \dot{S}_{gen} \geq 0 \ , \tag{89}$$

sometimes it is denoted as the *irreversibility*. It is an important engineering task to identify and quantify the irreversible work losses, and to reduce them by redesigning the system, or use of alternative processes. Loss analysis is an important part of technical thermodynamics that is featured in modern textbooks [8,23].

Entropy generation is due to friction, heat transfer over finite temperature differences, mixing, chemical reactions, and so forth. Full quantification of the entropy generation in nonequilibrium processes requires resolution of the process at all times, that is, solution of local transport equations (Navier-Stokes-Fourier, etc.). Nevertheless, already at the system level, loss analysis can lead to deeper insight into possibilities for process improvement.

The discussion of heat engines, refrigerators and heat pumps operating at steady state between two reservoirs, in particular of Carnot engines, is an important element of thermodynamic analysis. With the 1st and 2nd law in place, this is a special case of the above Equation (88), as is discussed next.

37. Heat Engines, Refrigerators, Heat Pumps

Engines operating at steady state between two reservoirs, one of them the environment, are shown in Figure 13, namely a heat engine (HE), a refrigerator (R), and a heat pump (HP). We discuss these engines with the combined 1st and 2nd law (88).

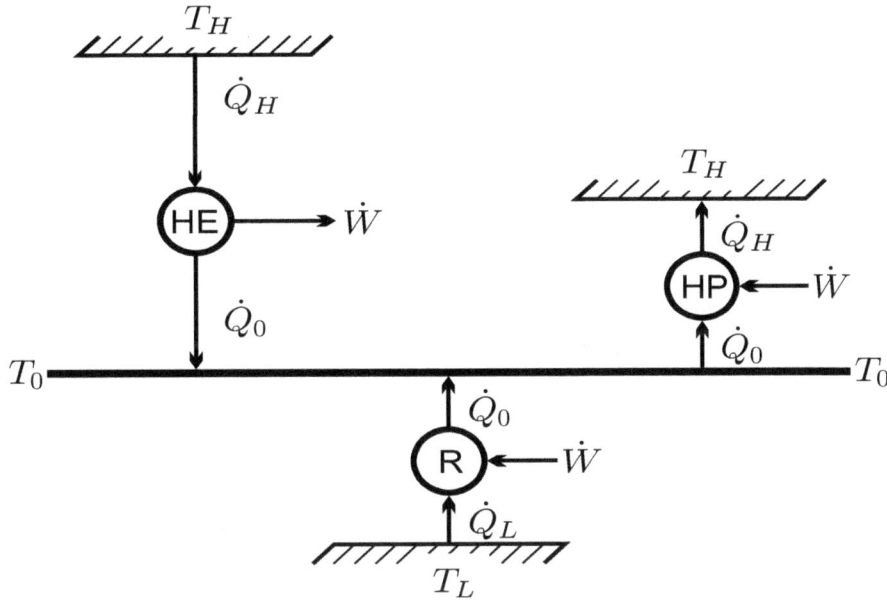

Figure 13. Heat engine (HE), refrigerator (R), and heat pump (HP) in contact with the environment at T_0.

For a heat engine that receives the heat $\dot{Q}_H$ from a hot reservoir at $T_H > T_0$, and rejects heat into the environment at T_0, the actual power produced is

$$\dot{W} = \left(1 - \frac{T_0}{T_H}\right) \dot{Q}_H - T_0 \dot{S}_{gen} > 0 \,, \tag{90}$$

where $T_0 \dot{S}_{gen} \geq 0$ is work loss to irreversibilities. A Carnot engine, named after Sadi Carnot (1796–1832), is a fully reversible engine, that is, it has no irreversible losses and provides the power

$$\dot{W}_{rev} = \left(1 - \frac{T_0}{T_H}\right) \dot{Q}_H > 0 \,. \tag{91}$$

The thermal efficiency is defined as the ratio of work produced (the gain) over heat input (the expense), $\eta = \dot{W}/\dot{Q}_H$, and we find the thermal efficiency of the Carnot engine as

$$\eta_C = \frac{\dot{W}_{rev}}{\dot{Q}_H} = 1 - \frac{T_0}{T_H} < 1 \,. \tag{92}$$

For a refrigerator that removes the heat $\dot{Q}_L$ from a cold space at $T_L < T_0$ and rejects heat into the environment at T_0, the power requirement is

$$\dot{W} = \left(1 - \frac{T_0}{T_L}\right) \dot{Q}_L - T_0 \dot{S}_{gen} < 0 \,, \tag{93}$$

where $T_0 \dot{S}_{gen} \geq 0$ is the extra work required to overcome irreversibilities. A fully reversible refrigerator, that is, a Carnot refrigerator, requires the power

$$\dot{W}_{rev} = \left(1 - \frac{T_0}{T_L}\right) \dot{Q}_L < 0 \,. \tag{94}$$

The coefficient of performance of a refrigerator is defined as the ratio of heat drawn from the cold (the gain) over work input (the expense), $\mathrm{COP}_R = \dot{Q}_L / |\dot{W}|$, and we find the coefficient of performance of the Carnot refrigerator as

$$\text{COP}_{R,C} = \frac{\dot{Q}_L}{|\dot{W}_{\text{rev}}|} = \frac{1}{\frac{T_0}{T_L} - 1} . \tag{95}$$

For a heat pump that supplies the heat $\dot{Q}_H$ to a warm space at $T_H > T_0$ and draws heat from the environment at T_0, the power requirement is

$$\dot{W} = \left(1 - \frac{T_0}{T_H}\right) \dot{Q}_H - T_0 \dot{S}_{gen} < 0 , \tag{96}$$

where $T_0 \dot{S}_{gen} \geq 0$ is the extra work required to overcome irreversibilities. A fully reversible heat pump, that is, a Carnot heat pump, requires the power

$$\dot{W}_{\text{rev}} = \left(1 - \frac{T_0}{T_H}\right) \dot{Q}_H < 0 . \tag{97}$$

The coefficient of performance of a heat pump is defined as the ratio between the heat provided (the gain) and the work input (the expense), $\text{COP}_{\text{HP}} = |\dot{Q}_H| / |\dot{W}|$, and we find the coefficient of performance of the Carnot heat pump as

$$\text{COP}_{\text{HP},C} = \frac{|\dot{Q}_H|}{|\dot{W}_{\text{rev}}|} = \frac{1}{1 - \frac{T_0}{T_H}} > 1 . \tag{98}$$

Due to irreversible losses, real engines always have lower efficiencies or coefficients of performance than the (fully reversible) Carnot engines operating between the same temperatures. While Carnot efficiencies cannot be reached—*all* real engines are irreversible—they serve as important benchmarks.

It must be emphasized that in the present approach the discussion of engines comes well after the 2nd law of thermodynamics is established. The classical Carnot-Clausius argument for finding the 2nd law puts engines front and center, which requires long discussion of processes and cycles before the 2nd law can be finally presented [2–5]. The present approach, where the 2nd law is derived from simple **Observations 1–5**, requires far less background and allows to introduce the 2nd law soon after the 1st law, so that both laws are available for the evaluation of all processes, cycles and engines right away, see also References [6–8]. Note that the above analysis of heat engine, refrigerator and heat pump does not require any details on the processes inside the engines.

38. Kelvin-Planck and Clausius Statements

The temperature difference between reservoirs provides the thermodynamic force that induces heat flux from hot to cold, due to the desire to equilibrate temperature. A heat engine converts a portion of this heat flux into work. That is, the nonequilbrium between reservoirs is essential for the process.

Not all heat received from the hot reservoir can be converted into work, some heat must be rejected to a colder reservoir. The Kelvin-Planck formulation of the second law states this as: *No steady state thermodynamic process is possible in which heat is completely converted into work.*

This statement is a direct consequence of the 1st and 2nd law. For a steady state process with just one heat exchange the laws require

$$-\frac{\dot{Q}_H}{T_H} = -\frac{\dot{W}}{T_H} = \dot{S}_{gen} \geq 0 , \tag{99}$$

hence heat and work must both be negative. Figure 14 shows the forbidden process, and also the—allowed—inverse process, the complete conversion of work into heat through friction. A typical example for the latter are resistance heaters in which electrical work is converted to heat through electric resistance (heat pump with $COP_{\text{RH}} = |\dot{Q}_H| / |\dot{W}| = 1$).

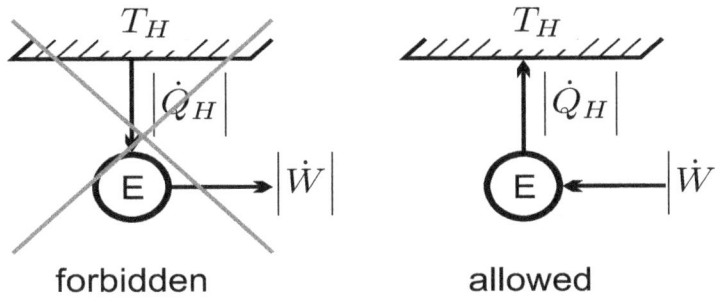

Figure 14. Heat cannot be completely converted into work, but work can be completely converted to heat.

Clausius' statement of the second law says that *heat will not go from cold to warm by itself.* This statement was used explicitly in our development of the 2nd law (**Observation 5**). Note that the two words "by itself" are important here—a heat pump system can transfer heat from cold to warm, but work must be supplied, so the heat transfer is not "by itself."

It is straightforward to show that both statements are equivalent [2,8].

39. Finding Equilibrium States

With the laws of thermodynamics now in place, we can use them to learn more about the equilibrium states that will be observed. For an isolated system, 1st and 2nd law reduce to

$$\frac{dE}{dt} = 0 \quad , \quad \frac{dS}{dt} = \dot{S}_{gen} \geq 0 , \tag{100}$$

with a constant mass m in the system. Since no work is exchanged, the system volume V must be constant as well. According to the second law, the state of the system will change (with $\dot{S}_{gen} > 0$) until the entropy has reached a maximum (when $\dot{S}_{gen} = 0$), where the process is restricted by having the initial mass, momentum and energy enclosed in the system. Starting with an arbitrary inhomogeneous initial state, the approach to equilibrium is a reorganization of the local properties of the system towards the final equilibrium state, which we will determine now for a single phase system.

Total mass, energy and entropy are obtained by integration over the full system,

$$m = \int_V \rho \, dV \; , \quad E = \int_V \rho \left(u + \frac{1}{2} \mathcal{V}^2 + g_n z \right) dV \; , \quad S = \int_V \rho s \, dV . \tag{101}$$

Here, ρ, T, $\mathcal{V}$, and $u\,(\rho, T)$, $s\,(\rho, T)$ are the *local* values of the thermodynamic properties, that is, $\rho = \rho\,(\vec{r})$, $T = T\,(\vec{r})$, $\mathcal{V}\,(\vec{r})$ and so forth, where $\vec{r}$ is the location in the volume V of the system, see Section 5. The gravitational acceleration g_n should not be confused with the Gibbs free energy g.

Often we are interested in systems that are globally at rest, where the overall momentum $\vec{M}$ vanishes, but we might consider also systems moving with a constant velocity $\vec{v}$, so that $\vec{M} = m\,\vec{v}$. Since all elements of the system have their own velocity $\vec{\mathcal{V}}\,(\vec{r})$, we find the total momentum by summing over the system,

$$\vec{M} = m\,\vec{v} = \int_V \rho \vec{\mathcal{V}} \, dV ; \tag{102}$$

here $\vec{\mathcal{V}}\,(\vec{r})$ is the local velocity vector with absolute value $\mathcal{V} = \sqrt{\vec{\mathcal{V}} \cdot \vec{\mathcal{V}}}$. As long as no forces act on the system, its momentum will be constant; total momentum vanishes for a system at rest in the observer frame, $\vec{M} = 0$.

The equilibrium state is the maximum of entropy S under the constraints of given mass m, momentum $\vec{M}$, and energy E. The best way to account for the constraints is the use of Lagrange multipliers Λ_ρ, $\vec{\Lambda}_M$ and Λ_E to incorporate the constraints and maximize not S but

$$\Phi = \int_V \rho s dV - \Lambda_\rho \left(\int_V \rho dV - m \right) - \vec{\Lambda}_M \cdot \left(\int_V \rho \vec{\mathcal{V}} dV - \vec{M} \right)$$
$$- \Lambda_E \left(\int_V \rho \left(u + \frac{1}{2}\mathcal{V}^2 + g_n z \right) dV - E \right) . \quad (103)$$

The maximization of Φ will give the local values of the thermodynamic equilibrium properties $\{\rho, T, \mathcal{V}\}$ in terms of the Lagrange multipliers, which then must be determined from the given values of $\{m, \vec{M}, E\}$.

For the solution of this problem, we employ the methods of variational calculus. For compact notation, we introduce the abbreviations $y(\vec{r}) = \{\rho, \vec{\mathcal{V}}, T\}$ and

$$X(y) = \rho \left[s - \Lambda_\rho - \vec{\Lambda}_M \cdot \vec{\mathcal{V}} - \Lambda_E \left(u + \frac{1}{2}\mathcal{V}^2 + g_n z \right) \right] . \quad (104)$$

The equilibrium state maximizes the integral $\int_V X(y) dV$. We denote the equilibrium state as $y_E(\vec{r})$ and consider small variations $\delta y(\vec{r})$ from the equilibrium state, so that $y = y_E + \delta y$. By means of a Taylor series we find

$$\int_V X(y) dV = \int_V \left[X(y_E) + \left(\frac{\partial X}{\partial y} \right)_{|E} \delta y + \frac{1}{2} \left(\frac{\partial^2 X}{\partial y \partial y} \right)_{|E} \delta y \delta y \right] dV \quad (105)$$

Since y_E maximizes the integral, the other terms on the right hand side must be negative for arbitrary values of the variation $\delta y(\vec{r})$, which implies $\left(\frac{\partial X}{\partial y} \right)_{|E} = 0$ and negative definiteness of the matrix $\left(\frac{\partial^2 X}{\partial y \partial y} \right)_{|E}$. We proceed with the evaluation of the first condition, and leave the discussion of the second conditions for Section 41.

The conditions for equilibrium read

$$\left(\frac{\partial X}{\partial y} \right)_{|E} = \left\{ \frac{\partial X}{\partial \rho}, \frac{\partial X}{\partial \vec{\mathcal{V}}}, \frac{\partial X}{\partial T} \right\}_{|E} = 0 . \quad (106)$$

These and the following relations are valid for the equilibrium values, $T_{|E}$, $\rho_{|E}$, $u_{|E}$, $\mathcal{V}_{|E}$, $s_{|E}$ and so forth. For better readability, the subscripts $|E$ referring to the equilibrium state are not shown. Evaluation yields

$$\frac{\partial X}{\partial \rho} = \left[s - \Lambda_\rho - \vec{\Lambda}_M \cdot \vec{\mathcal{V}} - \Lambda_E \left(u + \frac{1}{2}\mathcal{V}^2 + g_n z \right) \right] + \rho \left[\left(\frac{\partial s}{\partial \rho} \right)_T - \Lambda_E \left(\frac{\partial u}{\partial \rho} \right)_T \right] = 0 , \quad (107)$$

$$\frac{\partial X}{\partial \vec{\mathcal{V}}} = \rho \left[-\vec{\Lambda}_M - \Lambda_E \vec{\mathcal{V}} \right] = 0 . \quad (108)$$

$$\frac{\partial X}{\partial T} = \rho \left[\left(\frac{\partial s}{\partial T} \right)_\rho - \Lambda_E \left(\frac{\partial u}{\partial T} \right)_\rho \right] = 0 . \quad (109)$$

40. Homogeneous Equilibrium States

We proceed with evaluating the three conditions (107)–(109) to find the stable equilibrium state. For convenience, we begin with the middle Equation (108), which gives homogeneous velocity in equilibrium,

$$\vec{\mathcal{V}} = -\frac{\vec{\Lambda}_M}{\Lambda_E} \, . \tag{110}$$

For the case of a system at rest, where

$$0 = \vec{M} = \int_V \rho \vec{\mathcal{V}} \, dV = -\frac{\vec{\Lambda}_M}{\Lambda_E} \int_V \rho \, dV = -\frac{\vec{\Lambda}_M}{\Lambda_E} m \, , \tag{111}$$

this implies that in equilibrium all local elements are at rest, $\vec{\mathcal{V}} = \vec{\Lambda}_M = \vec{M} = 0$.

With the Gibbs equation, the last condition (109) becomes

$$\left(\frac{\partial s}{\partial T}\right)_\rho - \Lambda_E \left(\frac{\partial u}{\partial T}\right)_\rho = \frac{1}{T}\left(\frac{\partial u}{\partial T}\right)_\rho - \Lambda_E \left(\frac{\partial u}{\partial T}\right)_\rho = 0 \, . \tag{112}$$

It follows that in equilibrium the temperature is homogeneous, and equal to the inverse Lagrange multiplier,

$$T = \frac{1}{\Lambda_E} \, . \tag{113}$$

To evaluate the first condition, (107), we insert the above results for Λ_E, $\vec{\Lambda}_M$, $\vec{\mathcal{V}}$ and use again the Gibbs equation, which gives $\left(\frac{\partial s}{\partial \rho}\right)_T - \frac{1}{T}\left(\frac{\partial u}{\partial \rho}\right)_T = -\frac{p}{T\rho^2}$. After some reordering, we find

$$g + g_n z = u - Ts + \frac{p}{\rho} + g_n z = -T\Lambda_\rho \, , \tag{114}$$

where g is the Gibbs free energy, and g_n is gravitational acceleration. Thus, the sum of specific Gibbs free energy and specific potential energy, $g + g_n z$, is homogeneous in equilibrium, while density and pressure might be inhomogeneous.

In summary, maximizing entropy in the isolated system yields that the system is fully at rest, $\mathcal{V} = 0$, has homogeneous temperature, $T = 1/\Lambda_E$, and, in the gravitational field, has inhomogeneous density and pressure, given implicitly by $g(T,\rho) + g_n z = -T\Lambda_\rho$. The Lagrange multipliers must be determined from the constraints (101), their values depend on the size and geometry of the system.

Equilibrium states of systems in contact with the environment, for example, with prescribed boundary temperatures or pressures are determined similarly, this includes systems with several phases [8].

To gain insight into the influence of potential energy, we evaluate (114) for ideal gases and incompressible fluids. For an ideal gas, the Gibbs free energy is $g(\rho, T) = h(T) - T\left(\int_{T_0}^{T} \frac{c_v(T')}{T'} dT' - R \ln \frac{\rho}{\rho_0} + s_0\right)$. Using this in (114) and solving for density gives the barometric formula,

$$\rho = \rho^0 \exp\left[-\frac{g_n z}{RT}\right] \, , \tag{115}$$

where $\rho^0 = \rho_0 \exp\left[-\frac{\Lambda_\rho}{R} - \frac{h(T) - T\left(\int_{T_0}^{T} \frac{c_v(T')}{T'} dT' + s_0\right)}{RT}\right]$ is the density at reference height $z = 0$. The ideal

gas law gives the corresponding expression for pressure as

$$p = p^0 \exp\left[-\frac{g_n z}{RT}\right] \, , \tag{116}$$

where $p^0 = \rho^0 RT$ is the pressure at $z = 0$. This is the well known barometric formula [8]. Pressure variation of gases in the gravitational field is relatively small. In technical systems with a size of few metres, the variation is so small that one can assume homogeneous pressures. When climbing a mountain (say), the pressure variation is important, of course.

For incompressible fluids, $\rho = const.$, and internal energy and entropy depend only on temperature, so that the Gibbs free energy is $g(T, p) = u(T) + \frac{p}{\rho} - Ts(T)$. Using this in (114) and solving for pressure gives the hydrostatic pressure formula,

$$p = p^0 - \rho g_n z , \tag{117}$$

where $p^0 = \rho T \left[s(T) - u(T) / T - \Lambda_\rho \right]$ is the pressure at reference height $z = 0$. This is the well known hydrostatic pressure [8].

41. Thermodynamic Stability

The equilibrium state must be stable, which means that, indeed, it must be the maximum of the integral Φ (103). This requires that the second variation of Φ must be negative. In our case, where the integrand X depends only on y, this requires negative eigenvalues for the matrix of second derivatives $\partial^2 X / \partial y^2$ at the location of the maximum. With the help of the Gibbs equation, the second derivatives can be written as

$$
\begin{aligned}
\frac{\partial X}{\partial \rho^2} &= \left[\frac{1}{T} - \Lambda_E \right] \left[2 \left(\frac{\partial u}{\partial \rho} \right)_T + \rho \left(\frac{\partial^2 u}{\partial \rho^2} \right)_T \right] - \frac{1}{\rho T} \left(\frac{\partial p}{\partial \rho} \right)_T , \\
\frac{\partial^2 X}{\partial \vec{\mathcal{V}}^2} &= -\rho \Lambda_E , \\
\frac{\partial^2 X}{\partial T^2} &= \rho \left[\frac{1}{T} - \Lambda_E \right] \left(\frac{\partial^2 u}{\partial T^2} \right)_\rho - \frac{\rho}{T^2} \left(\frac{\partial u}{\partial T} \right)_\rho , \\
\frac{\partial^2 X}{\partial \rho \partial T} &= \frac{\partial^2 X}{\partial T \partial \rho} = \left[\frac{1}{T} - \Lambda_E \right] \left(\frac{\partial u}{\partial T} \right)_\rho , \\
\frac{\partial X}{\partial \rho \partial \vec{\mathcal{V}}} &= \frac{\partial X}{\partial \vec{\mathcal{V}} \partial \rho} = -\vec{\Lambda}_M - \Lambda_E \vec{\mathcal{V}} , \\
\frac{\partial X}{\partial T \partial \vec{\mathcal{V}}} &= \frac{\partial X}{\partial \vec{\mathcal{V}} \partial T} = 0 .
\end{aligned}
\tag{118}
$$

These must now be evaluated at the equilibrium state, $T = 1/\Lambda_E$ and $\vec{\mathcal{V}} = -\vec{\Lambda}_M/\Lambda_E$. All mixed derivatives vanish in equilibrium, hence the requirement reduces to negative values for the diagonal elements. With the definitions of isothermal compressibility $\kappa_T = -\frac{1}{v} \left(\frac{\partial v}{\partial p} \right)_T$ and the specific heat at constant volume $c_v = \left(\frac{\partial u}{\partial T} \right)_v$, the resulting conditions read

$$
\begin{aligned}
\frac{\partial X}{\partial \rho^2} \Big|_{eq} &= -\frac{1}{\rho T} \left(\frac{\partial p}{\partial \rho} \right)_T = -\frac{1}{\rho^2 T \kappa_T} < 0 , \\
\frac{\partial^2 X}{\partial \vec{\mathcal{V}}^2} \Big|_{eq} &= -\frac{\rho}{T} < 0 , \\
\frac{\partial^2 X}{\partial T^2} \Big|_{eq} &= -\frac{\rho}{T^2} \left(\frac{\partial u}{\partial T} \right)_\rho = -\frac{\rho}{T^2} c_v < 0 ;
\end{aligned}
\tag{119}
$$

With the mass density being positive, thermodynamic stability thus requires that isothermal compressibility, specific heat, and thermodynamic temperature are positive,

$$\kappa_T > 0 , \quad c_v > 0 , \quad T > 0 . \tag{120}$$

These conditions imply that the volume decreases when pressure is increased isothermally, and that the temperature rises when heat is added to the system. Once more we see that positivity of thermodynamic temperature guarantees dissipation of kinetic energy. The stability conditions (120) imply that $s\,(u, v)$ is a concave function, as shown in Appendix B.

42. Open Systems

So far we have considered only closed systems, which do not exchange mass. We shall now extend the discussion to systems which exchange mass with their surroundings. Figure 15 shows a generic *open system* with two inflows and two outflows. The amount of mass exchanged per unit time, the mass transfer rate or *mass flow*, is denoted by $\dot{m}$. The system also exchanges propeller and piston work, $\dot{W} = \dot{W}_{propeller} + \dot{W}_{piston}$, and heat, $\dot{Q} = \dot{Q}_1 + \dot{Q}_2$, with its surroundings, just as a closed system does.

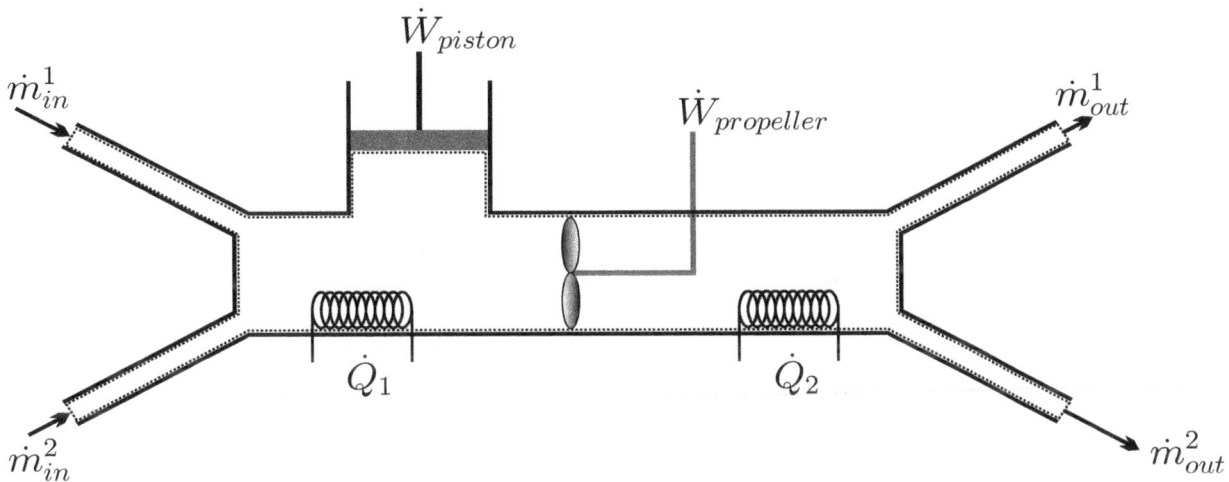

Figure 15. Open system with two inflows, two outflows and two heat sources. The dotted line indicates the system boundary.

States in open systems are normally inhomogeneous. One might think of a mass element entering the system of Figure 15 on the left. As the element travels through the system, it constantly changes its state: When it passes the heating, its temperature changes, when it passes the propeller its pressure and temperature change, and so on. Thus, at each location within the system one finds different properties. As discussed earlier, an inhomogeneous system is in a nonequilibrium state. In an open system the nonequilibrium is maintained through the exchange of mass, heat and work with the surroundings.

The equations for open systems presented below are those typically found in thermodynamics textbooks. These rely on a number of simplifying assumptions, such as use of average values for properties and ignore of viscous stresses across inlets and exits, which will only be mentioned, but will not be discussed in detail [8].

43. Balances of Mass, Energy and Entropy

Mass cannot be created or destroyed, that is mass is conserved. Chemical reactions change the composition of the material, but not its mass. In a closed system, the law of mass conservation states that the total mass m in the system does not change in time, that is, it simply reads $\frac{dm}{dt} = 0$. In an open system, where mass enters or leaves over the system boundaries, the conservation law for mass states that the change of mass in time is due to inflow—which increases mass—and outflow—which decreases system mass. In approximation, the mass flow can be written as

$$\dot{m} = \rho \mathcal{V} A \,, \tag{121}$$

where ρ and $\mathcal{V}$ are averages of mass density and velocity over the cross section A of the in/outflow boundary.

The rate of change in mass is due to the net difference of mass flows entering and leaving the system,

$$\frac{dm}{dt} = \sum_{in} \dot{m}_i - \sum_{out} \dot{m}_e \; . \tag{122}$$

The indices (i, e) indicate the values of the properties at the location where the respective flows cross the system boundary, that is their average values at the inlets and outlets, respectively.

The total energy E of an open system changes due to exchange of heat and work, and due to *convective energy transport* $\dot{E}$, that is energy carried in or out by the mass crossing the system boundary,

$$\dot{E} = \dot{m}e = \dot{m} \left(u + \frac{1}{2}\mathcal{V}^2 + g_n z \right) \; . \tag{123}$$

The power required to push mass over the system boundary is the force required times the velocity. The force is the local pressure (irreversible stresses are ignored) times the cross section, thus the associated *flow work* is

$$\dot{W}_{flow} = -\,(pA)\,\mathcal{V} = -\frac{p}{\rho}\dot{m} \; . \tag{124}$$

Work is done to the system when mass is entering, then $\dot{W}_{flow}$ must be negative. The system does work to push leaving mass out, then $\dot{W}_{flow}$ must be positive. Accordingly, flow work points opposite to mass flow, which is ensured by the minus sign in the equation.

Thus, in comparison to the energy balance for closed systems, the energy balance for the general open system of Figure 15 has additional contributions to account for convective energy transport and flow work, in condensed notation

$$\frac{dE}{dt} = \dot{Q} - \dot{W} + \sum_{in/out} \dot{E} - \sum_{in/out} \dot{W}^{flow} \; , \tag{125}$$

where the sums have to be taken over all flows crossing the system boundary.

Explicitly accounting for mass flows leaving and entering the system, and with enthalpy $h = u + p/\rho$, the 1st law—the balance of energy—for the general open system becomes

$$\frac{dE}{dt} = \sum_{in} \dot{m}_i \left(h + \frac{1}{2}\mathcal{V}^2 + g_n z \right)_i - \sum_{out} \dot{m}_e \left(h + \frac{1}{2}\mathcal{V}^2 + g_n z \right)_e + \dot{Q} - \dot{W} \; . \tag{126}$$

This equation states that the energy E within the system changes due to convective inflow and outflow, as well as due to heat transfer and work. Note that the flow energy includes the flow work required to move the mass across the boundaries (124). Moreover, there can be several contributions to work and heat transfer, that is $\dot{W} = \sum_j \dot{W}_j$ and $\dot{Q} = \sum_k \dot{Q}_k$.

All mass that is entering or leaving the system carries entropy. The entropy flow associated with a mass flow is simply $\dot{S} = \dot{m}s$, where s is the average specific entropy at the respective inlet or outlet. Adding the appropriate terms for inflow and outflow to the 2nd law (69) for closed systems yields the 2nd law—the balance of entropy—for open systems as

$$\frac{dS}{dt} = \sum_{in} \dot{m}_i s_i - \sum_{out} \dot{m}_e s_e + \sum_k \frac{\dot{Q}_k}{T_k} + \dot{S}_{gen} \quad \text{with} \quad \dot{S}_{gen} \geq 0 \; . \tag{127}$$

This equation states that the entropy S within the system changes due to convective inflow and outflow, as well as due to entropy transfer caused by heat transfer $(\dot{Q}_k/T_k)$ and entropy generation due to irreversible processes inside the system $(\dot{S}_{gen} \geq 0)$. If all processes within the system are reversible, the entropy generation vanishes $(\dot{S}_{gen} = 0)$. Recall that $\dot{Q}_k$ is the heat that crosses the system boundary where the boundary temperature is T_k.

44. One Inlet, One Exit Systems

A case of particular interest are steady-state systems with only one inlet and one exit, as sketched in Figure 16, for which the mass balance reduces to

$$\dot{m}_{in} = \dot{m}_{out} = \dot{m} .$$ (128)

There is just one constant mass flow $\dot{m}$ flowing through each cross section of the system.

For a steady state system, the corresponding forms for energy and entropy balance are

$$\dot{m} \left[h_2 - h_1 + \frac{1}{2} \left(V_2^2 - V_1^2 \right) + g_n (z_2 - z_1) \right] = \dot{Q}_{12} - \dot{W}_{12} ,$$ (129)

$$\dot{m} (s_2 - s_1) - \sum_k \frac{\dot{Q}_k}{T_k} = \dot{S}_{gen} \geq 0 .$$ (130)

It is instructive to study the equations for an infinitesimal step within the system, that is, for infinitesimal system length dx, where the differences reduce to differentials,

$$\dot{m} \left(dh + \frac{1}{2} dV^2 + g_n dz \right) = \delta \dot{Q} - \delta \dot{W} ,$$ (131)

$$\dot{m} ds - \frac{\delta \dot{Q}}{T} = \delta \dot{S}_{gen} .$$ (132)

Heat and power exchanged, and entropy generated, in an infinitesimal step along the system are process dependent, and as always we write $(\delta \dot{Q}, \delta \dot{W}, \delta \dot{S})$ to indicate that these quantities are not exact differentials. Use of the Gibbs equation in the form $T ds = dh - v dp$ allows to eliminate dh and $\delta \dot{Q}$ between the two equations to give an expression for power,

$$\delta \dot{W} = -\dot{m} \left(v dp + \frac{1}{2} dV^2 + g_n dz \right) - T \delta \dot{S}_{gen} .$$ (133)

The total power for the finite system follows from integration over the length of the system as

$$\dot{W}_{12} = -\dot{m} \int_1^2 \left(v dp + \frac{1}{2} dV^2 + g_n dz \right) - \int_1^2 T \delta \dot{S}_{gen} .$$ (134)

Since $T \delta \dot{S}_{gen} \geq 0$, we see—again—that irreversibilities reduce the power output of a power producing device (where $\dot{W}_{12} > 0$), and increase the power demand of a power consuming device (where $\dot{W}_{12} < 0$). Efficient energy conversion requires to reduce irreversibilities as much as possible.

When we consider (133) for a flow without work, we find Bernoulli's equation (Daniel Bernoulli, 1700–1782) for pipe flows as

$$v dp + \frac{1}{2} dV^2 + g_n dz = -\frac{1}{\dot{m}} T \delta \dot{S}_{gen} .$$ (135)

The Bernoulli equation is probably easier to recognize in its integrated form for incompressible fluids (where $v = \frac{1}{\rho} = const.$),

$$g_n (H_2 - H_1) = \frac{p_2 - p_1}{\rho} + \frac{1}{2} \left(V_2^2 - V_1^2 \right) + g_n (z_2 - z_1) = -\frac{1}{\dot{m}} \int_1^2 T \delta \dot{S}_{gen} .$$ (136)

Here, $H = \frac{p}{\rho g_n} + \frac{1}{2} \frac{V^2}{g_n} + z$ denotes hydraulic head. The right hand side describes loss of hydraulic head due to irreversible processes, in particular friction.

Finally, for reversible processes—where $\delta\dot{S}_{gen} = 0$ in (134)—we find the *reversible steady-flow work*

$$\dot{W}_{12}^{rev} = -\dot{m}\int_1^2\left(vdp + \frac{1}{2}dV^2 + g_ndz\right) . \tag{137}$$

For flows at relatively low velocities and without significant change of level the above relation can be simplified to

$$\dot{W}_{12}^{rev} = \dot{m}w_{12}^{rev} = -\dot{m}\int_1^2 vdp . \tag{138}$$

In a p-v-diagram, the *specific reversible flow work* w_{12}^{rev} is the area to the left of the process curve.

The heat exchanged in a reversible process in a steady-state, one inlet, one exit system follows from the integration of the second law (132) with $\delta\dot{S}_{gen} = 0$ as

$$\dot{Q}_{12}^{rev} = \int_1^2 \delta\dot{Q}^{rev} = \dot{m}\int_1^2 Tds = \dot{m}q_{12}^{rev} . \tag{139}$$

In a T-s-diagram, q_{12}^{rev} is the area below the process curve, just as in a closed system.

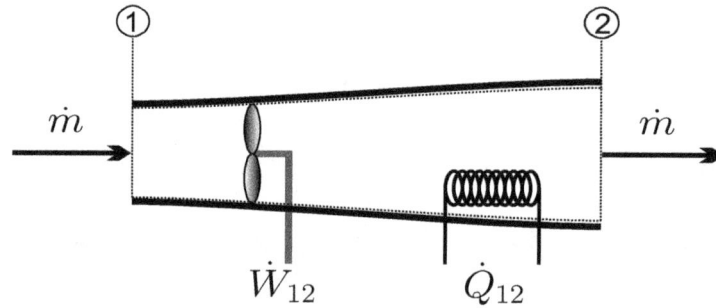

Figure 16. Typical one-inlet-one-exit system.

45. Entropy Generation in Mass Transfer

Friction in flows leads to loss of pressure and corresponding entropy generation. When we consider a simple flow with no work added or withdrawn, Equation (135) gives the entropy generated in dx as

$$\delta\dot{S}_{gen} = -\frac{\dot{m}}{T}\left(vdp + \frac{1}{2}dV^2 + g_ndz\right) . \tag{140}$$

The total entropy generated in a finite system is

$$\dot{S}_{gen} = -\dot{m}\int_{in}^{out}\frac{1}{T}\left(vdp + \frac{1}{2}dV^2 + g_ndz\right) . \tag{141}$$

For a system where kinetic and potential energy are unimportant, this reduces to

$$\dot{S}_{gen} = -\dot{m}\int_{in}^{out}\frac{v}{T}dp . \tag{142}$$

Once more, we interpret the entropy generation rate as the product of a flux, the mass flow $\dot{m}$, and a thermodynamic force, namely the integral over $-\frac{v}{T}dp$. Here, pressure in equilibrium is homogeneous, since gravitation is ignored. Deviation from homogeneous pressure is the thermodynamic force that induces a mass flow to equilibrate pressure.

Since specific volume v and thermodynamic temperature T are strictly positive, the force is proportional to the pressure difference, $-\int_{in}^{out}\frac{v}{T}dp \propto (p_{in} - p_{out})$. In order to obtain a positive entropy generation rate, linear irreversible thermodynamics suggests that the mass flow be proportional to the force, which is the case for

$$\dot{m} = \zeta A \left(p_{in} - p_{out}\right) = \zeta A \Delta p \ . \tag{143}$$

Here, A is the mass transfer area and $\zeta > 0$ is a positive transport coefficient that must be measured.

One particular example for this law is the Hagen-Poiseuille relation (Gotthilf Hagen, 1797–1884; Jean Poiseuille, 1797–1869) of fluid dynamics which gives the volume flow $\dot{V} = \dot{m}/\rho$ of a fluid with shear viscosity μ through a pipe of radius R and length L as

$$\dot{V} = \frac{\pi R^4}{8\mu L} \Delta p \ . \tag{144}$$

Another example for (143) is Darcy's law (Henry Darcy, 1803-1858) that describes flow through porous media. Then A is the cross section of the porous medium considered, and ζ is a coefficient of permeability.

Real processes are irreversible, and produce entropy. For a simple flow, the work loss to irreversibilities is

$$\dot{W}_{\text{loss}} = \int_{in}^{out} T \delta \dot{S}_{gen} \ . \tag{145}$$

Since $\delta \dot{S}_{gen} = -\frac{\dot{m}}{T} v dp$, for isothermal flow of an incompressible liquid, entropy generation and work loss are

$$\dot{S}_{gen} = \frac{\dot{V}}{T} \left(p_{in} - p_{out}\right) \ , \quad \dot{W}_{\text{loss}} = \dot{V} \left(p_{in} - p_{out}\right) \ , \tag{146}$$

where $\dot{V} = \dot{m} v$ is the volume flow.

For an ideal gas flow, we have instead

$$\dot{S}_{gen} = \dot{m} R \ln \frac{p_{in}}{p_{out}} \ , \quad \dot{W}_{\text{loss}} = -\dot{m} \int_1^2 v dp \ . \tag{147}$$

46. Global and Local Thermodynamics

From the very beginning of the discussion, we have emphasized entropy as a property describing equilibrium *and* nonequilibrium states. In that, it does not differ from other properties, such as mass density, momentum, or energy, which all are well defined in nonequilibrium. Nonequilibrium states typically are inhomogeneous, hence all properties must be defined locally, for example, as specific properties for the volume element dV.

Most thermodynamic systems, and certainly those of engineering interest, involve inhomogeneous states. While a lot can be learned about these systems by considering the thermodynamic laws for the system, for example, the global laws (122), (126) and (127), a complete understanding of a thermodynamic system requires the look inside, that is the complete resolution of local properties at all times, such as mass density $\rho\left(\vec{r}, t\right)$, velocity $\vec{V}\left(\vec{r}, t\right)$, temperature $T\left(\vec{r}, t\right)$. In transport theories such as *Fluid Dynamics* or *Heat Transfer* the local conservation laws are solved, their numerical solution is known as *Computational Fluid Dynamics* (CFD).

The topic of *Nonequilibrium Thermodynamics* is to identify the transport equations needed, and the constitutive equations required for their closure.

The Navier-Stokes-Fourier equations of classical thermo- and fluid dynamics are derived in the theory of *Linear Irreversible Thermodynamics* (LIT), which relies on the assumption of local thermodynamic equilibrium. A short outline of LIT is presented in Appendix C. The method combines the Gibbs equation with local conservation laws to find the balance law for entropy. Constitutive equations for stress tensor and heat flux are constructed such that the local entropy generation is always positive. Thus, the method provides not only transport equations, but also the accompanying

2nd law. The global laws as formulated above (122), (126) and (127) result from integration over system volume (with some simplifying assumptions).

Appendix C.4 also has a short discussion on boundary conditions for fluids and gases, where evaluation of the 2nd law suggests temperature jumps and velocity slip at the interface between a fluid and a wall.

Systems in local thermodynamic equilibrium, but global nonequilibrium, are relatively easy to describe, since the thermal and caloric equations of state, and the Gibbs equation, which are well-known from equilibrium thermodynamics, remain valid locally. Considering that equilibrium is approached over time, one will expect that local equilibrium states will be observed when the changes in the system, for example, the manipulation at system boundaries, are sufficiently slow, and gradients are sufficiently flat.

For systems with fast changes, and steep gradients, one will *not* encounter local equilibrium states, hence the equilibrium property relations cannot be used for their description. The question of determining the local property relations, and the transport equations, for systems in strong nonequilibrium is the subject of modern nonequilibrium thermodynamics. The answers depend on the school, and the material considered—a rarefied gas behaves differently from a visco-elastic fluid. Overarching frameworks are available, such as Extended Thermodynamics [20,21], or GENERIC (general equation for the nonequilibrium reversible-irreversible coupling) [22]. These include the equations of classical linear irreversible thermodynamics as proper limits, but go well beyond these. They typically include nonequilibrium property relations, often use an extended set of variables to describe the nonequilibrium state, and invariably include a formulation of the 2nd law, with nonequilibrium relations for entropy and its flux. Here is not the point to discuss this further, the cited books are good starting points for further inquiry.

47. Kinetic Theory of Gases

Instructive insights into the 2nd law can be found in Boltzmann's *Kinetic Theory of Gases*, which describes a gas as an ensemble of particles that move in space, and collide among themselves and with walls. An abridged and simplified overview of some aspects of the theory for ideal monatomic gases is presented in Appendix D. The Boltzmann equation describes the time-space evolution of the gas towards an equilibrium state from the microscopic viewpoint. The macroscopic conservation laws for mass, momentum and energy are obtained from suitable averages of the Boltzmann equation. The equations for local thermodynamic equilibrium can be obtained from suitable limits (small Knudsen number), in full agreement with LIT.

Boltzmann's celebrated H-theorem identifies a macroscopic quantity which has all properties of entropy, and, in our opinion is entropy, indeed. In particular, the H-function obeys a balance law with non-negative production, and in equilibrium it reduces to the equilibrium entropy of an ideal gas. The Boltzmann entropy is defined for arbitrary states, including *all* inhomogeneous nonequilibrium states far from local equilibrium, in agreement with our assumption on entropy throughout.

The underlying microscopic picture provides an interpretation for entropy, a quantity that arose first from purely phenomenological considerations. This interpretation is the topic of the next sections.

48. What is Entropy?

The arguments that gave us the second law and entropy as a property centered around the trend to equilibrium observed in any system left to itself (isolated system). Based on the derivation, the question *What is entropy?* can be answered simply by saying *it is a quantity that arises when one constructs an inequality that describes the trend to equilibrium.* Can there be a deeper understanding of entropy?

Before we try to answer, we look at internal energy—when the first law of thermodynamics was found, the concept of internal energy was new, and it was difficult to understand what it might describe. At that time, the atomic structure of matter was not known, and internal energy could not be interpreted—it appeared because it served well to describe the phenomena. Today we know more, and we understand internal energy as the kinetic, potential, and quantum energies of atoms and molecules on the microscopic level. Thus, while the concept of internal energy arose from the desire to describe phenomena, today it is relatively easy to understand, because it has a macroscopic analogue in mechanics.

Entropy also came into play to describe the phenomena, but it is a new quantity, without a mechanical analogue. A deeper understanding of entropy can be gained, as for internal energy, from considerations on the atomic scale. Within the framework of his *Kinetic Theory of Gases*, Ludwig Boltzmann (1844–1905) found the microscopic interpretation of entropy (see Appendix D).

Macroscopically, a state is described by only a few macroscopic properties, for example, temperature, pressure, volume. Microscopically, a state is described through the location and momentum of all atoms within the system. The microscopic state is constantly changing due to the microscopic motion of the atoms, and there are many microscopic states that describe the same macroscopic state. If we denote the total number of all microscopic states that describe the same macroscopic state by Ω, then the entropy of the macroscopic state according to Boltzmann is

$$S = k_B \ln \Omega . \tag{148}$$

The constant $k_B = \bar{R}/A = 1.3804 \times 10^{-23} \frac{J}{K}$ is the Boltzmann constant, which can be interpreted as the gas constant per particle; $A = 6.022 \times 10^{23} \frac{1}{mol}$ is the Avogadro constant.

The growth of entropy in an isolated system, $\frac{dS}{dt} \geq 0$, thus means that the system shifts to macrostates which have larger numbers of microscopic realizations. As we will see, equilibrium states have particularly large numbers of realizations, and this is why they are observed.

49. Ideal Gas Entropy

To make the ideas somewhat clearer, we consider the expansion of a gas when a barrier is removed, see Section 35. This is a particularly simple case, where the internal energy, and thus the distribution of energy over the particles, does not change. Hence, we can ignore the distribution of thermal energy over the particles, and the exchange of energy between them.

We assume a system of N gas particles in a volume V. The volume of a single particle is v_0, and in order to be able to compute the number Ω, we "quantize" the accessible volume V into $n = V/v_0$ boxes that each can accommodate just one particle. Note that in a gas most boxes are empty. Due to their thermal energy, the atoms move from box to box. The number of microstates is simply given by the number of realizations of a state with N filled boxes and $(n - N)$ empty boxes, which is

$$\Omega (N, V) = \frac{n!}{N! (n - N)!} . \tag{149}$$

By means of Stirling's formula $\ln x! = x \ln x - x$ (for $x \gg 1$) the entropy (148) for this state becomes

$$S (N, V) = k_B \left[-N \ln \frac{N}{n} - (n - N) \ln \left(1 - \frac{N}{n} \right) \right] . \tag{150}$$

Now we can compute the change of entropy with volume. For this, we consider the same N particles in two different volumes, $V_1 = n_1 v_0$ and $V_2 = n_2 v_0$. The entropy difference $S_2 - S_1 = S (N, V_2) - S (N, V_1)$ between the two states is

$$S_2 - S_1 = k_B \left[N \ln \frac{n_2}{n_1} + n_1 \ln \left(1 - \frac{N}{n_1} \right) - n_2 \ln \left(1 - \frac{N}{n_2} \right) + N \ln \frac{\left(1 - \frac{N}{n_2} \right)}{\left(1 - \frac{N}{n_1} \right)} \right] . \tag{151}$$

In an ideal gas the number of possible positions n is much bigger than the number of particles N, that is $\frac{N}{n_1} \ll 1, \frac{N}{n_2} \ll 1$. Taylor expansion yields the entropy difference to leading order as

$$S_2 - S_1 = k_B N \ln \frac{n_2}{n_1} = mR \ln \frac{V_2}{V_1} , \tag{152}$$

where we reintroduced volume ($V_{1,2} = n_{1,2}v_0$), and introduced the mass as $m = MN/A$; $R = \bar{R}/M$ is the gas constant. This is just the change of entropy computed in Section 35.

50. Homogeneous vs. Inhomogeneous States

It is instructive to compare the number of realizations for the two cases, for which we find

$$\frac{\Omega_2}{\Omega_1} = \exp \frac{S_2 - S_1}{k} = \exp \left(N \ln \frac{V_2}{V_1} \right) = \left(\frac{V_2}{V_1} \right)^N . \tag{153}$$

For a macroscopic amount of gas, the particle number N is extremely large (order of magnitude $\sim 10^{23}$), so that already for a small difference in volume the ratio of microscopic realization numbers is enormous. For instance for $V_2 = 2V_1$, we find $\frac{\Omega_2}{\Omega_1} = 2^N$.

Microscopic states change constantly due to travel of, and collisions between, particles. Each of the Ω microstates compatible with the given macrostate is observed with the same probability, $1/\Omega$. The Ω_1 microstates in which the gas is confined in the volume V_1 are included in the Ω_2 microstates in which the gas is confined in the larger volume V_2. Thus, after removal of the barrier, there is a finite, but extremely small probability of $P = \frac{\Omega_1}{\Omega_2} = \left(\frac{V_1}{V_2} \right)^N$ to find all gas particles in the initial volume V_1. This probability is so small that the expected waiting time for observing a return into the original volume exceeds the lifetime of the universe by many orders of magnitude. If we do not want to wait that long for the return to initial state, we have to push the gas back into the initial volume, which requires work.

In generalization of the above, we can conclude that it is quite unlikely that a portion V_ν of the volume is void of particles. The corresponding probability is $P_\nu = \left(\frac{V - V_\nu}{V} \right)^N$. The average volume available for one particle is $\bar{V} = \frac{V}{N}$, and when $V_\nu = \nu \bar{V}$ we find, for the large particle numbers in an macroscopic amount of gas, $P_\nu = \left(1 - \frac{\nu}{N} \right)^N \simeq e^{-\nu}$. Thus, as long as V_ν is bigger than the average volume for a single particle, so that $\nu > 1$, the probability for a void is very small. Moreover, strongly inhomogeneous distributions are rather unlikely, since the number of homogeneous distributions is far larger in number. Hence, we observe homogeneous distributions in equilibrium.

A closer look at equilibrium properties reveals small local fluctuations of properties, for example, mass density, which are induced by the thermal motion of particles. The equilibrium state is stable, that is these random disturbances decay in time, so that in average the equilibrium state is observed. For macroscopic systems the fluctuations are so small that they can be ignored. Nevertheless, fluctuations in density lead to light scattering, which can be used to determine transport coefficients such as viscosity and heat conductivity from equilibrium states [20]. Since blue light is more likely to be scattered in density fluctuations of the atmosphere, the sky appears blue.

51. Entropy and Disorder

Often it is said that entropy is a measure for disorder, where disorder has a higher entropy. One has to be rather careful with this statement, since order, or disorder, are not well-defined concepts. To shed some light on this, we use the following analogy—the ordered state of an office is the state where all papers, folders, books and pens are in their designated shelf space. Thus, they are confined to a relatively small initial volume of the shelf, V_1. When work is done in the office, all these papers, folders, books and pens are removed from their initial location, and, after they are used, are dropped somewhere in the office—now they are only confined to the large volume of the office, V_2. The actions of the person working in the office constantly change the microstate of the office (the precise location of that pen ... where is it now?), in analogy to thermal motion.

At the end of the day, the office looks like a mess and needs work to clean up. Note, however, that the final state of the office—which appears to be so disorderly—is just *one* accessible microstate, and therefore it has the same probability as the fully ordered state, where each book and folder is at its designated place on the shelf. A single microstate, for example, a particular distribution of office material over the office in the evening, has no entropy. Entropy is a macroscopic property that counts the number of all possible microstates, for example, all possible distributions of office material.

A macroscopic state which puts strong restrictions on the elements has a low entropy, for example, when all office material is in shelves behind locked doors. When the restrictions are removed—the doors are unlocked—the number of possible distributions grows, and so does entropy. Thermal motion leads to a constant change of the distribution within the inherent restrictions.

To our eye more restricted macroscopic states—all gas particles only in a small part of the container, or all office material behind closed doors—appear more orderly, while less restricted states generally appear more disorderly. Only in this sense one can say that entropy is a measure for disorder.

In the office, every evening the disordered state differs from that of the previous day. Over time, one faces a multitude of disordered states, that is the disordered office has many realizations, and a large entropy. In the end, this makes cleaning up cumbersome, and time consuming.

Our discussion focussed on spatial distributions where the notion of order is well-aligned with our experience. The thermal contribution to entropy is related to the distribution of microscopic energy e_m over the particles, where e_m is the microscopic energy per particle. In *Statistical Thermodynamics* one finds that in equilibrium states the distribution of microscopic energies between particles is exponential, $A \exp\left[-\frac{e_m}{kT}\right]$. The factor A must be chosen such that the sum over all particles gives the internal energy, $U = \sum_m A e_m \exp\left[-\frac{e_m}{k_B T}\right]$. One might say that the exponential itself is an orderly function, so that the equilibrium states are less disordered than nonequilibrium states. Moreover, for lower temperatures the exponential is more narrow, the microscopic particle energies are confined to lower values, and one might say that low temperature equilibrium states are more orderly than high temperature equilibrium states. And indeed, we find that entropy grows with temperature, that is colder systems have lower entropies.

52. Summary

Looking back at the above, it is clear that we have not established any new thermodynamics, but provided our perspective on entropy and the 2nd law. Throughout the discussion entropy is established as a property for any state, be it in equilibrium and nonequilibrium. While this is standard in all theories on nonequilibrium thermodynamics, and in kinetic theory of gases, one finds many discussions of thermodynamics that define entropy only for equilibrium states. Restriction of entropy to equilibrium state is an unnecessary assumption, that reduces applicability of thermodynamics, but can easily be avoided.

Engineering applications of thermodynamics invariably have to account for inhomogeneous nonequilibrium states, hence a clear description of entropy as nonequilibrium quantity is required, and is, indeed, used. The global balance laws used in engineering textbooks follow from the assumption of local thermodynamic equilibrium, which allows to use the equilibrium property relations locally. The same assumption gives the Navier-Stokes-Fourier equations, that is, the partial differential equations describing all local process details. Global and local descriptions are equivalent.

Entropy, and all other properties such as density, energy, and so forth, are also meaningful for extreme nonequilibrium states, which are not in local equilibrium. The associated property relations and transport equations might differ considerably from those based on the local equilibrium hypothesis, but this does not imply that energy or entropy lose their meaning.

Positivity of thermodynamic temperature guarantees dissipation of work and kinetic energy. This is best seen in the stability analysis, where motion of volume elements, that is, kinetic energy, is included. In the frame where system momentum vanishes, the local velocity will vanish in equilibrium, and this equilibrium resting state is stable only if thermodynamic temperature is positive.

Entropy and the 2nd law were introduced based on five intuitive observations that are in agreement with daily experience. There is no need of any discussion of thermodynamic cycles and engines to introduce entropy and the 2nd law. This greatly simplifies access to the subject—both for teaching and studying thermodynamics—since all thermodynamic cycles and engines are discussed only after the laws of thermodynamics are established.

For teaching thermodynamics, I use a variant—with some shortcuts—of the approach developed here [8]. This allows fast and meaningful access to the thermodynamic laws as early in the course as possible, so that all applications can rely on 1st and 2nd law from the beginning.

Acknowledgments: The origins of this contribution go back to my stay in 2015 as visitor at the Aachen Institute for Advanced Study in Computational Engineering Science (AICES) at RWTH Aachen University, Germany. I wish to express my sincere thanks to AICES for the hospitality, and interesting discussions. In particular I thank Manuel Torrilhon for deep exchanges on a wide range of topics related to nonequilibrium thermodynamics. I gratefully acknowledge Livio Gibelli (Edinburgh), Karl Heinz Hoffmann (Chemnitz), and Peter Van (Budapest) who read versions of the manuscript and gave me valuable feedback and suggestions for improvements and clarification. Larger sections of this contribution are taken, in somewhat modified form, from my textbook on thermodynamics [8], and I wish to thank my publisher, Springer Heidelberg, for the excellent collaboration over the years, and their permission to use the material.

Appendix A. Caratheodory's Approach

Our approach towards entropy and the 2nd law is based on the observation of phenomena. Attempts for building thermodynamics axiomatically were presented by C. Caratheodory [9,19], and more recently by Lieb and Yngvason [12,13].

Restricting his attention to adiabatic processes and equilibrium states only, Caratheodory formulated the 2nd law in the statement that *in each neighbourhood of an arbitrary initial state there exist states that cannot be reached by adiabatic processes*. While initial and end states are equilibrium states, the processes that connect these states are allowed to be irreversible (work only). This statement is, of course, compatible with the 2nd law (44), which for adiabatic processes reduces to $\frac{dS}{dt} = \dot{S}_{gen} \geq 0$ —states with smaller entropy than the initial state cannot be reached adiabatically.

Using sophisticated reasoning within multivariable calculus, Caratheodory proceeds to find the Gibbs equation where thermodynamic temperature is identified as a material independent integrating factor.

Compared to our approach, entropy is defined only for equilibrium states and, since only adiabatic processes are considered, heat—and therefore entropy flux as heat over temperature—does not appear in the 1st and 2nd law. Therefore, the full entropy balance (44), which describes irreversible processes with heat and work interaction, cannot be found within this framework.

Appendix B. Concavity of Equilibrium Entropy

We show that the stability conditions (120) are equivalent to the statement that equilibrium entropy $s(u,v)$ is a concave function of its arguments. The requirement for concavity is that the matrix of second derivatives is negative definite. A symmetric matrix $A = \begin{Bmatrix} a & b \\ b & c \end{Bmatrix}$ is negative definite if $x \cdot A \cdot x < 0$ for arbitrary vectors $x = \{x_1, x_2\}$. This is the case for $a < 0$, $c - \frac{b^2}{a} < 0$, which implies $c < 0$. Hence, concavity of $s(u,v)$ is given for

$$\left(\frac{\partial^2 s}{\partial u \partial u}\right)_v < 0 \quad , \quad \left(\frac{\partial^2 s}{\partial u \partial u}\right)_v - \frac{\left(\frac{\partial^2 s}{\partial v \partial u}\right)^2}{\left(\frac{\partial^2 s}{\partial v \partial v}\right)_u} < 0 \, . \tag{A1}$$

The reformulation of these conditions into the stability conditions (120) is a beautiful application of multi-variable calculus.

The first derivatives of entropy are obtained from the Gibbs equation $Tds = du + pdv$ as

$$\left(\frac{\partial s}{\partial u}\right)_v = \frac{1}{T} \quad , \quad \left(\frac{\partial s}{\partial v}\right)_u = \frac{p}{T} \, . \tag{A2}$$

From this we have, with the definition of specific heat (19),

$$\left(\frac{\partial^2 s}{\partial u \partial u}\right)_v = \left(\frac{\partial}{\partial u}\frac{1}{T}\right)_v = -\frac{1}{T^2}\left(\frac{\partial T}{\partial u}\right)_v = -\frac{1}{T^2 c_v} < 0 \, , \tag{A3}$$

To bring the other second derivatives into a compact form requires repeated use of the identity $\left(\frac{\partial x}{\partial z}\right)_y = -\left(\frac{\partial x}{\partial y}\right)_z \left(\frac{\partial y}{\partial z}\right)_x$. For the mixed derivative we find

$$\frac{\partial^2 s}{\partial v \partial u} = \left(\frac{\partial\left(\frac{1}{T}\right)}{\partial v}\right)_u = -\frac{1}{T^2}\left(\frac{\partial T}{\partial v}\right)_u = -\frac{1}{T^2}\left[-\left(\frac{\partial T}{\partial u}\right)_v\left(\frac{\partial u}{\partial v}\right)_T\right] = \frac{1}{T^2 c_v}\left(\frac{\partial u}{\partial v}\right)_T . \tag{A4}$$

For the second volume derivative we find at first

$$\left(\frac{\partial^2 s}{\partial v \partial v}\right)_u = \left(\frac{\partial\left(\frac{p}{T}\right)}{\partial v}\right)_u = -\frac{p}{T^2}\left(\frac{\partial T}{\partial v}\right)_u + \frac{1}{T}\left(\frac{\partial p}{\partial v}\right)_u = \frac{p}{c_v T^2}\left(\frac{\partial u}{\partial v}\right)_T - \frac{1}{T}\left(\frac{\partial p}{\partial u}\right)_v\left(\frac{\partial u}{\partial v}\right)_p . \tag{A5}$$

For further simplification, we consider energy as a function of (T, v), with the differential $du = c_v dT + \left(\frac{\partial u}{\partial v}\right)_T dv$, from which we find the partial derivatives

$$\left(\frac{\partial u}{\partial v}\right)_p = c_v\left(\frac{\partial T}{\partial v}\right)_p + \left(\frac{\partial u}{\partial v}\right)_T \quad , \quad \left(\frac{\partial u}{\partial p}\right)_v = \frac{1}{\left(\frac{\partial p}{\partial u}\right)_v} = c_v\left(\frac{\partial T}{\partial p}\right)_v . \tag{A6}$$

We use the above, and (54), to find

$$\begin{aligned}\left(\frac{\partial^2 s}{\partial v \partial v}\right)_u &= \frac{1}{T^2 c_v}\left[p - T\left(\frac{\partial p}{\partial T}\right)_v\right]\left(\frac{\partial u}{\partial v}\right)_T - \frac{1}{T}\left(\frac{\partial T}{\partial v}\right)_p\left(\frac{\partial p}{\partial T}\right)_v \\ &= -\frac{1}{T^2 c_v}\left(\frac{\partial u}{\partial v}\right)_T\left(\frac{\partial u}{\partial v}\right)_T + \frac{1}{T}\left(\frac{\partial p}{\partial v}\right)_T ,\end{aligned}$$

so that

$$\left(\frac{\partial^2 s}{\partial v \partial v}\right)_u - \frac{\left(\frac{\partial^2 s}{\partial v \partial u}\right)^2}{\left(\frac{\partial^2 s}{\partial u \partial u}\right)_v} = \frac{1}{T}\left(\frac{\partial p}{\partial v}\right)_T = -\frac{1}{v T \kappa_T} < 0 \, .$$

Here, we have used the stability conditions (120), which state that $T > 0$, $\kappa_T = -\frac{1}{v}\left(\frac{\partial v}{\partial p}\right)_T > 0$. Hence concave equilibrium entropy $s(u, v)$ follows from stability of the equilibrium state as expressed in (120).

Appendix C. Local Formulation of Nonequilibrium Thermodynamics

Appendix C.1. Global and Local Balance Laws

In the body of the paper, we have discussed thermodynamics of systems, only occasionally looking inside the systems, when states are inhomogeneous. For full resolution of what happens inside a system, we need to formulate the conservation laws for mass, momentum and energy, and the non-conservation law for entropy, for each point in space, that is as partial differential equations. For doing so, it is best to first discuss the general structure of balance laws in global and local form, and then specify for the individual quantities to be balanced (mass, momentum, energy, entropy).

We consider a system of fixed volume V_0. The outer surface of the system is denoted by ∂V_0, and n_i denotes the normal vector on the boundary, pointing outwards. Following Reference [5], we use index notation for vectors and tensors.

A global balance law considers the change in time of a global quantity

$$\Psi = \int_{V_0} \rho \psi dV , \qquad (A7)$$

where ψ is the mass specific property to the the global property Ψ.

The change in time $d\Psi/dt$ can be effected by: (a) a convective flux, that is the amount of Ψ that is transferred in or out of the system when mass crosses the system boundary,

$$\dot{\Psi} = - \oint_{\partial V_0} \rho \psi \mathcal{V}_i n_i dA ; \qquad (A8)$$

here, dA is a surface element of the boundary ∂V_0 and $-\rho \mathcal{V}_i n_i dA$ is the amount of mass crossing dA during a time interval dt.

(b) a non-convective flux with local flux vector φ_i, so that the overall amount flowing over the system boundary is

$$\dot{\Phi} = - \oint_{\partial V_0} \varphi_i n_i dA ; \qquad (A9)$$

(c) a production Π inside the system with local production density π,

$$\Pi = \int_{V_0} \pi dV ; \qquad (A10)$$

and (d) a supply Θ with local supply density $\rho\theta$ to the bulk of the system

$$\Theta = \int_{V_0} \rho\theta dV . \qquad (A11)$$

The difference between supply and production is that a supply can, at least in principle, be controlled from the outside, while a production cannot be controlled, and is due to the processes inside the system.

Combining the above into the balance law gives

$$\frac{d\Psi}{dt} = \dot{\Psi} + \dot{\Phi} + \Pi + \Theta . \qquad (A12)$$

With the Gauss divergence theorem, $\oint_{\partial V_0} \gamma_i n_i dA = \int_{V_0} \frac{\partial \gamma_k}{\partial r_k} dV$, the flux terms can be converted into volume integrals, and since the system volume is fixed, the time derivative can be moved into the integral, so that the balance assumes the form

$$\int_{V_0} \left(\frac{\partial \rho \psi}{\partial t} + \frac{\partial (\rho \psi \mathcal{V}_k + \varphi_k)}{\partial r_k} - \pi - \rho\theta \right) dV = 0 .$$

The integral must vanish for arbitrary system volumes, hence the integrand must vanish as well. As a result, we obtain the general form of a balance law in local formulation,

$$\frac{\partial \rho \psi}{\partial t} + \frac{\partial (\rho \psi \mathcal{V}_k + \varphi_k)}{\partial r_k} = \pi + \rho\theta . \qquad (A13)$$

Appendix C.2. Local Conservation Laws

For the balance of mass, we have $\psi = 1$. Mass is conserved ($\pi = 0$), and can only be transferred by convection, that is there is neither a non-convective flux ($\varphi_k = 0$), nor a supply ($\theta = 0$). Hence, the local mass balance reads

$$\frac{\partial \rho}{\partial t} + \frac{\partial}{\partial r_k}[\rho V_k] = 0 \, . \tag{A14}$$

For momentum, we have $\psi = V_i$. Momentum is conserved ($\pi = 0$), it can be transferred to the system by convection, and by forces $\varphi_k = -t_{ik}n_i$ acting on the system boundary; body forces G_i, such as gravity, serve as a supply ($\theta = G_i$):

$$\frac{\partial [\rho V_i]}{\partial t} + \frac{\partial}{\partial r_k}[\rho V_i V_k - t_{ik}] = \rho G_i \, . \tag{A15}$$

Here, t_{ij} is the symmetric stress tensor defined such that $t_{ik}n_i$ is the force on a fluid surface element with normal vector n_i; in equilibrium the stress tensor reduces to the pressure, $t_{ik|Eq} = -p\delta_{ik}$.

For energy, we have $\psi = u + \frac{1}{2}V^2$. Also energy is conserved ($\pi = 0$). It can be transferred to the system by convection, by non-convective heat flux q_k, and by the power of the surface forces $-t_{ik}n_k V_i$; the supply is due to the power $\rho G_i V_i$ of body forces. Hence the energy balance reads

$$\frac{\partial}{\partial t}\left[\rho\left(u + \frac{1}{2}V^2\right)\right] + \frac{\partial}{\partial r_k}\left[\rho\left(u + \frac{1}{2}V^2\right)V_k + q_k - t_{ik}V_i\right] = \rho G_i V_i \, . \tag{A16}$$

The above Equations (A14)–(A16) are valid for any fluid or gas. Integrating the mass balance (A14) and the energy balance (A16) over the (time dependent) system volume V results in the system conservation laws (122) and (126), with the mass flow over an open section A (i.e., an inflow or outflow) of the boundary as

$$\dot{m} = \int_A \rho \, |V_k n_k| \, dA \, , \tag{A17}$$

and total heat and work given by

$$\dot{Q} = -\oint_{\partial V} q_i n_i dA \, , \quad \dot{W} = -\int_{\partial V,\text{solid}} t_{ik}V_k n_i dA \, ; \tag{A18}$$

here, ∂V is the boundary of the system, and (∂V, solid) is the solid part of the system boundary, where no mass can cross. The derivation requires some assumptions, including that properties at in/outflows can be replaced by averages, and that the body force has a time-independent potential, $G_i = -\frac{\partial(gnz)}{\partial r_i}$.

In the form given above, the conservation laws (A14)–(A16) do not form a closed system of equations for the variables (ρ, V_i, T), since they contain internal energy u, heat flux q_i and stress tensor t_{ij}, which must be related to the variables by means of constitutive relations—this will be discussed in the next section.

For the constitutive theory based on the assumption of local thermodynamic equilibrium, it is most convenient to rewrite the conservation laws with the material time derivative $\frac{D}{Dt} = \frac{\partial}{\partial t} + V_k\frac{\partial}{\partial r_k}$. After some manipulation, including use of mass balance to simplify momentum and energy balance, and momentum balance (after scalar product with V_i) to simplify the energy balance, the result can be written as

$$\begin{aligned} \frac{D\rho}{Dt} + \rho\frac{\partial V_k}{\partial r_k} &= 0 \, , \\ \rho\frac{DV_i}{Dt} - \frac{\partial t_{ik}}{\partial r_k} &= \rho G_i \, , \\ \rho\frac{Du}{Dt} + \frac{\partial q_k}{\partial r_k} &= -t_{ik}\frac{\partial V_i}{\partial x_k} \, . \end{aligned} \tag{A19}$$

The last equation is the balance of internal energy, which is not a conservation law, due to the term on the right which describes production of internal energy due to frictional heating.

Appendix C.3. Linear Irreversible Thermodynamics

The question of finding constitutive equations for u, q_i, t_{ij}, and entropy s, and the construction of the entropy balance, is the main question in *Nonequilibrium Thermodynamics*. For strong nonequilibrium, this is a rather difficult topic, with many competing, or complimentary, ideas. The approach for the case of *local thermodynamic equilibrium*, however, is now well established as *Linear Irreversible Thermodynamics* (LIT) [18].

The basic idea is to assume that the property relations for equilibrium are valid locally, for any element of the fluid, that is we have the thermal and caloric equations of state, $p(T, \rho)$, $u(T, \rho)$ and the Gibbs equation, $T ds = du - \frac{p}{\rho^2} d\rho$ valid in a volume element dV located at $\vec{r}$.

For the construction of the local balance law for entropy—the 2nd law—we recall that the Gibbs equation was derived for a closed system, that is, it is valid for a material element. Therefore its material time derivative assumes the form

$$T \frac{Ds}{Dt} = \frac{Du}{Dt} - \frac{p}{\rho^2} \frac{D\rho}{Dt} , \tag{A20}$$

which is the base for the construction of the entropy balance. Replacing the time derivatives of ρ, u by means of the conservation laws, multiplying with $\frac{\rho}{T}$, and an integration by parts, yields

$$\rho \frac{Ds}{Dt} + \frac{\partial}{\partial r_k} \left[\frac{q_k}{T} \right] = \sigma , \tag{A21}$$

with

$$\sigma = q_k \frac{\partial \frac{1}{T}}{\partial r_k} + \frac{1}{T} t_{\langle ik \rangle} \frac{\partial \mathcal{V}_{\langle i}}{\partial r_{k \rangle}} + \frac{1}{T} \left(p + \frac{1}{3} t_{rr} \right) \frac{\partial \mathcal{V}_k}{\partial r_k} . \tag{A22}$$

Here, $t_{\langle ik \rangle}$ denotes the trace free and symmetric part of the stress tensor, and t_{rr} is its trace.

Equation (A21) is the local balance of entropy with the non-convective entropy flux $\phi_k = \frac{q_k}{T}$, and the entropy production density σ. Integration over the system volume yields the system balance (69) that was derived under the assumption of local thermal equilibrium, with the overall entropy generation rate $\dot{S}_{gen} = \int_{V_0} \sigma dV$.

The requirement of non-negative entropy generation, $\dot{S}_{gen} \geq 0$, for all choices of the system volume V_0 implies non-negative values for the production density, that is, $\sigma \geq 0$. This is a strong condition on the constitutive equations for q_i, t_{ik}. We recognize σ in (A22) as the product of the thermodynamic fluxes $F_A = \left\{ q_i, t_{\langle ik \rangle}, \left(p + \frac{1}{3} t_{rr} \right) \right\}$ and thermodynamic driving forces, viz. the space derivatives of velocity $\mathcal{V}_i$ and temperature T. Note that the forces describe deviations from equilibrium states, which have homogeneous temperature and velocity.

LIT assumes a linear relationship between fluxes and forces, such that corresponding forces and fluxes have the same tensorial degree (Curie principle) [18]. The result is the stress tensor of the Navier-Stokes equations, and Fourier's law for the heat flux,

$$\left(p + \frac{1}{3} t_{rr} \right) = \nu \frac{\partial \mathcal{V}_k}{\partial r_k} \quad , \quad t_{\langle ik \rangle} = 2\mu \frac{\partial \mathcal{V}_{\langle i}}{\partial r_{k \rangle}} \quad , \quad q_i = -\kappa \frac{\partial T}{\partial r_i} . \tag{A23}$$

Here, $\nu > 0$ is the bulk viscosity, $\mu > 0$ is the shear viscosity, and $\kappa > 0$ is the heat conductivity; all three must be measured as functions of (ρ, T). The signs ensure that the entropy generation is always positive. Note that the viscosities must be positive since thermodynamic temperature is positive—once more we see that positive temperature guarantees mechanical stability.

The entropy generation rate becomes

$$\sigma = \frac{\kappa}{T^2} \frac{\partial T}{\partial r_i} \frac{\partial T}{\partial r_k} + \frac{2\mu}{T} \frac{\partial \mathcal{V}_{\langle i}}{\partial r_{k\rangle}} \frac{\partial \mathcal{V}_{\langle i}}{\partial r_{k\rangle}} + \frac{\nu}{T} \frac{\partial \mathcal{V}_j}{\partial r_j} \frac{\partial \mathcal{V}_k}{\partial r_k} \geq 0. \tag{A24}$$

In summary, LIT derives the Navier-Stokes-Fourier equations that are routinely used in thermodynamics, fluid mechanics, and heat transfer. Moreover, LIT provides the expression for the local entropy balance, which opens the door to full analysis of thermodynamic losses—recall that $\dot{W}_{loss} = T_0 \dot{S}_{gen} = T_0 \int \sigma dV$.

The entropy generation rate (A24) is quadratic in gradients of temperature and velocity. This indicates that processes with small gradients have very small entropy generation, with the limit of reversible processes for vanishing gradients.

Appendix C.4. Jump and Slip at Boundaries

It is a common assumption that a fluid at a wall assumes the temperature and velocity of the wall. To re-examine boundary conditions for fluids, we consider a wall-fluid boundary with a normal vector n_k, pointing from the wall into the fluid. For this section properties of the wall are denoted with a superscript W, while properties with a superscript F denote fluid properties *directly at the wall*.

The wall does not allow fluid to pass, hence the normal velocity of the fluid relative to the wall vanishes,

$$\left(\mathcal{V}_k^F - \mathcal{V}_k^W \right) n_k = 0. \tag{A25}$$

Since momentum and energy are conserved, their non-convective fluxes must be continuous, that is

$$- t_{ik}^F n_k = - t_{ik}^W n_k \quad, \quad q_k^F n_k - t_{ik}^F \mathcal{V}_i^F n_k = q_k^W n_k - t_{ik}^W \mathcal{V}_i^W n_k. \tag{A26}$$

It must be expected that the interaction of fluid and wall is irreversible, so that the entropy flux into the fluid is larger than the flux out of the wall. With the surface entropy production $\bar{\sigma} \geq 0$, we write

$$\frac{q_k^F n_k}{T^F} = \frac{q_k^W n_k}{T^W} + \bar{\sigma}. \tag{A27}$$

To obtain boundary conditions for the fluid, we eliminate the stress and heat flux in the wall from the conservation conditions, to find

$$\bar{\sigma} = \frac{q_k^F n_k}{T^F} - \frac{q_k^W n_k}{T^W} = \left(\frac{1}{T^F} - \frac{1}{T^W} \right) q_k^F n_k + \left(\mathcal{V}_i^F - \mathcal{V}_i^W \right) \frac{t_{ik}^F n_k - n_i t_{jk}^F n_j n_k}{T^W} \geq 0. \tag{A28}$$

Again we require non-negative production at all states, and, employing the LIT constitutive expressions, find expressions for the temperature jump and the velocity slip,

$$T^F - T^W = \lambda_T \frac{\partial T}{\partial r_k} n_k \quad, \quad \mathcal{V}_i^F - \mathcal{V}_i^W = \lambda_{\mathcal{V}} \left(\frac{\partial \mathcal{V}_{\langle i}}{\partial r_{k\rangle}} - \frac{\partial \mathcal{V}_{\langle j}}{\partial r_{k\rangle}} n_i n_j \right) n_k. \tag{A29}$$

Here, $\lambda_T > 0$ and $\lambda_{\mathcal{V}} > 0$ are the jump and slip length, respectively, which must be obtained from experiments.

For most engineering applications, λ_T and $\lambda_{\mathcal{V}}$ are so small that they can be set to zero, which yields the common boundary conditions that the fluid sticks to the wall, and has its boundary temperature, $\mathcal{V}_i^F = \mathcal{V}_i^W$, $T^F = T^W$. The above derivation shows, however, that jump and slip are the natural expectation, since they imply irreversibility at the wall. In particular for rarefied gases, jump and slip might be marked effects [17]. Note that in nonequilibrium systems temperature jumps are expected at thermometer boundaries, so that a thermometer might not show the actual temperature of the fluid.

Appendix D. Elements of Kinetic Theory of Gases

Much on entropy and the 2nd law can be learned from the kinetic theory of monatomic gases, which we briefly discuss in this Appendix. Kinetic Theory is a fascinating and deep subject, and it should be clear that we can at best give a flavor of its most salient results. For deeper insight, we must refer the reader to the literature, for example, References [15–17].

Appendix D.1. Microscopic Description of a Gas

A gas consists of a huge number—in the order of 10^{23}—interacting particles α whose physical state is described by their locations $\mathbf{x}^\alpha = \{x_1^\alpha, x_2^\alpha, x_3^\alpha\}$ and their velocities $\mathbf{c}^\alpha = \{c_1^\alpha, c_2^\alpha, c_3^\alpha\}$ at any time t. The (micro-) state of the gas is given by the complete set of the $\{\mathbf{x}^\alpha, \mathbf{c}^\alpha\}$, and each particle can be described through its trajectory in the 6-dimensional phase space spanned by $\mathbf{x}$ and $\mathbf{c}$.

Thus, to describe the gas one could establish the equation of motion for each particle, and then had to solve a set of $\sim 10^{23}$ coupled equations. Clearly this is not feasible, and therefore kinetic theory chooses to describe the state of the gas on the micro-level through the phase density, or distribution function, $f(\mathbf{x}, t, \mathbf{c})$ which is defined such that $N_{\mathbf{x},\mathbf{c}} = f(\mathbf{x}, t, \mathbf{c})\, d\mathbf{x}d\mathbf{c}$ gives the number of particles that occupy a cell of phase space $d\mathbf{x}d\mathbf{c}$ at time t. In other words, $N_{\mathbf{x},\mathbf{c}}$ is the number of particles with velocities in $\{\mathbf{c}, \mathbf{c} + d\mathbf{c}\}$ located in the interval $\{\mathbf{x}, \mathbf{x} + d\mathbf{x}\}$ at time t.

With this definition, a certain level of inaccuracy is introduced, since now the state of each particle is only known within an error of $d\mathbf{x}d\mathbf{c}$. The phase density $f(\mathbf{x}, t, \mathbf{c})$ is the central quantity in kinetic theory, since the state of the gas is (almost) completely known when f is known.

We consider particles of mass m. When we integrate mf over velocity, we obtain the mass density ρ. We frequently have to integrate over the full velocity space, and in order to condense notation we write one integral sign without limits,

$$\rho = m \iiint_{-\infty}^{\infty} f d\mathbf{c} = m \int f d\mathbf{c} \, . \tag{A30}$$

The momentum density of the gas is obtained by averaging particle momentum,

$$\rho \mathcal{V}_i = m \int c_i f d\mathbf{c} \, . \tag{A31}$$

The peculiar velocity $C_i = c_i - \mathcal{V}_i$ of the particles gives the particle speed as measured by an observer moving with the gas at the local velocity $\mathcal{V}_i$, that is, an observer in the rest-frame, hence the first moment of f over C_i vanishes,

$$0 = m \int C_i f d\mathbf{c} \, . \tag{A32}$$

The kinetic energy of a particle is given by $\frac{m}{2}c^2$, so that the energy density of the gas is given by (e denotes the specific energy)

$$\rho e = \frac{m}{2} \int c^2 f d\mathbf{c} = \frac{m}{2} \int \left(C^2 + 2\mathcal{V}_k C_k + \mathcal{V}^2\right) f dc = \rho u + \frac{\rho}{2}\mathcal{V}^2 \, , \tag{A33}$$

where

$$\rho u = \frac{m}{2} \int C^2 f d\mathbf{c} \tag{A34}$$

is the internal, or thermal, energy of the gas, and $\frac{\rho}{2}\mathcal{V}^2$ is the kinetic energy of its macroscopic motion. Thus, the internal energy of an ideal monatomic gas is the kinetic energy of its particles as measured in the rest-frame.

Pressure tensor p_{ij} and heat flux q_i are the fluxes of momentum and energy,

$$p_{ij} = -t_{ij} = m \int C_i C_j f d\mathbf{c} \quad \text{and} \quad q_i = \frac{m}{2} \int C^2 C_i f d\mathbf{c} \, . \tag{A35}$$

Note that the stress tensor of fluid dynamics is just the negative pressure tensor. Pressure is the trace of the pressure tensor,

$$p = \frac{1}{3} p_{kk} = \frac{1}{3} m \int C^2 f d\mathbf{c} = \frac{2}{3} \rho u ,$$

(A36)

From the ideal gas law $p = \rho R T$ follows the relation between energy and temperature as

$$u = \frac{3}{2} R T , \quad \text{or} \quad T = \frac{1}{3} \frac{m}{\rho R} \int C^2 f d\mathbf{c} .$$

(A37)

Temperature in kinetic theory is usually defined by the above relation for all situations, including strong deviations from equilibrium.

The atoms in a monatomic gas have three translational degrees of freedom (the three components of velocity), and each degree of freedom contributes $\frac{1}{2} R T$ to the specific internal energy, or $\frac{1}{2} R$ to the specific heat $c_v = \left(\frac{\partial u}{\partial T} \right)_\rho$.

Appendix D.2. Equilibrium and the Maxwellian Distribution

A simple argument going back to Maxwell allows us to find the velocity distribution in equilibrium. Equilibrium is a state where no changes will occur when the gas is left to itself, and this will imply that the gas is homogeneous, that is, displays no gradients in any quantity (external forces such as gravity are ignored), and the phase density is isotropic, that is independent of the direction $v_i = C_i / C$. The following argument considers the gas in the rest-frame, where $V_i = 0$.

An arbitrary atom picked from the gas will have the velocity components $C_k, k = 1, 2, 3$, and the probability to find the component in direction r_k within the interval $[C_k, C_k + dC_k]$ is given by $\Pi (C_k) dC_k$. Note that, due to isotropy, the probability function Π is the same for all components. Then, the probability to find a particle with the velocity vector $\{C_1, C_2, C_3\}$ is given by

$$F (C) \, dC_1 dC_2 dC_3 = \Pi (C_1) \, \Pi (C_2) \, \Pi (C_3) \, dC_1 dC_2 dC_3$$

(A38)

where $F (C) = f (C) / (\rho / m)$ depends only on the absolute value of velocity, C, since the probability must be independent of direction. Thus, F and Π are related as

$$F (C) = \Pi (C_1) \, \Pi (C_2) \, \Pi (C_3) .$$

(A39)

Taking the logarithmic derivative of this equation with respect to C_1 we see

$$\frac{\partial}{\partial C_1} \ln F (C) = \frac{\partial}{\partial C_1} [\ln \Pi (C_1) + \ln \Pi (C_2) + \ln \Pi (C_3)]$$

(A40)

or

$$\frac{1}{C} \frac{F' (C)}{F (C)} = \frac{1}{C_1} \frac{\Pi' (C_1)}{\Pi (C_1)} = -2\gamma .$$

(A41)

Since the left and the right side of this equation depend on different variables, γ must be a constant, and integration gives an isotropic Gaussian distribution,

$$F = \frac{f}{n} = A \exp \left[-\gamma C^2 \right] \quad \text{and} \quad \Pi (C_k) = \sqrt[3]{A} \exp \left[-\gamma C_k^2 \right] ,$$

(A42)

where A is a constant of integration. The two constants γ and A follow from the conditions that the phase density must reproduce mass and energy density, that is

$$\rho = m \int f d\mathbf{c} \quad \text{and} \quad \rho u = \frac{3}{2} \rho R T = \frac{m}{2} \int C^2 f d\mathbf{c} .$$

(A43)

The resulting equilibrium phase density is the Maxwellian distribution

$$f_M = \frac{\rho}{m} \frac{1}{\sqrt{2\pi RT}^3} \exp\left[-\frac{C^2}{2RT}\right] . \tag{A44}$$

Appendix D.3. Kinetic Equation and Conservation Laws

The evolution of the phase density f in space time is given by the Boltzmann equation, which describes the change of f due to free flight, external forces G_k, and binary collisions between particles. Instead of discussing the full Boltzmann equation, we consider a kinetic model equation, known as the BGK model, where the Boltzmann collision term is replaced by a relaxation term,

$$\frac{\partial f}{\partial t} + c_k \frac{\partial f}{\partial r_k} + G_k \frac{\partial f}{\partial c_k} = -\frac{1}{\tau}(f - f_M) . \tag{A45}$$

Here, τ is the mean free time, that is, the average time a particle travels freely between two collisions with other particles, and f_M is the local Maxwell distribution. The right hand side describe the change of the distribution due to collisions, which move the distribution function closer towards the Maxwellian. If the collision frequency $1/\tau$ is relatively large, than we expect that the local state is close to the Maxwellian at all times, which implies local equilibrium. The discussion further below will give more details on this particular limit.

While not exact, the BGK model shares the main features of the Boltzmann equation, and is used in the present brief sketch because of its simplicity.

To simplify notation, we write moment of the phase density as

$$\rho \langle \psi \rangle = m \int \psi f d\mathbf{c} , \tag{A46}$$

where ψ is any function of $(\mathbf{r}, t, \mathbf{c})$, so that, for example, $\rho = \rho \langle 1 \rangle$, $\rho v_i = \rho \langle c_i \rangle$, $\rho e = \frac{\rho}{2} \langle c^2 \rangle$, and so forth. The evolution equation for $\langle \psi \rangle$, the equation of transfer, is obtained by multiplying (A45) with $\psi(\mathbf{r}, t, \mathbf{c})$, and subsequent integration,

$$\frac{\partial \rho \langle \psi \rangle}{\partial t} + \frac{\partial \rho \langle \psi c_k \rangle}{\partial r_k} = \rho \left\langle \frac{\partial \psi}{\partial t} \right\rangle + \rho \left\langle c_k \frac{\partial \psi}{\partial r_k} \right\rangle + \rho \left\langle G_k \frac{\partial \psi}{\partial c_k} \right\rangle + S_\psi . \tag{A47}$$

The production term S_ψ is defined as

$$S_\psi = -\frac{1}{\tau} \int \psi (f - f_M) d\mathbf{c} , \tag{A48}$$

with no productions for the conserved quantities mass, momentum and energy,

$$S_1 = S_{c_i} = S_{c^2} = 0 , \tag{A49}$$

that is the Boltzmann equation, and the BGK model, guarantee conservation of mass, momentum, and energy.

For $\psi = \left\{1, c_i, \frac{1}{2}c^2\right\}$, (A47) and the above definitions of macroscopic properties imply the conservation laws for mass, momentum and energy (A14)–(A16).

Appendix D.4. Entropy and 2nd Law in Kinetic Theory

In the equation of transfer (A47) we chose, following Boltzmann [15–17],

$$\psi = -k_B \ln \frac{f}{y} \tag{A50}$$

with the Boltzmann constant k_B and another constant y. We introduce

$$\eta = -k_B \int f \ln \frac{f}{y} d\mathbf{c} \;, \quad \Phi_\kappa = -k_B \int c_k f \ln \frac{f}{y} d\mathbf{c} \tag{A51}$$

and

$$\Sigma = S_{-k_B \ln \frac{f}{y}} = \frac{k_B}{\tau} \int \ln \frac{f}{y} (f - f_M) \, d\mathbf{c} = \frac{k_B}{\tau} \int \ln \frac{f}{f_M} (f - f_M) \, d\mathbf{c} \;, \tag{A52}$$

so that the corresponding transport equation for η reads

$$\frac{\partial \eta}{\partial t} + \frac{\partial \Phi_k}{\partial r_k} = \Sigma \;. \tag{A53}$$

Closer examination shows that the collision term (A52) cannot be negative, and Σ vanishes in equilibrium, so that

$$\Sigma \geq 0 \;. \tag{A54}$$

This feature of the BGK model reproduces the behavior of the full Biolzmann collision term. Thus, η always has a positive production which vanishes in equilibrium. Accordingly, η can only grow in an isolated system, where no flux over the surface is allowed ($\oint_{\partial V} \Phi_k n_k dA = 0$), and reaches its maximum in equilibrium, where $\Sigma = 0$. This property of η and in particular the definite sign of Σ are known as the H-theorem.

In our view, the H-theorem is equivalent to the second law of thermodynamics, the entropy law, which was introduced above on purely phenomenological grounds. Then, η is the entropy density of the gas, and we write

$$\eta = \rho s \;, \tag{A55}$$

where s denotes the specific entropy. Φ_κ as given in (A51)$_2$ is the entropy flux, and the non-convective entropy flux $\phi_k = \Phi_k - \rho s v_k$ can be computed according to

$$\phi_\kappa = -k_B \int C_k f \ln \frac{f}{y} d\mathbf{c} \;. \tag{A56}$$

This gives the second law in the form

$$\rho \frac{Ds}{Dt} + \frac{\partial \phi_k}{\partial x_k} = \Sigma \geq 0 \;; \tag{A57}$$

Σ is the entropy production (entropy generation rate).

The specific entropy of the monatomic gas in equilibrium follows by evaluating (A51)$_1$ with the Maxwellian (A44) as

$$s = R \ln \frac{T^{3/2}}{\rho} + s_0 \quad \text{where} \quad s_0 = R \left[\frac{3}{2} + \ln \left(m R^{3/2} y \sqrt{2\pi}^3 \right) \right] \;. \tag{A58}$$

This result stands in agreement with classical thermodynamics.

Phase density and Boltzmann equation are valid for arbitrary nonequilibrium states, hence the above expressions for entropy, entropy flux, and entropy generation, and the balance law (A57) hold for arbitrary nonequilibrium states. There is no restriction of Boltzmann entropy to equilibrium states!

The Boltzmann equation is constructed such that entropy η has a strictly non-negative production rate $\Sigma \geq 0$. This is not in agreement with the behavior of microscopic mechanical systems. The instantaneous reversal of the microscopic velocities of all particles should send the gas back to its initial state, which would imply destruction of entropy (Loschmidt's reversibility paradox). Also, any mechanical system will, after some time, return to (microscopic) states arbitrarily close to its initial state, with an entropy close to that of the initial state, which also implies destruction of entropy (Zermelo's recurrence paradox).

The reversibility paradox is resolved by observing that the overwhelming majority of possible microscopic initial conditions (for the same macroscopic state) will result in processes with non-negative entropy generation. Microscopic states that result from sudden inversion of all particle velocities belong to the comparatively rather small number of possible initial conditions that lead to entropy destruction. Indeed, the relative number of such initial conditions must be so small that these can be ignored—after all, observation of macroscopic processes shows non-negative entropy generation. Also, it must be noted that it is impossible to realize the sudden reversal of all microscopic velocities.

The recurrence paradox is resolved by observing that the recurrence time is so long (much longer than the lifetime of the universe for systems with many particles) that entropy destruction would not be observed during our lifetimes.

In summary, we can state that the Boltzmann equation describes the probable behavior of gases, with non-negative generation of entropy. For a deeper discussion of these points we must refer the reader to References [1,15–17].

Appendix D.5. Kinetic Theory in the Continuum Limit

To align the idea of local thermodynamic equilibrium with kinetic theory, we use the Chapman-Enskog method to determine an approximate phase density from the kinetic equation. The approach presented is justified for flows where the Knudsen number is sufficiently small [16,17]. The Knudsen number is defined as the ratio of average mean free path $\lambda = \tau \sqrt{RT}$ of the gas and the relevant length scale of a process, as $\mathrm{Kn} = \frac{\tau \sqrt{RT_0}}{L}$, where we use $\sqrt{RT_0}$ with a meaningful reference temperature T_0 as a measure of average particle speed in the restframe.

In dimensionless form, with scales based on L, T_0, the BGK collison term appears as $\frac{1}{\mathrm{Kn}} (f - f_M)$, and evaluation of the Boltzmann equation can be performed in the limit of small Knudsen numbers. Instead, one can work with the dimensional equation, and introduce a formal scaling parameter ε that plays the role of the Knudsen number in taking limits. The parameter will be set to unity at the end of the scaling procedure. Hence, we write the kinetic equation with the convective time derivative as

$$\frac{Df}{Dt} + C_k \frac{\partial f}{\partial r_k} + G_k \frac{\partial f}{\partial c_k} = -\frac{1}{\varepsilon} \frac{1}{\tau} (f - f_M) \ . \tag{A59}$$

The idea of the Chapman-Enskog expansion is to write f as a series in ε (or, in the Knudsen number Kn, if the dimensionless equation is used),

$$f = f^{(0)} + \varepsilon f^{(1)} + \varepsilon^2 f^{(2)} + \cdots \tag{A60}$$

The expansion terms $f^{(\alpha)}$ are determined from inserting the series into the Boltzmann equation, and evaluation of the contributions at the different powers of ε. Here, we are only interested in the first two contributions. Inserting the ansatz into (A59), and equating terms of equal powers in ε^{-1} and ε^0 yields at first

$$0 = -\frac{1}{\tau} \left(f^{(0)} - f_M \right) \quad , \quad \frac{Df^{(0)}}{Dt} + C_k \frac{\partial f^{(0)}}{\partial r_k} + G_k \frac{\partial f^{(0)}}{\partial c_k} = -\frac{1}{\tau} f^{(1)} \ . \tag{A61}$$

Hence, the leading order term is the local Maxwellian, $f^{(0)} = f_M$, and the first order correction $f^{(1)}$ is determined from the Maxwellian. Using the conservation laws for eliminating the ensuing time derivatives, one finds

$$f^{(1)} = -f_M \tau \left[\frac{C_{\langle k} C_{l \rangle}}{\theta} \frac{\partial v_{\langle k}}{\partial r_{l \rangle}} + C_k \left(\frac{C^2}{2\theta} - \frac{5}{2} \right) \frac{1}{\theta} \frac{\partial \theta}{\partial r_k} \right] \ . \tag{A62}$$

When the first order approximation of the distribution function, $f^{(0)} + \varepsilon f^{(1)}$, is inserted into (A35) to determine stress tensor and heat flux to first order, one finds just the laws of Navier-Stokes and Fourier,

$$t_{ij}^{(1)} = -p\delta_{ij} + 2p\tau \frac{\partial v_{\langle i}}{\partial r_{k\rangle}} \quad \text{and} \quad q_i^{(1)} = -\frac{5}{2}p\tau \frac{\partial \theta}{\partial r_i} , \tag{A63}$$

where viscosity and thermal conductivity for the BGK model are identified as

$$\mu = p\tau \quad \text{and} \quad \kappa = \frac{5}{2}p\tau . \tag{A64}$$

We note that the BGK model gives an incorrect Prandtl number, $\text{Pr} = \frac{5}{2}\frac{\mu}{\kappa} = 1$, while the full Boltzmann equation yields the proper value of $\text{Pr} \simeq \frac{2}{3}$. Moreover, the monatomic gas has no bulk viscosity.

According to the CE expansion, the first order solution for the phase density can be written as

$$f = f_M (1 + \varepsilon\varphi) , \tag{A65}$$

and the corresponding expressions for entropy and its flux follow from first order expansion of the logarithm, $\ln(1 + \varepsilon\varphi) \simeq \varepsilon\varphi$. For the entropy we obtain to first order in ε

$$\rho s = -k \int f_M (1 + \varepsilon\varphi) \left(\ln \frac{f_M}{y} + \varepsilon\varphi \right) d\mathbf{c} = -k \int f_M \ln \frac{f_M}{y} d\mathbf{c} . \tag{A66}$$

Since the first order contributions vanish, this is just the entropy in equilibrium (24). We conclude that to first order in ε the gas can be described locally as if it were in equilibrium. This result gives justification to the hypothesis of local thermodynamic equilibrium.

The corresponding first order entropy flux is obtained as

$$\phi_\kappa = -k \int C_k f_M \varepsilon\varphi \left(\ln \frac{f_M}{y} + 1 \right) d\mathbf{c} . \tag{A67}$$

Since $\ln f_M + 1 = -C^2/2\theta + F(\rho, T)$, and $\int C_k f_M \varphi d\mathbf{c} = 0$, we obtain, again after setting the formal parameter $\varepsilon = 1$, an approximation for the entropy flux to first order in ε,

$$\phi_\kappa = \frac{1}{T} \frac{m}{2} \int C_k C^2 f_M \varphi d\mathbf{c} = \frac{q_k}{T} . \tag{A68}$$

Also this relation between heat flux and entropy flux is well-known in classical thermodynamics. The corresponding entropy generation rate Σ is the same as in LIT, given in (A24).

Appendix D.6. Entropy as $S = k \ln \Omega$

Boltzmann's choice (A50) allows to relate entropy to microscopic properties, and to interpret entropy as a measure for disorder [5,17].

We consider a gas of N particles in the volume V, and compute its total entropy as

$$H = \int_V \eta d\mathbf{r} = -k \int_V \int f \ln \frac{f}{y} d\mathbf{c} d\mathbf{r} . \tag{A69}$$

The number of particles in a phase cell is

$$N_{\mathbf{r},\mathbf{c}} = f d\mathbf{r} d\mathbf{c} = fY , \tag{A70}$$

where $Y = d\mathbf{r}d\mathbf{c}$ is the size of the cell. By writing sums over phase cells instead of integrals, and with $N = \sum_{\mathbf{r},\mathbf{c}} N_{\mathbf{r},\mathbf{c}}$ follows

$$H = -k \sum_{\mathbf{r},\mathbf{c}} N_{\mathbf{r},\mathbf{c}} \ln \frac{N_{\mathbf{r},\mathbf{c}}}{yY} = -k \sum_{\mathbf{r},\mathbf{c}} N_{\mathbf{r},\mathbf{c}} \ln N_{\mathbf{r},\mathbf{c}} + kN \ln yY \,. \tag{A71}$$

Use of Stirling's formula, $\ln N! = N \ln N - N$, yields

$$H = -k \sum_{\mathbf{r},\mathbf{c}} (\ln N_{\mathbf{r},\mathbf{c}}! + N_{\mathbf{r},\mathbf{c}}) + kN \ln yY = k \ln \frac{N!}{\prod_{\mathbf{r},\mathbf{c}} N_{\mathbf{r},\mathbf{c}}!} + kN \ln \frac{yY}{N} \,. \tag{A72}$$

Choosing $y = N/Y$ finally allows to write

$$H = k \ln \Omega \tag{A73}$$

where

$$\Omega = \frac{N!}{\prod_{\mathbf{r},\mathbf{c}} N_{\mathbf{r},\mathbf{c}}!} \tag{A74}$$

is the number of possibilities to distribute N particles into the cells of phase space, so that the cell at $(\mathbf{r}, \mathbf{c})$ contains $N_{\mathbf{r},\mathbf{c}}$ particles.

Equation (A73) relates the total gas entropy H to the number of possibilities to realize the state of the gas. The growth of entropy, which is imperative in an isolated system, therefore corresponds to an increasing number of possibilities to realize the state. Since a small number of possibilities refers to an ordered state, and a large number to disorder, we can say that the H-theorem states that disorder must grow in an isolated process. Accordingly, entropy is often interpreted as a measure for disorder.

The relation (A73) is generally accepted as being valid not only for monatomic ideal gases—where it originated—but for any substance. However, the evaluation for other substances can be quite difficult, or impossible, since it requires a detailed understanding of the microscopic details of phase space, accessible states, and so forth, in order to determine Ω properly.

Finally we note that we used the symbols η, H to denote Boltzmann entropies, but that we consider these to be the actual entropies of the gas, that is $\eta = \rho s, H = \int \eta dV = \int \rho s dV = S$.

References

1.   Müller, I. *A History of Thermodynamics*; Springer: Berlin, Germany, 2007.
2.   Cengel, Y.; Boles, M. *Thermodynamics: An Engineering Approach*, 5th ed.; McGraw-Hill: Boston, MA, USA, 2005.
3.   Moran, M.J.; Shapiro, H.N. *Fundamentals of Engineering Thermodynamics*, 6th ed.; John Wiley & Sons: New York, NY, USA, 2006.
4.   Borgnakke, C.; Sonntag, R.E. *Fundamentals of Thermodynamics*, 7th ed.; John Wiley & Sons: New York, NY, USA, 2008.
5.   Müller, I.; Müller, W.H. *Fundamentals of Thermodynamics and Applications: With Historical Annotations and Many Citations from Avogadro to Zermelo*; Springer: Berlin, Germany, 2009.
6.   Baehr, H.D.; Kabelac, S. *Thermodynamik: Grundlagen und technische Anwendungen*, 15th ed.; Springer: Berlin, Germany, 2012.
7.   Bhattacharjee, S. *Thermodynamics: An Interactive Approach*; Prentice Hall: New York, NY, USA, 2014.
8.   Struchtrup, H. *Thermodynamics and Energy Conversion*; Springer: Heidelberg, Germany, 2014.
9.   Caratheodory, C. Untersuchungen über die Grundlagen der Thermodynamik. *Math. Ann.* **1909**, *67*, 355–386. [CrossRef]
10.   Callen, H.B. *Thermodynamics and an Introduction to Thermostatistics*, 2nd ed.; John Wiley & Sons: New York, NY, USA, 1985.
11.   Uffink, J. Bluff Your Way in the Second Law of Thermodynamics. *Stud. Hist. Phil. Mod. Phys.* **2001**, *32*, 305–394. [CrossRef]

12. Lieb, E.H.; Yngvason, J. The physics and mathematics of the second law of thermodynamics. *Phys. Rep.* **1999**, *310*, 1–96. [CrossRef]

13. Lieb, E.H.; Yngvason, J. A fresh look at entropy and the second law of thermodynamics. *Phys. Today* **2000**, *53*, 32. [CrossRef]

14. Thess, A. *The Entropy Principle: Thermodynamics for the Unsatisfied*; Springer: Berlin, Germany, 2011.

15. Cercignani, C.*Theory and Application of the Boltzmann Equation*; Scottish Academic Press: Edinburgh, UK, 1975.

16. Kremer, G.M. *An Introduction to the Boltzmann Equation and Transport Processes in Gases*; Springer: Berlin, Germany, 2010.

17. Struchtrup, H. *Macroscopic Transport Equations for Rarefied Gas Flows Approximation Methods in Kinetic Theory*; Springer: Berlin, Germany, 2005.

18. de Groot, S.R.; Mazur, P. *Non-Equilibrium Thermodynamics*; Dover: New York, NY, USA, 1984.

19. Müller, I. *Thermodynamics*; Pitman: Boston, MA, USA, 1985.

20. Müller, I.; Ruggeri,T. *Rational Extended Thermodynamics (Springer Tracts in Natural Philosophy)*, 2nd ed.; Springer: New York, NY, USA, 1998.

21. Jou, D.; Casas-Vasquez, J.; Lebon, G. *Extended Irreversible Thermodynamics*, 4th ed.; Springer: New York, NY, USA, 2009.

22. Öttinger, H.C. *Beyond Equilibrium Thermodynamics*; John Wiley & Sons: Hoboken, NJ, USA, 2005.

23. Bejan, A. *Advanced Engineering Thermodynamics*, 4th ed.; John Wiley & Sons: Hoboken, NJ, USA, 2016.

24. de Podesta, M. Rethinking the Kelvin. *Nature Phys.* **2016**, *12*, 104. [CrossRef]

25. Fischer, J.; Ullrich, J. The new system of units. *Nature Phys.* **2016**, *12*, 4–7. [CrossRef]

3

Discrepancy between Constant Properties Model and Temperature-Dependent Material Properties for Performance Estimation of Thermoelectric Generators

Prasanna Ponnusamy [1,*]**, Johannes de Boor** [1] **and Eckhard Müller** [1,2,*]

[1]   Institute of Materials Research, German Aerospace Center (DLR), D-51170 Köln, Germany;
       Johannes.deboor@dlr.de

[2]   Institute of Inorganic and Analytical Chemistry, Justus Liebig University Gießen, D-35392 Gießen, Germany

*   Correspondence: prasanna.ponnusamy@dlr.de (P.P.); Eckhard.mueller@dlr.de (E.M.)

Abstract: The efficiency of a thermoelectric (TE) generator for the conversion of thermal energy into electrical energy can be easily but roughly estimated using a constant properties model (CPM) developed by Ioffe. However, material properties are, in general, temperature (T)-dependent and the CPM yields meaningful estimates only if physically appropriate averages, i.e., spatial averages for thermal and electrical resistivities and the temperature average (TAv) for the Seebeck coefficient (α), are used. Even though the use of α_{TAv} compensates for the absence of Thomson heat in the CPM in the overall heat balance, we find that the CPM still overestimates performance (e.g., by up to 6% for PbTe) for many materials. The deviation originates from an asymmetric distribution of internally released Joule heat to either side of the TE leg and the distribution of internally released Thomson heat between the hot and cold side. The Thomson heat distribution differs from a complete compensation of the corresponding Peltier heat balance in the CPM. Both effects are estimated quantitatively here, showing that both may reach the same order of magnitude, but which one dominates varies from case to case, depending on the specific temperature characteristics of the thermoelectric properties. The role of the Thomson heat distribution is illustrated by a discussion of the transport entropy flow based on the $\alpha(T)$ plot. The changes in the lateral distribution of the internal heat lead to a difference in the heat input, the optimum current and thus of the efficiency of the CPM compared to the real case, while the estimate of generated power at maximum efficiency remains less affected as it is bound to the deviation of the optimum current, which is mostly <1%. This deviation can be corrected to a large extent by estimating the lateral Thomson heat distribution and the asymmetry of the Joule heat distribution. A simple guiding rule for the former is found.

Keywords: TEG performance; device modeling; temperature profile; constant properties model; Fourier heat; Thomson heat; Joule heat

1. Introduction

Thermoelectric generator (TEG) materials convert a certain fraction of the heat passed through them into useful electrical power, as the charge carriers (holes/electrons) absorb the thermal energy and move from the hot side to the cold side, carrying entropy [1,2]. The transport entropy flux related to the convective heat transport is given by αj, with the Seebeck coefficient $\alpha(T)$ and current density j. Typically, a thermoelectric (TE) module consists of a series of pn leg pairs (thermocouples), electrically connected in series and thermally in parallel [3]. In steady-state conditions, the exact performance of the TEG is obtained by solving the thermoelectric heat balance equation [4] for the temperature profile $T(x)$. In 1D, it reads

$$\kappa(T)\frac{\partial^2 T}{\partial x^2} + \frac{d\kappa}{dT}\left(\frac{\partial T}{\partial x}\right)^2 - jT\frac{d\alpha}{dT}\frac{\partial T}{\partial x} = -\rho(T)j^2 \tag{1}$$

where the thermal conductivity κ, the electrical resistivity ρ and α are the three main temperature-dependent thermoelectric properties. Here, $\frac{\partial}{\partial x}\cdot\left\{\kappa(T)\frac{\partial T}{\partial x}\right\} = \kappa(T)\frac{\partial^2 T}{\partial x^2} + \frac{d\kappa}{dT}\left(\frac{\partial T}{\partial x}\right)^2$ corresponds to the (negative) divergence of the Fourier heat flux, i.e., its local change; $jT\frac{d\alpha}{dT}\frac{\partial T}{\partial x}$ corresponds to the local Thomson heat absorption driven by the change of the convective entropy flux αj related to the temperature dependence of $\alpha(T)$, and $\rho(T)j^2$ corresponds to the local Joule heat dissipation. With a typical TE material, with $\kappa(T)$ falling with T and the amount of $\alpha(T)$ rising with T, Thomson heat will be released and Fourier heat flow will grow from the hot to the cold side along a TE leg in TEG operation, where the current flow is driven by the thermo-voltage generated by the leg. Equation (1) is a second-order non-linear partial differential equation, which can be solved using numerical methods like finite element methods (FEMs) [5,6], finite volume methods (FVMs) [7–9] or finite difference methods (FDMs) [6]. However, these solution methods are costly and time-consuming.

On the other hand, when assuming constant properties of the TE properties, an approximate solution can be found analytically, as suggested by Ioffe [1]. This solution by the constant property model (CPM) involves a discrepancy from the exact results due to the underlying simplification. Moreover, the choice of the averaged constant properties to be obtained from the actual temperature-dependent data is not straightforward. As can be seen from Equation (1), the Thomson heat vanishes when the Seebeck coefficient (and with that the convective entropy flux αj) remains constant. Various models corrected the CPM to compensate for this "missing Thomson heat" [10–16] have been proposed. Meanwhile, Sandoz et al. [17] attempted to explain the use of the T-averaged Seebeck coefficient in predicting exact power in the CPM mathematically, but did not recognize the importance of the asymmetry in heat distribution for the prediction of efficiency.

In a previous study [18], on the physically appropriate choice of averages in the CPM, we highlighted that spatial averages (SpAv) for resistivities (electrical and thermal) and temperature averaging (TAv) for the Seebeck coefficient are essential for a meaningful CPM estimate. However, there is still a remaining deviation due to unconsidered local redistribution of internal heat release or absorption and of thermal conduction in the CPM, which is linked to a change in the T profile $T(x)$ [1,12,18,19]. Here, we will analyze the individual heat contributions exemplarily for six representative thermoelectric materials that we considered previously [18] plus PbTe [20], as this is one of the best TE materials in practice and shows an especially large deviation between CPM and exact results.

Initially, the effect of the T dependence of each of the TE properties, α, ρ and κ, leading to locally shifted heat release and transport over the TE element, for performance estimation, is studied separately. Calculated maximum efficiency in the full temperature-dependent case, η_{max}, is compared with tailored model materials, in order to separate and quantify the individual contributions. Model materials are defined by setting one or two of the three TE properties constant at its respective average while keeping the other properties T dependent. Next, we explain the physical origin of a relevant part of the discrepancy between CPM results and the real situation using a schematic plot of the convective entropy flux derived from an $\alpha(T)$ graph, alongside showing that the net Peltier/Thomson heat is correctly considered by the CPM when appropriate temperature averaging is used for $\alpha(T)$. Marked areas in the entropy flux diagram quantify the exchange of Peltier and Thomson heat, and with that, a correction for the related deviation in CPM efficiency estimation, $d\eta_{max} = \frac{\eta_{max} - \eta_{max}^{CPM}}{\eta_{max}^{CPM}}$, is suggested and demonstrated.

2. Methods, Results and Discussion

2.1. Role of the T Dependence of Material Properties in Performance Estimation

Since a generalized temperature dependence study for all types of T dependence is quite elaborate, a comparative study based on seven well-known and representative TE materials [20–24] was conducted. To understand the role of the T dependence of each of α, ρ and κ in performance estimation, the calculated maximum efficiencies when all properties are considered as T dependent (referred to as "real case" or "exact" from now on) were compared with the calculated efficiencies of model materials. These model materials have the same T dependence as the real materials for one or two of the three thermoelectric transport properties, while the remaining properties are kept constant; these materials are denoted as two temperature-dependent property (2TD) materials and 1TD materials, respectively. The constants used to define the model materials were obtained using the spatial averages (SpAv; for electrical and thermal resistivity) at a current density corresponding to the maximum efficiency of the real material and the temperature average (TAv; for the Seebeck coefficient). The SPAv and TAv of a T-dependent quantity p for a hot side temperature T_h and a cold side temperature T_c are given by [1,12,18,25]

$$p_{TAv} = \bar{p} = \frac{1}{\Delta T} \int_{T_c}^{T_h} p(T)dT \tag{2}$$

$$p_{SpAv} = \langle p \rangle = \frac{1}{L} \int_0^L p(T(x))dx \tag{3}$$

where $\Delta T = T_h - T_c$ and L are the length of the TE leg. The exact efficiency using T-dependent properties was obtained using the 1D solution algorithm developed in [18] by calculating

$$P = V \cdot I, \\ \text{where } V = V_o - R_{in}I, V_o = \alpha \Delta T \text{ and} \tag{4}$$

$$\eta = P/Q_{in} \tag{5}$$

Here, P is the output power, V is the net output voltage which is given by the Seebeck voltage generated, $V_o = \int_{T_c}^{T_h} \alpha(T)dT$, minus the voltage drop due to internal resistance $R_i = \frac{\rho_{SpAv}L}{A}$, where A is the area of the TE leg and $\rho_{SpAv} = \frac{1}{L}\int_0^L \rho(T(x))dx$. $I = jA$ is the current passing through the TE material due to the generated voltage. The efficiency (η) is given by the ratio of output power to the input heat flow (Q_{in}) as in Equation (5), where Q_{in} is given by

$$Q_{in} = -\kappa_h \cdot A \cdot \frac{dT}{dx}_h + I \cdot \alpha_h \cdot T_h \tag{6}$$

Q_{in} consists of the Fourier heat flow $-\kappa_h \cdot A \cdot \frac{dT}{dx}_h$ (including the fraction of Joule and Thomson heat contributions released in the leg which is flowing to the hot side) plus the Peltier heat ($I \cdot \alpha_h \cdot T_h$) absorbed at the hot side. The suffix h indicates the hot side values, i.e., $\kappa_h = \kappa(T_h)$ and $\alpha_h = \alpha(T_h)$. As the spatial averages depend on $T(x)$, which in turn varies with current, they were formed pre-assuming the optimum current of the real materials. For brevity, the efficiency was also calculated at the optimum current of the real material. The optimum current in the numerical calculation was obtained by finding the current where $\frac{d\eta}{dI}$ becomes zero.

The relative deviation (RD) of the calculated maximum efficiency between the 2TD model materials and the real materials, $\delta\eta_{max}^{model} = \frac{\eta_{max} - \eta_{max}^{model}}{\eta_{max}}$, is shown in Figure 1a. Here, and in the following, for brevity, we will use δ and d to denote a relative and absolute deviation, respectively. The comparison shows how strongly each of the contributing T dependences alone would shift efficiency. Obviously, the T dependence of ρ will affect the calculated efficiency to a lower extent than

$\alpha(T)$ and $\kappa(T)$ will do for some materials (middle section of Figure 1a); the asymmetry of Joule heat generation mostly plays a minor role. However, this does not hold for all materials and it does not mean that the RD between the CPM and a real material due to asymmetric distribution of Joule heat, $\delta\eta_{\text{maxJ}} = \frac{d\eta_{\text{max}}}{d\dot{Q}_J^{\text{h}}}\delta\dot{Q}_J^{\text{h}}$, would be insignificant, as all of the three identified effects will act simultaneously when comparing the CPM and the real case. Although the effects of the T dependence of $\alpha(T)$ and $\kappa(T)$ are much larger for some materials, they often partly cancel each other. A comparison of the real Joule heat partial T profiles in Figure 1b shows a considerable asymmetry, in correlation to the deviations in the $\rho(T) = const.$ case for SnSe and PbTe (Figure 1a, mid); however, the RD contribution related to the profiles in Figure 1b is larger as they contain an asymmetry due to the asymmetry of axial heat conduction linked to $\kappa(T)$, in addition to the asymmetry of Joule heat generation which alone is represented by Figure 1a. Calculation of the partial T profiles is explained in Appendix A.2. It should be noted that unlike for $\alpha(T)$, where the absence of the T dependence means an absence of Thomson heat, the absence of the T dependence of ρ just means that there is no local asymmetry in Joule heat generation, whereas the amount of Joule heat that appears remains unchanged. Both symmetrically or asymmetrically released Joule heat will contribute, together with Thomson heat, to the effect of a T dependence of $\kappa(T)$ that consists in shifting the distribution of the inner reversible and irreversible heat towards the hot and cold sides. Accordingly, the magnitude of the effect of a T dependence of $\kappa(T)$ will scale with the total amount of inner heat.

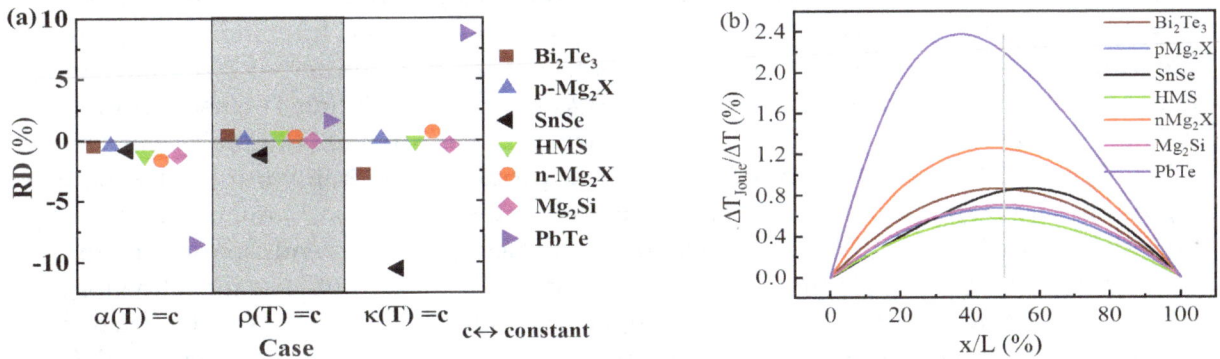

Figure 1. (a) Comparison of the relative deviation of the calculated maximum efficiency of 2TD (two temperature-dependent property) model materials (one of the thermoelectric properties kept constant,) to their real counterpart for the example materials, (b) T profile bending caused by Joule heat for example materials. Distinct asymmetry is observed particularly for PbTe and SnSe, correlated to maximum offset values in the middle part of Figure (a).

When α or κ is kept constant, there can be large discrepancies, as seen from the scatter in the left and right section of Figure 1a. Switching off Thomson heat results in a change from non-constant to constant convective entropy flux linked to a different partition of reversible (Peltier + Thomson-bound) heat to both sides of the leg. When setting $\kappa(T) = const.$, net Fourier heat transmitted does not change as the thermal resistance of the leg is fixed by the definition of the SpAv. Rather, the observed differences are merely due to a changed lateral distribution of Thomson and Joule heat. Comparing this to Figure 2a reveals that a large RD for $\kappa(T) = const.$ correlates to strongly non-linear T profiles linked to $\kappa(T)$ (see T profiles for $j = 0$); see also Appendix A.2, Figure A1a, where SnSe, Bi$_2$Te$_3$ and PbTe have significantly different κ_{h} and κ_{c} and Figure A2a, showing that the weight of Joule and Thomson heat to Q_{in} is comparably large for these materials.

The dominating effect of the T dependence of κ and α on the estimated performance is also seen by comparing the T profiles of the model cases with the real temperature profile of n-type Mg$_2$(Si,Sn) (referred to as n-Mg$_2$X), Figure 2b. All profiles are calculated for the optimum current for maximum efficiency of the real material. Here, in addition to the 2TD materials, 1TD materials were also involved.

$\alpha(T)$ and $\kappa(T)$ play a dominating role in the shaping of the temperature profile, which is reflected by the closeness of the $\alpha(T) \neq const.$, $\kappa(T) \neq const.$ case to the real material.

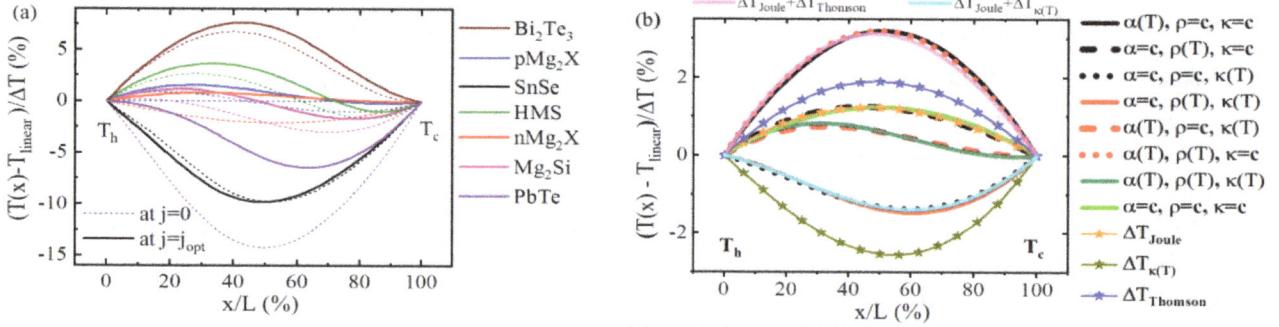

Figure 2. (a) Bending of T profiles for the real materials at $j = 0$ (dotted lines) and $j = j_{opt}$ (solid lines), normalized to ΔT, (b) T profile bending for the 1TD and 2TD model materials in comparison to the full T-dependent case and the constant properties case, along with the individual contributions to the fully T-dependent profile for an n-Mg$_2$(Si,Sn) TE leg with $T_h = 723$ K and $T_c = 383$ K.

The effects of the 2TD cases on the overall inflowing Fourier heat and thus on the efficiency of n-Mg$_2$X from Figure 1a (red dots) can be discussed in terms of the hot side slopes of the corresponding temperature profiles (red lines) in Figure 2b when comparing between cases with the same $\kappa(T)$. The downward $\frac{dT}{dx}_h$ for the 2TD material with $\alpha(T) = const.$(red solid line) indicates an increase in the inflowing Fourier due to missing Thomson heat, compared to the actual case (dark green line). Simultaneously, but only partly compensated in the Q_{in} balance by missing Thomson heat, less Peltier heat is absorbed at the hot side and therefore the efficiency is overestimated (Figure 1a left side, red dot). The 2TD $\kappa(T) = const.$(red dotted line) deforms the T profile considerably but hardly increases the heat input (Equation (6)) compared to the real material, as the SpAv of $\kappa(T)$ maintains an unchanged thermal resistance of the TE leg. We can conclude that replacing the T dependence of $\alpha(T)$ and $\kappa(T)$ by adequate constants will, although significantly changing the T profile, influence the inflowing heat and thus efficiency to a much lower extent due to compensating effects. The RD of CPM efficiency in effect arises mainly from a redistribution of internal Joule and Thomson heat due to considerable deformation of the T profile by neglecting the T dependence of $\kappa(T)$ and $\alpha(T)$ and local redistribution of reversible heat generation as a consequence of neglect of the T dependence of the convective entropy flux.

When comparing the 1TD and 2TD model materials, additionally a shift of the SpAv values of ρ and κ as a consequence of different T profiles, as well as coupling effects among the individual contributions, play a role, but only to a very minor extent, as proven by the close coincidence of their profiles to combinations of the individual partial T profiles of the real material, see Figure 2b (pink and cyan lines). The latter represent the physical contributions to the real temperature profile, ΔT_{Joule}, $\Delta T_{Thomson}$ and $\Delta T_{\kappa(T)}$ and are plotted by symbols and lines in Figure 2b. They sum up, together with the linear part, $T_{lin}(x) = T_h - x\frac{\Delta T}{L}$, to the total temperature profile

$$T(x) = T_{lin}(x) + \Delta T_{Joule}(x) + \Delta T_{Thomson}(x) + \Delta T_{\kappa(T)}(x) \tag{7}$$

The procedure to calculate the partial profiles is described in Appendix A.3.

From the close coincidence of combinations of the real partial T profiles to the T profiles of the 1TD and 2TD model materials, as evident from Figure 2b, we can conclude that the contributions from each of the effects (Thomson heat, Joule heat, T dependence of κ) to the total $T(x)$ behave in good approximation and are independent and additive (a small note on this is given in the Appendix A.1.). The reason for the overall weak cross-coupling between the contributing effects is the small amplitude of the partial T profiles ΔT_{Joule}, $\Delta T_{Thomson}$, $\Delta T_{\kappa(T)}$ compared to the overall ΔT but also the fact that $\Delta T_{Thomson}$ and $\Delta T_{\kappa(T)}$ often partially compensate. Therefore, the T profiles of a real material and the

CPM may also be quite close to each other for some materials. It is evident that the shape of $\alpha(T)$ and $\kappa(T)$ affects the temperature profile much more than that of $\rho(T)$ but this does not mean that the asymmetry of Joule heat distribution between the hot and cold side would contribute insignificantly to the difference of the inflowing heat between the CPM case and a real material. The redistribution of Joule heat affects the maximum efficiency to a relevant extent along with the redistribution of Thomson heat. Thus, we can split the RD of the maximum efficiency according to the physical origin—redistribution of Peltier–Thomson heat and Joule heat—as $\delta\eta_{max} = \frac{\eta_{max} - \eta_{max}^{CPM}}{\eta_{max}^{CPM}} = \delta\eta_{max\pi\tau} + \delta\eta_{maxJ}$.

Depending on the slope ratio of $\kappa(T)$ and $\alpha(T)$, the efficiency discrepancy due to Joule heat asymmetry, $\delta\eta_{maxJ}$, will vary considerably between different materials and may change sign from case to case, as observed in [18].

Now, let us proceed to understand in more detail how the absence of Thomson heat in the CPM will affect the efficiency calculation. We will see that it is partially and usually not entirely compensated by the difference in Peltier heat between a real material and its CPM approximation.

2.2. Peltier–Thomson Heat Balance and the Resulting Uncertainty in CPM Efficiency

Consider a TE material with constant κ and a linearly increasing $\alpha(T)$ curve (which is typical for a TE material below the peak zT temperature), as schematically shown in Figure 3. In a TE material under current flow, the convective entropy flux is given by $\dot{s}(T) = j\alpha(T)$. Hence, in a TE leg with a current flow I, the convective entropy flow $\dot{S}(T) = I\alpha(T)$ is directly linked to the temperature dependence of the Seebeck coefficient.

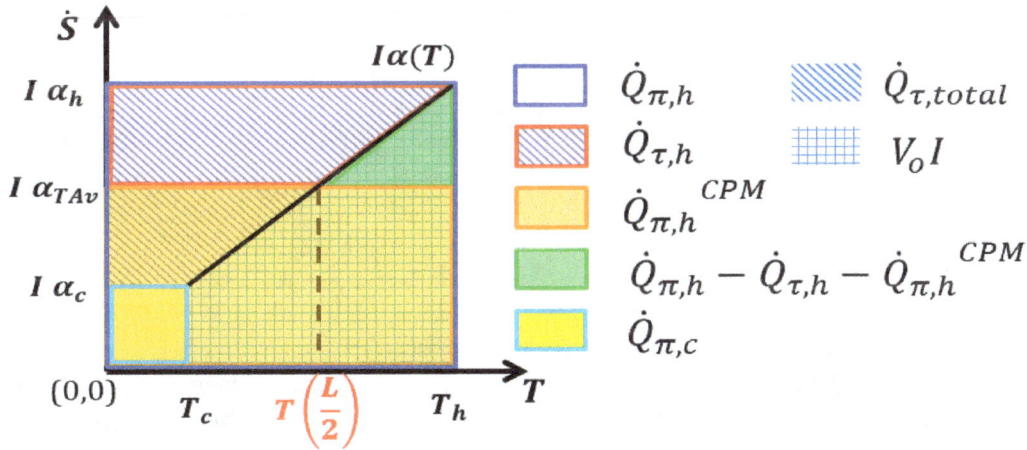

Figure 3. Schematic representation of reversible heat exchange in a TE leg for a linear $\alpha(T)$ curve (black line) in a plot of a convective 1D entropy flow with a constant current I. According to the relation $\dot{Q}_\pi = I\alpha T$, areas in the $\dot{S}(T)$ diagram represent certain amounts of (flowing or exchanged) Peltier (including Thomson) heat. The dark blue and light blue rectangles—in- and outflowing Peltier heat; trapezium above the $\dot{S}(T)$ curve—Thomson heat (marked with slant lines); trapezium below the $\dot{S}(T)$ curve (marked in checked lines) —gross electrical power generated ($V_0 I$); red trapezium—Thomson heat flowing to the hot side; orange rectangle—hot side Peltier heat (CPM). The green triangle indicates part of the difference in the amount of absorbed Peltier heat at the hot side between the actual and the CPM cases that is not compensated in the real material by backflowing Thomson heat $\dot{Q}_{\tau,h}$.

Peltier heat absorbed at the hot side (T_h) in the real case is given by $\dot{Q}_{\pi,h} = I\alpha_h T_h$, while at the cold side, it is $\dot{Q}_{\pi,c} = I\alpha_c T_c$. Areas in the diagram of Figure 3 represent certain amounts of Peltier and Thomson heat but also generated electric power. This allows a schematic comparison of reversible heat exchange in a T-dependent material to its CPM approximation. The difference in the Peltier heat balance, $I(\alpha_h T_h - \alpha_c T_c)$, is given by the difference of the light and dark blue line-marked areas. It is composed of the area below the $\dot{S}(T)$ curve (marked in checked lines) given by $P_0 = IV_0 = I\int_{T_c}^{T_h} \alpha(T)dT$, which is the gross produced electrical power (which includes Joule heat). The area to the left from the $\dot{S}(T)$ curve (indicated by slant lines) is

$$\int_{I\alpha_c}^{I\alpha_h} T d\dot{S} = I \int_{T_c}^{T_h} T \frac{d\alpha}{dT} dT = I \int_{T_c}^{T_h} \tau dT = \dot{Q}_\tau \tag{8}$$

where $\tau = T \frac{d\alpha}{dT}$ is the Thomson coefficient. This area represents the net Thomson heat generated in the TE leg, $\dot{Q}_\tau$, which is directly linked to the variation of the convective entropy flow over the leg. The reversible heat balance

$$\dot{Q}_{\pi,h} - \dot{Q}_{\pi,c} = \dot{Q}_\tau + P_0 \tag{9}$$

shows that the loss of Peltier heat in the sample equals released Thomson heat plus produced gross electrical power. $\dot{Q}_\tau$ and P_0 are counted here as positive when going out of the system. Part of the Thomson heat will flow back, as a contribution to the overall Fourier heat flow, to the hot side. For simplification we assume that Thomson heat that is released at any point in the leg will flow out to the closer side. This is physically not strict but sufficient to qualitatively illustrate the relevant effect of undercompensation of the difference in Peltier heat exchanged at the hot side in a real material compared to the CPM by Thomson heat flowing back to the hot side, i.e., compensation of $dQ_{\pi,h} = \dot{Q}_{\pi,h} - \dot{Q}_{\pi,c}^{CPM} = IT_h(\alpha_h - \overline{\alpha})$ by $\dot{Q}_{\tau,h} = I \int_{\alpha_{\tau,ex}}^{\alpha_h} T d\alpha$. The relevant question on the Seebeck value $\alpha_{\tau,ex}$, from which the integration gives the correct amount of $\dot{Q}_{\tau,h}$ (and its corresponding temperature $T_{\tau,ex}$ with $\alpha_{\tau,ex} = \alpha(T_{\tau,ex})$), will be touched on below.

In the CPM, the Peltier heat at the hot side is given by $I\overline{\alpha}T_h$, while at the cold side it is $I\overline{\alpha}T_c$, where $\overline{\alpha} = \alpha_{TAv}$ is the temperature average of $\alpha(T)$ (see Equation (2)). Therefore, the following equation holds:

$$\dot{Q}_{\pi,h}^{CPM} - \dot{Q}_{\pi,c}^{CPM} = I\overline{\alpha}(T_h - T_c) = I \int_{T_c}^{T_h} \alpha(T) dT \tag{10}$$

i.e., Peltier heat is completely balanced by electrical production.

From Equations (9) and (10), it is obvious that globally the explicit absence of Thomson heat in the CPM is taken care of correctly by the use of temperature averaged $\overline{\alpha}$ in the CPM, i.e.,

$$\dot{Q}_{\pi,h} - \dot{Q}_{\pi,c} - \dot{Q}_\tau = \dot{Q}_{\pi,h}^{CPM} - \dot{Q}_{\pi,c}^{CPM} = I\overline{\alpha}\Delta T = P_0 \tag{11}$$

With this choice of $\overline{\alpha}$ as the CPM value, the gross power generated is exactly the same in the CPM as in the real material, at the same current. On the other hand, it implies that, typically, considerably less Peltier heat is absorbed at the hot side in the CPM case than in reality, whereas back-flowing Thomson heat partly compensates the actually higher Peltier heat intake. Figure 3 visualizes with the green triangle that this compensation is incomplete, i.e., $d\dot{Q}_{\pi\tau,h} = d\dot{Q}_{\pi,h} - \dot{Q}_{\tau,h} > 0$. Accordingly, more Thomson heat is leaving at the cold side. It is evident that this holds not only for a linear but also for a left- or right-hand bowed Seebeck curve.

In a less typical case with strongly asymmetric heat conduction, i.e., $\kappa(T)$ strongly increasing with T, or if $\alpha(T)$ forms a significant maximum, this typical tendency could reverse, but mostly it leads to underestimation of the inflowing heat in the CPM case Q_{in}^{CPM} and hence to overestimation of the efficiency by the CPM. With p-Mg$_2$X, a particular example is given in Appendix A.3.2 (Figure A2c) where, with $\alpha(T)$ weakly changing between T_c and T_h but peaking inside, this compensation can also be almost perfect, or, as for SnSe (Figures 4, 5b and 6), overcompensation may even occur.

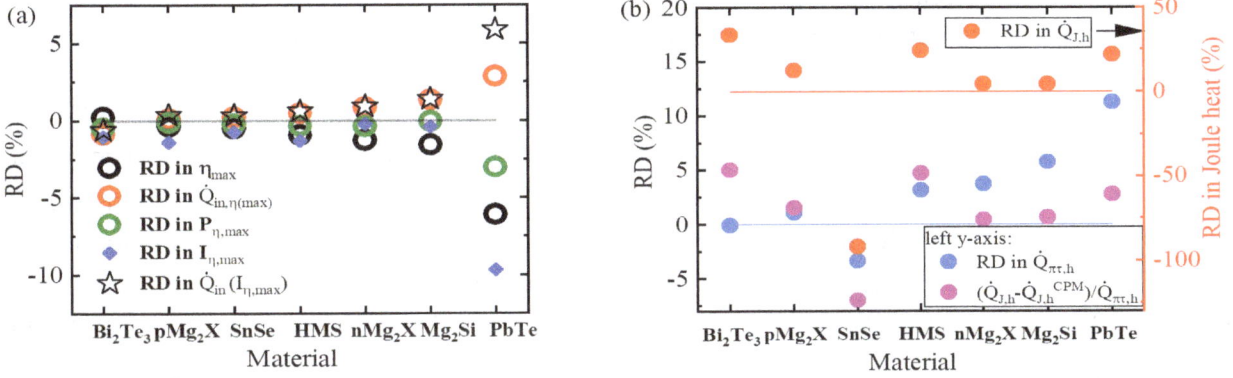

Figure 4. Calculated relative deviation (RD) of (**a**) the maximum efficiency, $\delta\eta_{max}$, heat input, $\delta\dot{Q}_{in}$, power at maximum efficiency, $\delta P_{\eta_{max}}$, and optimum current, $\delta I_{opt,\eta}$; additionally, $\delta\dot{Q}_{in}$ when neglecting $\delta I_{opt,\eta}$ (black stars), (**b**) Joule heat, $\delta\dot{Q}_{J}^{h}$, reversible heat, $\delta\dot{Q}_{\pi\tau}^{h}$, (see Equation (12)) and, for direct comparison, also $d\dot{Q}_{J}^{h}/\dot{Q}_{\pi\tau}^{h}$.

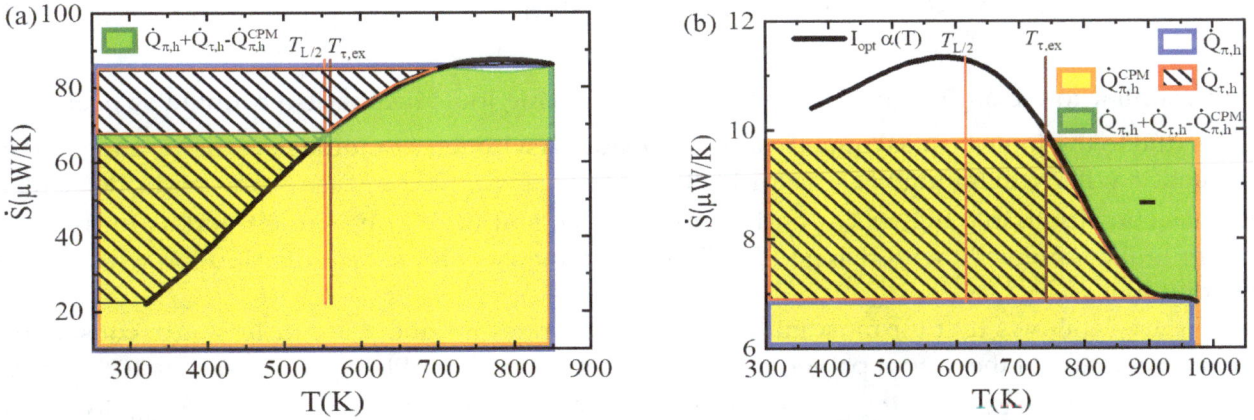

Figure 5. Plot of the convective 1D entropy flow at constant current I for (**a**) PbTe and (**b**) SnSe. Relevant areas are marked to determine the uncompensated Peltier–Thomson heat $d\dot{Q}_{\pi\tau}^{h}$ (green area). Note that the $\frac{L}{2}$ temperature and the temperature $T_{\tau,ex}$ according to the extremum of $\Delta T_{Thomson}(x)$ may be located quite far apart (**b**) whereas $T_{\tau,ex}$ is very close to the crossing point of $\alpha(T)$ to $\bar{\alpha}$.

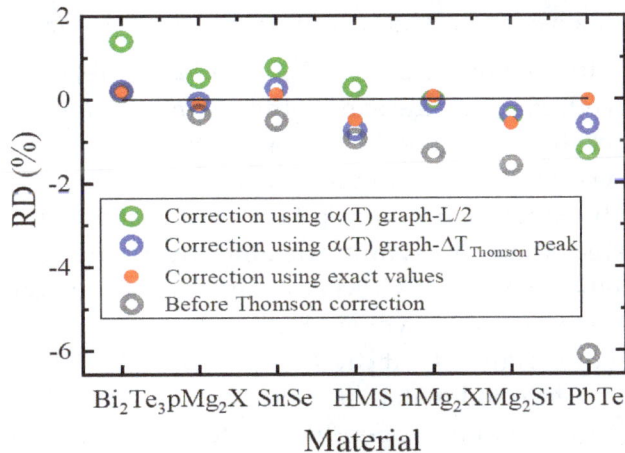

Figure 6. RD in maximum efficiency, $\delta\eta_{max}^{corr}$, corrected with respect to $d\dot{Q}_{\pi\tau}^{h,\,I=const}$ ($T_{\tau,ex}$ according to the peak of the exact Thomson profile; blue), $d\dot{Q}_{\pi\tau}^{h}$ (exact numerical calculation; red; compare also Equation (12)) and a first guess by the $\frac{L}{2}$ position.

Overall, the efficiency deviation between the real and CPM cases would be negligible if $Q_{in} = Q_{in}{}^{CPM}$. For a rising $\alpha(T)$ curve, which is the typical case applied for most of the established TE materials, the Peltier–Thomson part, $\dot{Q}_{\pi\tau}{}^{h,CPM}$, of $\dot{Q}_{in}$ will remain lower than the real $\dot{Q}_{\pi\tau}{}^{h}$. Thus, the efficiency is often overestimated by the CPM. Furthermore, a shift in $I_{\eta,opt}{}^{CPM}$ against the true $I_{\eta,opt}$ has to be taken into consideration due to a change in the current-dependent contributions to $\dot{Q}_{in}$. The usually higher intake of reversible heat at the hot side in the real case, $\dot{Q}_{\pi\tau}{}^{h}$, compared to the CPM ($\delta Q_{\pi\tau,h} > 0$) results in a steeper curve $\dot{Q}_{in}(I)$ than $\dot{Q}_{in}{}^{CPM}(I)$. Efficiency, as defined by $\eta(I) = P/\dot{Q}_{in}$, will accordingly have a lower slope in reality than for the CPM, equivalent to a lower maximum position $I_{opt,\eta}$. Thus, usually, the CPM will overestimate the optimum current, $\delta I_{opt,\eta} = \frac{I_{opt,\eta} - I_{opt,\eta}^{CPM}}{I_{opt,\eta}^{CPM}} < 0$, and hence will overestimate output power at maximum efficiency ($\delta P_{\eta max} < 0$), which adds to the overestimate of maximum efficiency: $\delta\eta_{max} = \delta P_{\eta max} - \delta\dot{Q}_{in}$, amplifying the effect of $\delta\dot{Q}_{in}$ (see Figure 4a). Hence, for a quantitative analysis, we have to consider three contributions to the (absolute) deviation of $\dot{Q}_{in}$

$$d\dot{Q}_{in} = d\dot{Q}_{\pi\tau}{}^{h} - d\dot{Q}_{J}{}^{h} = d\dot{Q}_{\pi\tau}{}^{h,\,I=const} - d\dot{Q}_{J}{}^{h,\,I=const} + \left.\frac{\partial\dot{Q}_{in}}{\partial I}\right|_{I_{opt,\eta}} dI_{opt,\eta} \qquad (12)$$

where, similar to the outflowing Thomson heat, outflowing Joule heat is also counted as positive and $d\dot{Q}_{J}{}^{h}$ is due to the Joule heat asymmetry at the hot side. Asymmetry of Joule heat distribution and heat conduction will, with falling $\kappa(T)$, as for PbTe and SnSe, favor heat release to the cold side. This will likewise contribute to a higher $\dot{Q}_{in}$ and steeper $\dot{Q}_{in}(I)$, amplifying the same trend as for reversible heat, or will counteract it with rising $\kappa(T)$. Thus, asymmetry of Joule heat distribution will add to the mispoint in $I_{opt,\eta}{}^{CPM}$.

Figure 4a shows that for most materials, $I_{opt,\eta}$ changes for about 1% or less and, consequently, also the deviation of the output power, remain small. However, for PbTe, $\delta I_{opt,\eta}$ reaches 10%. Then the deviation of output power, $\delta P_{\eta max}$, may grow in absolute amount to be as large as $\delta\dot{Q}_{in}$, doubling its effect. Whereas the contribution to $\delta\dot{Q}_{in}$, due to $\delta I_{opt,\eta}$ usually remains insignificant, it becomes relevant for PbTe where it compensates half of $d\dot{Q}_{in}$ related to the distribution of inner heat at an unchanged current, $d\dot{Q}_{\pi\tau}{}^{h}(I_{opt,\eta}) - d\dot{Q}_{J}{}^{h}(I_{opt,\eta})$, see Equation (12) and black stars in Figure 4a.

The RD of hot side Joule heat, $\delta\dot{Q}_{J}{}^{h}$, and Peltier/Thomson heat, $\delta\dot{Q}_{\pi\tau}{}^{h}$ with $\dot{Q}_{\pi\tau}{}^{h} = \dot{Q}_{\pi}{}^{h} - \dot{Q}_{\tau}{}^{h}$, are shown in Figure 4b. $\delta\dot{Q}_{J}{}^{h}$ reaches quite significant nominal values (SnSe), mainly due to the low magnitude of $\dot{Q}_{J}{}^{h}$ itself. For direct comparison to $\delta\dot{Q}_{\pi\tau}{}^{h}$, the (absolute) deviation $d\dot{Q}_{J}{}^{h}$ related to $\dot{Q}_{\pi\tau}{}^{h}$ is plotted and shows that both effects reach the same order of magnitude. Typically, both contributions partly compensate. Furthermore, no general behavior can be observed in their mutual relation over the materials, as in some cases clearly one effect dominates, in others the other.

As seen from Figure 1b, usually, more Joule heat is released to the hot side than to the cold side in a real material, whereas there are symmetric amounts in the CPM case. This contributes to an underestimation of the efficiency in the CPM case, $\delta\eta_{maxJ} > 0$. On the other hand, as explained, the Peltier–Thomson balance tends to an overestimation, $\delta\eta_{max\pi\tau} < 0$, thus, both effects counteract and partially compensate. From Figure 4a, it can be seen that the CPM overestimates the efficiency compared to the real case for all selected materials except Bi_2Te_3, which has an exceptionally higher κ_h compared to the cold side (Figure A1a in Appendix A.2) together with high Joule release (Figure 4b) and almost compensation of the Peltier–Thomson balance. Thus, the Joule contribution dominates, leading to an underestimation of the efficiency. Additionally, SnSe behaves somewhat differently from the general trend, with a falling $\alpha(T)$ curve (Appendix A.2 Figure A1b) and the over-resistivity at the cold side (Appendix A.2 Figure A1c). Moreover, κ_h is much lower than κ_c. As an effect, Joule heat is preferentially led to the cold side; consequently, hot side Joule heat is greatly overestimated in the CPM (Figure 4b), but as the relative contribution of Joule heat to Q_{in} is small (Figure A2a), the resulting trend towards the overestimation of performance in the CPM remains moderate. On the other hand, as seen from Appendix A.3.2 Figure A1b, Thomson heat is absorbed in the leg as $\alpha(T)$ for SnSe is a falling curve and is mainly bound to the hot side. As seen from Figure 4b, for SnSe, the hot side Peltier–Thomson heat will, unlike for most of the other materials, be overestimated by the CPM. However, the resulting underestimation of efficiency in the CPM will be overcompensated by the counteracting Joule heat distribution.

The first four materials in our list (see Figure 4a) show a minor discrepancy of the CPM with reality. Although Joule heat asymmetry is contributing comparably, from case to case, the dominating source of discrepancy is mostly the uncompensated Peltier heat according to Equation (11)). It is particularly relevant in the cases of n-Mg_2X, Mg_2Si and PbTe, which have larger Thomson contributions (Figure A2a), leading to larger discrepancies of the CPM efficiency estimate.

2.3. Refining the CPM Efficiency Estimate

Having identified the effects causing a systematic uncertainty in the CPM efficiency estimation, they can be accordingly corrected.

We want to analyze how this can be done practically for the Thomson contribution, $\delta \dot{Q}_{\pi\tau}^h$, by calculating the uncompensated Peltier heat at the hot side. Therefore, we discuss the approach for example materials with dissimilar $\alpha(T)$ characteristics.

The values of $\dot{Q}_{\pi,h}$ and $\dot{Q}_{\pi,h}^{CPM}$ are known from T_h, α_h and $\bar{\alpha}$, for a given current, where, as a first approximation, $I_{opt,\eta}^{CPM}$ is used. We have seen that the Thomson heat flowing to the hot side is strictly calculated from the partial T profile $\Delta T_{Thomson}(x)$ by $\dot{Q}_{\tau,h} = -\kappa_h \cdot \frac{d\Delta T_{Thomson}}{dx}_h$. We apply this route to form a reference for an approximate estimation to be developed and, because of this, we omit a numerical calculation of exact T profiles. As derived from Equation (10), we obtain the uncompensated Peltier–Thomson heat from $d\dot{Q}_{\pi\tau}^h = \dot{Q}_{\pi,h} + \dot{Q}_{\tau,h} - \dot{Q}_{\pi,h}^{CPM}$. Neglecting any deviation of current, this can be illustrated in the $\alpha(T)$ diagram based on our interpretation of areas by amounts of reversible heat, see Figure 3. Thus, we aim for a good approximation of the green marked area in Figure 3 by an appropriateand simple approximation. The problem splits into two aspects: finding the temperature $T_{\tau,ex}$ above which the inner Thomson heat is conducted to the hot side and finding a close approximation of the integral. As $\alpha(T)$ may be quite different (see Figure A1b), we meet various situations, represented by different $\Delta T_{Thomson}(x)$ temperature profiles (Figure A2b), among them typical ones with a single maximum according to Thomson heat flowing out to both sides, but also less typical ones with a single minimum (Thomson heat flowing in from both sides) or even two extrema (for Bi_2Te_3) where Thomson heat is released to the cold side but absorbed from the hot side. A rule to treat all of the cases likewise is needed. Figure 5a,b and Figure A2c,d accordingly show scenarios where $\alpha(T)$ contains almost linear intervals along with strongly bowed ones, where $\alpha(T)$ is monotonous or contains a maximum, where α_h and $\bar{\alpha}$ are far from each other or close together or where $\alpha(T)$ crosses the $\bar{\alpha}$ horizontal once or twice. The position of the extrema (maxima or minima) of $\Delta T_{Thomson}(x)$ is marked in each diagram by a brown line. Accordingly, the area corresponding to the uncompensated heat might be more complex than is shown in Figure 3, e.g., see Figure 5a. The area to the left of the $\alpha(T)$ curve to the α-axis from this point up to the hot side α_h (marked by a red border) represents $\dot{Q}_{\tau,h}$. The fact that the respective area also contains negatively counted parts when $\alpha(T)$ goes through a maximum is also taken into account. Accordingly, the upper slim boat-shaped area in Figure 5a counts as negative; symbolically, it is mirrored in the green area.

However, in such a case, the integration can be simplified, switching from the hot to the cold side, as $\dot{Q}_{\tau,h} = \dot{Q}_\tau - \dot{Q}_{\tau,c}$ and with Equation (10), $\dot{Q}_\tau = \dot{Q}_{\pi,h} - \dot{Q}_{\pi,c} - P_0$. Note that if there are two extrema of $\Delta T_{Thomson}(x)$, then we have two $T_{\tau,ex}$ values where the Thomson heat between both can be neglected as it cancels out completely. Only the intervals outside, $(T_c; T_{\tau,ex})$ or $(T_{\tau,ex}; T_h)$, have to be considered. Among both intervals, the side has to be chosen where $\alpha(T)$ is a monotonous function in the relevant temperature interval, where it is closer to linearity, and possibly where $T_{\tau,\text{ex}}$ is closer to T_h or T_c.

Applying Equation (11) accordingly to the chosen interval, the integration for $\dot{Q}_\tau$ can be substituted by one for P_0, e.g., for the cold side:

$$\dot{Q}_{\tau,\text{c}} = \dot{Q}_{\pi,T_{\tau,\text{ex}}} - \dot{Q}_{\pi,\text{c}} - \int_{T_\text{c}}^{T_{\tau,\text{ex}}} \alpha\, dT \tag{13}$$

This facilitates practical execution as $\alpha(T)$ is mostly known as a low-order polynomial, thus integration could be done analytically.

If the Thomson T profile is not known, half of the leg length, $\frac{L}{2}$, can be taken as a first guess of the position for the calculation of $\dot{Q}_{\tau,\text{h}}$. The corresponding temperature is marked in the diagrams. This can be a quite good estimate when the Thomson T profile is close to symmetric, as for PbTe (see Figure A2b), but may fail greatly when Thomson heat is strongly asymmetric, as for SnSe. On the contrary, an entropy consideration of Thomson heat in the TE leg (see Appendix A.4.) leads to a rule of thumb for $T_{\tau,\text{ex}}$ that is

$$\alpha(T_{\tau,\text{ex}}) \approx \overline{\alpha} \tag{14}$$

Indeed, it applies well for all example materials involved here. With this rule, approximation of $\dot{Q}_{\tau,\text{h}}$ is facilitated considerably, as just a crossing point of $\alpha(T)$ with its TAv has to be found.

Figure 6 shows the remaining efficiency deviation, $\delta\eta_{\max}^{\text{corr}}$, corrected by the uncompensated Peltier–Thomson heat calculated from the $\alpha(T)$ graph using the $\frac{L}{2}$ position, using $T_{\tau,\text{ex}}$ according to the extremum (maximum) position of $\Delta T_{\text{Thomson}}(x)$ but neglecting the current deviation $\delta I_{\text{opt},\eta}$, as well as corrected by the exact deviation $d\dot{Q}_{\pi\tau}^\text{h} = \dot{Q}_{\pi,\text{h}} - \dot{Q}_{\tau,\text{h}} - \dot{Q}_{\pi,\text{h}}^{\text{CPM}}$. The efficiency estimate by the CPM is greatly improved when the $\Delta T_{\text{Thomson}}(x)$ extremum position is used (red dots).

Only occasionally, e.g., when $\alpha(T)$ is close to linear, the $\frac{L}{2}$ position works well for correction but fails for most materials as it does not take into account the asymmetry of heat sources and heat conduction. Similarly, models suggesting half of the Thomson heat on either side for correcting the CPM results [14–16,26,27] will mostly not work sufficiently. The correction employing the $\Delta T_{\text{Thomson}}(x)$ peak position is close to the exact numerical correction for most materials as this position considers the asymmetry exactly. The difference between both cases is merely due to the change of the optimum current which is as yet unconsidered by the graphical correction. The remaining discrepancy is due to Joule heat asymmetry.

Whereas we have used exact numerical calculations to demonstrate the principle of the Thomson correction method and to show that the rule $\alpha(T_{\tau,\text{ex}}) \approx \overline{\alpha}$ holds well, the suggested practical procedure for the correction of $d\dot{Q}_{\pi\tau}^{\text{h, I=const}}$ described here, which is based on an analysis of the physical effects behind the deviation of CPM performance estimates, is limited to basic algebraic operations which can be instantaneously calculated by any table calculation software.

3. Conclusions

From the study of 2TD and 1TD model materials with one or two selected properties among α, ρ and κ set as constant, which results in both redistribution of heat between the hot and cold side of the element and the change of spatial averages, we see that in some examples, large deviations in efficiency $\delta\eta_{\max}^{\text{model}}$ arise as a consequence of considerable modification of the T profile. In comparison to the efficiency deviation between the CPM and real materials $\delta\eta_{\max}$ which conserve the spatial property averages and are mostly below 2%, this shows that a change of spatial averages due to an arbitrary modification of the T profile may contribute a strong shift to the efficiency estimate. Thus, conservation of the leg's thermal and electrical resistance is essential for a valid efficiency estimate. However, the shift mainly remains low if only $\rho(T)$ is switched to constant. Nevertheless, it cannot be concluded from this that the temperature dependence of the electrical resistivity plays a minor role in the efficiency estimation by the CPM. The 2TD and 1TD model materials lead to quite good approximations of the partial T profiles $\Delta T_{\text{Joule}}(x)$, $\Delta T_{\text{Thomson}}(x)$ and $\Delta T_{\kappa(T)}(x)$.

It is shown that the deviation of a CPM-based efficiency estimate, $\delta\eta_{\text{max}}$, is not just due to the absence of Thomson heat in the CPM, as the choice of the temperature average of $\alpha(T)$ as a CPM parameter mainly compensates for the absence of Thomson heat. Rather, the discrepancy in efficiency determination in the CPM is shown to be, to a major extent, due to the excess unaccounted heat at the hot side in the CPM $\delta\dot{Q}_{\text{in}}$, which usually leads to overestimation of performance, and, to a minor extent, due to a shift of the optimum current $\delta I_{\text{opt},\eta}$ and, consequently, of the produced electrical power at maximum efficiency, $\delta P_{\eta_{\text{max}}}$. In most cases, the change of the optimum current is small. In materials with rising $\alpha(T)$, less of the released Thomson heat flows back to the hot side than would compensate for the reduced hot side Peltier heat absorption assumed by the CPM. This systematic undercompensation tends towards a higher actual heat intake at the hot side compared to the CPM, thus overestimating efficiency when the CPM is used. Asymmetry of Joule heat usually has an opposite influence but is overcompensated in most cases.

In order to correct for the Peltier–Thomson heat-related deviation $\delta\dot{Q}_{\pi\tau}^{\text{h}}$, a graphical illustration in terms of convective entropy flow based on the $\alpha(T)$ curve is given. It confirms that the rule for the splitting of Thomson heat to the sides $\alpha(T_{\tau,\text{ex}}) \approx \overline{\alpha}$, which results from an entropy consideration, holds well. This enables a valid approximation of $\delta\dot{Q}_{\pi\tau}^{\text{h}}$ with a simple algebraic procedure, omitting the exact numerical calculation of the temperature profile. Although a considerable deformation of the T profile caused by the T dependence of $\kappa(T)$ is observed, it will affect the deviation between the real situation and its CPM approximation simply via a local shift of the thermal and electrical resistivity but will not explicitly contribute to the inflowing heat balance $\dot{Q}_{\text{in}}$.

In summary, the performance of a TE material does not only depend on its averaged material parameters but also on local asymmetry of Thomson and Joule heat, driven by the T dependence of the TE properties. In particular, Thomson heat can show highly asymmetric distribution. Thus, TE device efficiency can be varied beyond the averaged properties, represented by a figure of merit.

Author Contributions: Conceptualization, E.M. and P.P.; methodology, E.M. and P.P.; software, P.P.; validation, P.P., E.M. and J.d.B.; formal analysis and investigation, P.P., E.M. and J.d.B.; resources, P.P.; data curation, P.P.; writing—original draft preparation, P.P.; writing—review and editing, P.P., E.M. and J.d.B.; visualization, P.P. and E.M.; supervision, E.M. and J.d.B.; project administration and funding acquisition, E.M. and J.d.B. All authors have read and agreed to the published version of the manuscript.

Acknowledgments: We would like to gratefully acknowledge the endorsement from the DLR Executive Board Member for Space Research and Technology and the financial support from the Young Research Group Leader Program. P.P would like to acknowledge the German Academic Exchange Service, DAAD (Fellowship No. 247/2017) for financial support.

Appendix A

Appendix A.1. Note from Section 2.1

A very good approximation of the actual T profile and hence the SpAv of ρ and κ in accordance with a real material can be calculated in a straightforward way from the T-dependent properties without using an iterative solution [18] for $T(x)$. This may considerably simplify the estimation of appropriate SpAvs as CPM property values. $\Delta T_{\text{Joule}}(x)$ can be obtained analytically from the CPM case, $\Delta T_{\kappa(T)}$ from a integration of the Fourier equation and $\Delta T_{\text{Thomson}}$ and from a 1TD $\alpha(T)$ model by a single integration.

Appendix A.2. Material Data and Boundary Conditions

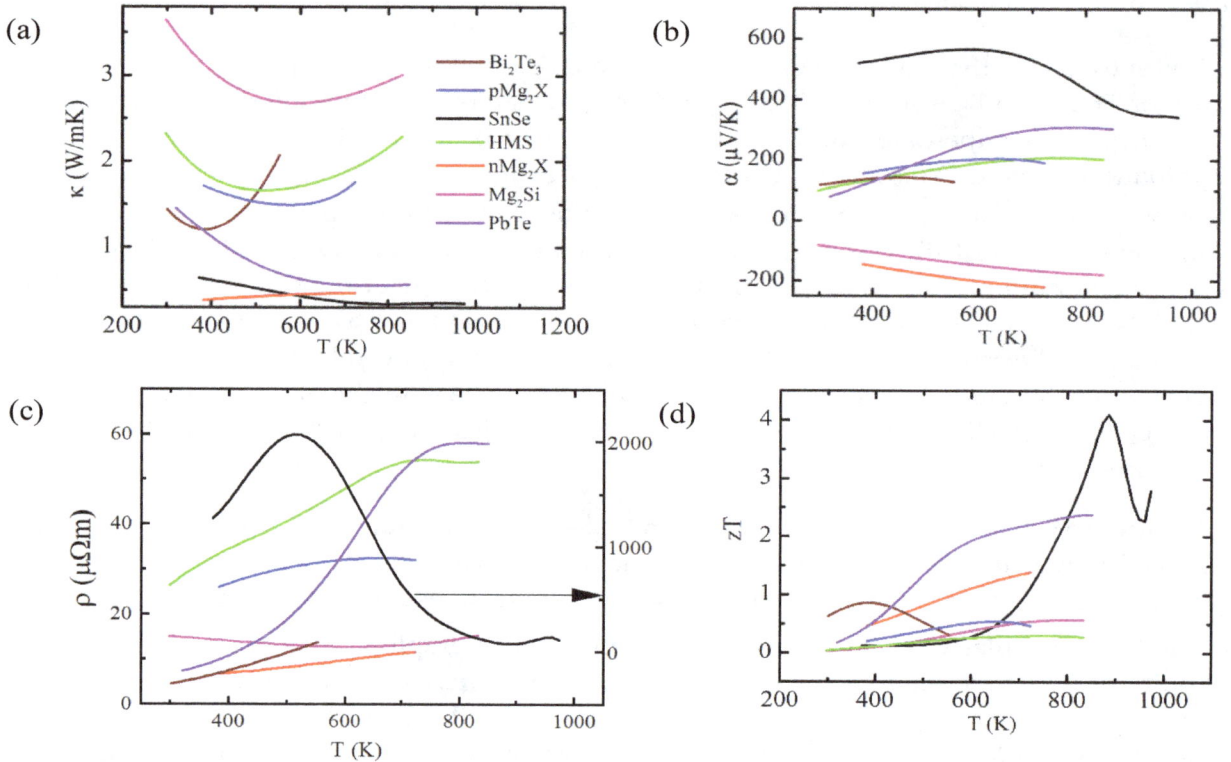

Figure A1. Temperature-dependent thermoelectric material properties of representative material classes: (**a**) thermal conductivity, (**b**) Seebeck coefficient, (**c**) electrical resistivity and (**d**) figure of merit. Since SnSe has much higher resistivity, the scale for it is given on the right y-axis. All the raw experimental data taken from the literature [20–24] were fitted with appropriate polynomials (usually 3rd or 4th order). For SnSe, a 9th order polynomial fit was used owing to the complex T dependence and hence shows an unusually high zT_{max}. However, this does not affect the physics discussed and hence these fitted data were used throughout the manuscript.

Table A1. Temperature range of analysis for all materials of Figure A1.

Material	Temperature Range of Analysis
p-Mg_2(Si,Sn)	723 K to 383 K
n-Mg_2(Si,Sn)	723 K to 383 K
HMS	833 K to 298 K
Mg_2Si	833 K to 298 K
p-Bi_2Te_3	553 K to 301 K
SnSe	973 K to 373 K
PbTe	850 K to 320 K

Appendix A.3. Additional Information

Appendix A.3.1. Finding Individual Contributions to the Total T Profile

The partial T profiles are each found by equating $\kappa(T)\frac{\partial^2 T}{\partial x^2}$ in Equation (1) to each of the other corresponding terms, assuming isothermal boundary conditions and fixing all coefficients in the equation according to the total T profile $T(x)$. Thus, solving for the respective partial T profile reduces to a double integration, where the first step provides the total amount of each partial heat contribution to the thermal balance.

As the partial T profiles can have opposite signs in amplitude and partially compensate for many of the common TE materials (however, not always), the T profiles of a real material and the CPM may be quite close, as in the example of n-type Mg_2X, Figure 2b.

Appendix A.3.2. Contributions to $\dot{Q}_{in}$

As both Joule and Thomson heat, after appearing inside the leg, will flow out, physically, as Fourier heat, we have to consider in this discussion the pure Fourier heat $Q_{F,h} = K\Delta T$ (with $K = \langle\kappa^{-1}\rangle^{-1}A/L$), which is merely related to the thermal resistance of the leg and is constant along the leg, separately from the Joule- and Thomson-related contributions. Accordingly, $\dot{Q}_{in}$ is composed of

$$\dot{Q}_{in} = \dot{Q}_{F,h} + \dot{Q}_{\pi,h} - \dot{Q}_{\tau,h} - \dot{Q}_{J,h} \tag{A1}$$

The real Joule- and Thomson-related contributions, $-\dot{Q}_{\tau,h}$ and $-\dot{Q}_{J,h}$, to the inflowing hot side heat are calculated by splitting the overall temperature profile $T(x)$ into additive partial T profiles, each related to one of the individual physical contributions. Partial Thomson T profiles of example materials are plotted in Figure A2b. Evaluating $-\kappa_h \cdot \left(\frac{d}{dx}\Delta T_{Thomson}\right)_h$ and $-\kappa_h \cdot \left(\frac{d}{dx}\Delta T_{Joule}\right)_h$ from the partial T profiles gives $\dot{Q}_{\tau,h}$ and $\dot{Q}_{J,h}$, respectively.

Figure A2a shows the relative contribution of each heat to $\dot{Q}_{in}$: $\frac{\dot{Q}_{F,h}}{\dot{Q}_{in}}, \frac{\dot{Q}_{\pi,h}}{\dot{Q}_{in}}, -\frac{\dot{Q}_{J,h}}{\dot{Q}_{in}}, -\frac{\dot{Q}_{\tau,h}}{\dot{Q}_{in}}$. This comparison reveals that Joule and Thomson heat contribute about 1–5% to $\dot{Q}_{in}$, usually flowing out, with their contributions being roughly of the same order. Figure A2a also shows the fraction of Thomson heat and Joule heat distributed to the hot side ($\frac{\dot{Q}_{\tau,h}}{\dot{Q}_\tau}$ and $\frac{\dot{Q}_{J,h}}{\dot{Q}_J}$).

In order to illustrate example situations of the distribution of Peltier and Thomson heat along the leg, $\alpha(T)$ graphs for p-Mg_2X and Bi_2Te_3 are given in Figure A2c,d, respectively. Due to the bowed shape of the $\alpha(T)$ graph and relatively close values of α_h to α_c for p-Mg_2X, the difference between $\dot{Q}_{\pi,h}$ and $\dot{Q}_{\pi,h}^{CPM}$ is almost negligible, but $\dot{Q}_{\tau,h}$ amounts to more than twice the amount of $\dot{Q}_{\pi,h} - \dot{Q}_{\pi,h}^{CPM}$. Nevertheless, this did not affect the efficiency deviation $\delta\eta_{max}$ too much, as $\dot{Q}_{\tau,h}$ is quite small in absolute terms. In the case of Bi_2Te_3, $\dot{Q}_{\pi,h}^{CPM}$ is even higher than $\dot{Q}_{\pi,h}$ again due to the curved shape of $\alpha(T)$, affecting the position of α_{TAv}. However, $\dot{Q}_{\tau,h}$ almost completely compensates for this Peltier heat difference, keeping the influence on the efficiency deviation negligible.

Appendix A.4. Thomson Heat Distribution and Entropy

With the TEG leg, we discuss the entropy flow in a reversible system of Peltier heat transport and Thomson heat exchange which is running on a non-equilibrium temperature background mainly fixed by the continuous flow of Fourier heat. As released Thomson heat will be transported as Fourier heat but is small in relation to the Fourier heat background (see Figure A2a), which is driven by the temperature difference and the thermal resistance of the TE leg, we will treat the variation of the temperature profile by the conducted Thomson heat as insignificant for the following consideration.

In the steady state, the entropy of the system remains constant; there is a continuous entropy production by the dissipative heat transport from hot to cold and the balancing continuous entropy export by transmitted Fourier heat (plus a negligible fraction arising from outflowing Joule heat). Assuming ideal outer current leads with $\alpha = 0$, there is no other entropy exchange at the hot and cold sides.

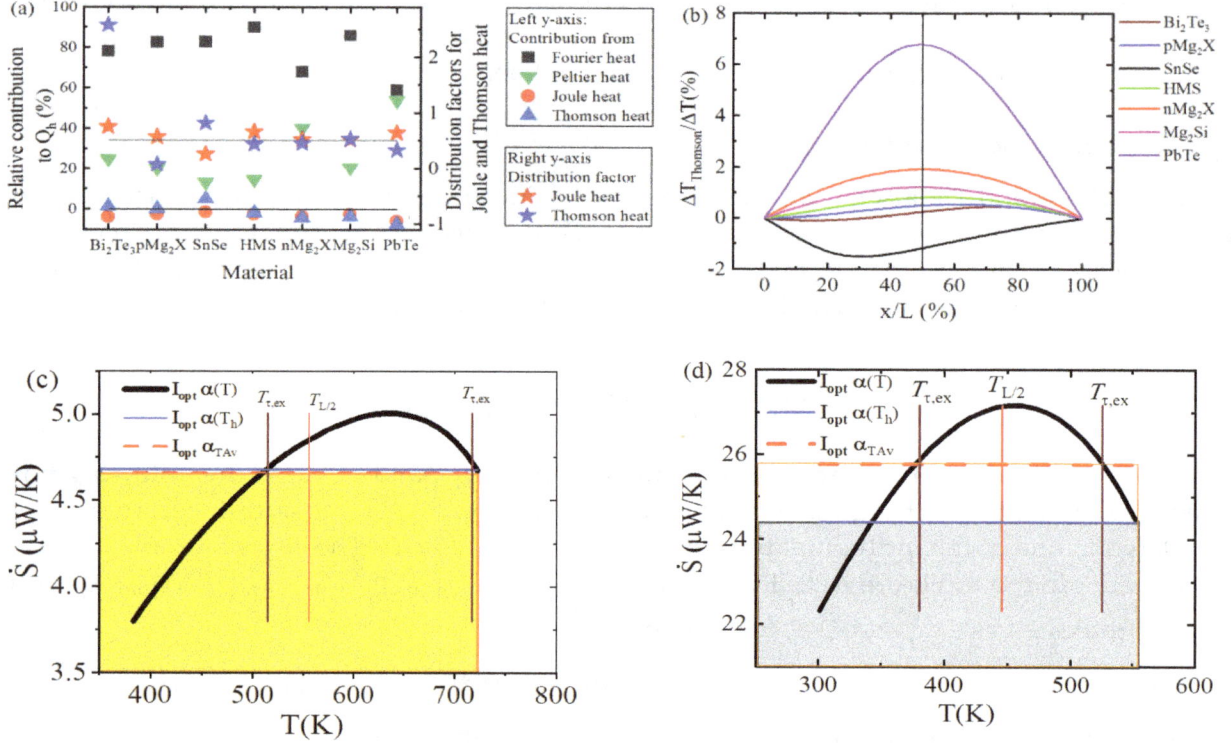

Figure A2. (a) Ratio of individual heat contributions to $\dot{Q}_{in}$ (Equation (A1)) calculated from the corresponding partial temperature profiles (for comparison, all quantities are counted as positive when flowing into the element) (left y-axis), and distribution factors (right y-axis) for Thomson and Joule heat. (b) Thomson T profiles for all example materials (c) $\dot{S}(T)$ diagram for p-Mg2X showing the area between $I\alpha(T_h)$ and $I\alpha_{TAv}$ (corresponding to the Peltier heat difference between the CPM and real case), which is very small due to the shape of $\alpha(T)$. The position of the first peak in the Thomson partial T profile is marked as a brown vertical line. (d) $\dot{S}(T)$ diagram for Bi$_2$Te$_3$, where $\alpha_{TAv} > \alpha(T_h)$. Hence, $\dot{Q}_{\pi,h}^{CPM}$ is higher than $\dot{Q}_{\pi,h}$.

In the CPM, we have a constant convective entropy flow $\bar{\alpha}I$ throughout the element, equal to the absorbed and released entropy rate $\bar{\alpha}I$ by absorption and release of Peltier heat at the terminals. In a real material, the absorbed entropy rate $\alpha_h I$ equals the convective entropy flow at the hot side, and, likewise, the amount of $\alpha_c I$ at the cold side. The variation of α along the leg drives local Thomson heat production $d\dot{Q}_\tau = T\frac{d\alpha}{dT}IdT = TId\alpha$, contributing an entropy flow increment $d\dot{S} = Id\alpha$. Thomson heat flows to the hot and cold sides and the related total entropy exchange is $(\alpha_h - \alpha_c)I = \Delta\alpha I$. It distributes by the fraction x_h to the hot and cold sides:

$$\Delta\dot{S}_{\tau,h} = x_h(\alpha_h - \alpha_c)I \text{ and } \Delta\dot{S}_{\tau,c} = (1 - x_h)(\alpha_h - \alpha_c)I. \tag{A2}$$

Driven by the gradient of the partial Thomson temperature profile, all Thomson heat released at one side of a maximum (or minimum) of this profile will be exchanged to this side of the leg. With the temperature $T_{\tau,ex}$ of this position and its Seebeck coefficient $\alpha_{\tau,ex} = \alpha(T_{\tau,ex})$, the shares of the entropy exchange which are bound to each of the sides are

$$\Delta\dot{S}_{\tau,h} = (\alpha_h - \alpha_{\tau,ex})I \text{ and } \Delta\dot{S}_{\tau,c} = (\alpha_{\tau,ex} - \alpha_c)I. \tag{A3}$$

Multiplying both by the respective temperature of the side yields total Thomson heat:

$$\dot{Q}_\tau = T_h(\alpha_h - \alpha_{\tau,ex})I + T_c(\alpha_{\tau,ex} - \alpha_c)I = \{T_h\alpha_h - T_c\alpha_c - \alpha_{\tau,ex}(T_h - T_c)\}I = \Delta\dot{Q}_\pi - I\alpha_{\tau,ex}\Delta T. \tag{A4}$$

Comparing Equation (A4) with the energy balance of reversible heat $\dot{Q}_\tau = \Delta \dot{Q}_\pi - IV_0$, we can conclude that

$$\alpha_{\tau,ex}\Delta T = V_0 = \overline{\alpha}\Delta T, \text{ thus } \alpha_{\tau,ex} = \overline{\alpha} \tag{A5}$$

This gives us a rule for the temperature intervals over which the Thomson heat is flowing to either side of the leg. Consequently, Thomson heat has to be integrated from the crossing point of the curve of the Seebeck coefficient $\alpha(T)$ with its temperature average $\overline{\alpha}$. As a reversible approximation, this result is approximate and not strict as we have neglected here that dissipative processes are involved when Thomson heat is conducted to the leg sides. Below we will analyze these changes and find that these are small, and thus the rule stated here on the position of $\alpha_{\tau,ex}$, although not strict, is a good guide for estimates of the distribution of Thomson heat. Indeed, as observed by comparison to exact numerical calculations, this rule is almost perfectly fulfilled for all the example materials.

Within this reversible approximation, the Thomson heat flowing to the hot side is obtained as $\dot{Q}_{\tau,h} = T_h(\alpha_h - \overline{\alpha})I$. This would be equivalent to a complete compensation of the Peltier heat difference between reality and the CPM, i.e., the vanishing axial redistribution of reversible heat which is consistent with the simplifying assumption that the Thomson heat flowing to the outside is transmitted free of dissipation, i.e., equivalent to reversible heat. Here, the (additional) T gradient related to the flow of Thomson heat is neglected, whereas an underlying T profile related to an independent heat flow (here, the background of Fourier heat transfer) does, in effect, not contribute to its dissipation. We will see below that this happens as Thomson heat flowing to different sides will contribute almost compensating shares to the entropy balance. What is neglected here is that the Thomson heat itself when flowing to the ends of the leg will dissipate, according to the slight shift of the inner T profile it is causing. Above, this T offset was separated and called the partial T profile due to Thomson heat, $\Delta T_{Thomson}(x)$. Additionally, this omission will contribute to a weak deviation from the position rule $\alpha_{\tau,ex} = \overline{\alpha}$.

The dissipative part of the entropy transport to the sides of the leg is related to the T drop or step-up between the location where an increment of Thomson heat $d\dot{Q}_\tau$ is released and the side temperature, T_h or T_c. The entropy increment is released over a segment of the leg with the T increment dT is $d\dot{S} = Id\alpha = \frac{d\dot{Q}_\tau}{T}$. With the transfer to the cold side, for example, the transmitted increment of Thomson heat $d\dot{Q}_\tau$ increases its entropy up to $d\dot{S}_c = \frac{d\dot{Q}_\tau}{T_c}$, and the according entropy gain is

$$d\Delta\dot{S}_c = d\dot{S}_c - d\dot{S} = \frac{d\dot{Q}_\tau}{T_c} - \frac{d\dot{Q}_\tau}{T} = \frac{d\dot{Q}_\tau}{TT_c}(T - T_c) = d\dot{S}\frac{T - T_c}{T_c} \tag{A6}$$

Summing over all Thomson heat flowing to that side, we have

$$\Delta\dot{S}_c = \frac{1}{T_c}\int_{I\alpha_c}^{I\overline{\alpha}}(T - T_c)d\dot{S} = \frac{I}{T_c}\int_{\alpha_c}^{\overline{\alpha}}(T - T_c)d\alpha = \frac{I}{T_c}\int_{T_c}^{T_{\overline{\alpha}}}T\frac{d\alpha}{dT}dT - I(\overline{\alpha} - \alpha_c) \tag{A7}$$

Multiplying with the cold side temperature, $\Delta\dot{Q}_{\tau,c} = T_c\Delta\dot{S}_c = \int_{T_c}^{T_{\overline{\alpha}}}T\frac{d\alpha}{dT}dT - I(\overline{\alpha} - \alpha_c)T_c$ gives us the amount of Thomson heat that is just the difference from the Peltier–Thomson heat balance of the CPM, $\Delta\dot{Q}_{\tau,c} = \dot{Q}_{\tau,c} - \left(\dot{Q}_{\pi,c}^{CPM} - \dot{Q}_{\pi,c}\right)$, i.e., the part that we have identified as uncompensated Peltier–Thomson heat in a real material. Note that $\dot{Q}_{\pi,c}^{CPM}$ contains merely completely reversible exchange of Peltier heat. Thus, the incomplete compensation of the Peltier–Thomson heat balance can be understood as an effect of the partly dissipative character of the exchange of the Thomson heat in a real system when conducted to the side. Accordingly, with the same consideration for the hot side, with $dS_h = \frac{d\dot{Q}_\tau}{T_h}$, we obtain

$$\Delta \dot{S}_h = d\dot{S}_h - d\dot{S} = \frac{d\dot{Q}_\tau}{T_h} - \frac{d\dot{Q}_\tau}{T} = \frac{I}{T_h} \int_{\overline{\alpha}}^{\alpha_h} (T - T_h) d\alpha = \frac{I}{T_h} \int_{T_{\overline{\alpha}}}^{T_h} T \frac{d\alpha}{dT} dT - I(\alpha_h - \overline{\alpha}) \qquad (A8)$$

i.e., $\Delta \dot{S}_h$ gives a negative contribution to the entropy balance. This sounds contradictory to the second law of thermodynamics but it is not, as the Thomson heat is not really flowing from a lower to a higher temperature but, when released, reduces the T gradient of the underlying background of flowing Fourier heat, thus reducing the Fourier heat flow by the amount of "upstreaming" Thomson heat.

The hot and cold side entropy changes together give

$$\Delta \dot{S} = \Delta \dot{S}_h + \Delta \dot{S}_c = \frac{I}{T_h} \int_{\overline{\alpha}}^{\alpha_h} (T - T_h) d\alpha + \frac{I}{T_c} \int_{\alpha_c}^{\overline{\alpha}} (T - T_c) d\alpha = \frac{I}{T_c} \int_{T_c}^{T_{\overline{\alpha}}} T \frac{d\alpha}{dT} dT +$$
$$\frac{I}{T_h} \int_{T_{\overline{\alpha}}}^{T_h} T \frac{d\alpha}{dT} dT - I(\alpha_h - \alpha_c). \qquad (A9)$$

With $\frac{1}{T_c} \int_{\alpha_c}^{\overline{\alpha}} T d\alpha \widetilde{>} \overline{\alpha} - \alpha_c$ and $\frac{1}{T_h} \int_{\overline{\alpha}}^{\alpha_h} T d\alpha \widetilde{<} \alpha_h - \overline{\alpha}$ we get $\frac{I}{T_c} \int_{T_c}^{T_{\overline{\alpha}}} T \frac{d\alpha}{dT} dT + \frac{I}{T_h} \int_{T_{\overline{\alpha}}}^{T_h} T \frac{d\alpha}{dT} dT \approx I(\alpha_h - \alpha_c)$ and thus $\Delta \dot{S} \approx 0$. Hence, assuming $\alpha_{\tau,ex} = \overline{\alpha}$, the entropy balance of the inner Thomson heat transfer as an offset of a much larger background Fourier heat flow is almost zero. This indeed confirms our approach to deduce a rule for the local distribution of Thomson heat based on a reversible approximation, i.e., assuming $\Delta \dot{S} \approx 0$, but also shows that the rule is not completely strict.

References

1. Zeier, W.G.; Schmitt, J.; Hautier, G.; Aydemir, U.; Gibbs, Z.M.; Felser, C.; Snyder, G.J.J.N.R.M. Engineering half-Heusler thermoelectric materials using Zintl chemistry. *Nat. Rev. Mater.* **2016**, *1*, 1–10. [CrossRef]

2. Goupil, C.; Seifert, W.; Zabrocki, K.; Müller, E.; Snyder, G.J.J.E. Thermodynamics of thermoelectric phenomena and applications. *Entropy* **2011**, *13*, 1481–1517. [CrossRef]

3. Snyder, G.J.; Toberer, E.S. Complex thermoelectric materials. In *Materials For Sustainable Energy: A Collection of Peer-Reviewed Research and Review Articles from Nature Publishing Group*; World Scientific: Singapore, 2011; pp. 101–110.

4. Rowe, D.M. *Thermoelectrics Handbook: Macro to Nano*; CRC Press: Boca Raton, FA, USA, 2005.

5. Antonova, E.E.; Looman, D.C. Finite elements for thermoelectric device analysis in ANSYS. In Proceedings of the ICT 2005. 24th International Conference on Thermoelectrics, Clemson, SC, USA, 19–23 June 2005; pp. 215–218.

6. Goupil, C. *Continuum Theory and Modeling of Thermoelectric Elements*; John Wiley & Sons: Hoboken, NJ, USA, 2015.

7. Kim, C.N. Development of a numerical method for the performance analysis of thermoelectric generators with thermal and electric contact resistance. *Appl. Therm. Eng.* **2018**, *130*, 408–417. [CrossRef]

8. Hogan, T.; Shih, T. Modeling and characterization of power generation modules based on bulk materials. *Thermoelectr. Handb. Macro Nano* **2006**. [CrossRef]

9. Oliveira, K.S.; Cardoso, R.P.; Hermes, C.J. Two-Dimensional Modeling of Thermoelectric Cells. In Proceedings of the International Refrigeration and Air Conditioning Conference, West Lafayette, IN, USA, 14–17 July 2014.

10. Kim, H.S.; Liu, W.; Ren, Z. The bridge between the materials and devices of thermoelectric power generators. *Energy Environ. Sci.* **2017**, *10*, 69–85. [CrossRef]

11. Kim, H.S.; Liu, W.; Ren, Z.J.J.o.A.P. Efficiency and output power of thermoelectric module by taking into account corrected Joule and Thomson heat. *J. Appl. Phys.* **2015**, *118*, 115103. [CrossRef]

12. Ryu, B.; Chung, J.; Park, S. Thermoelectric efficiency has three Degrees of Freedom. *arXiv* **2018**, arXiv:1810.11148.

13. Sunderland, J.E.; Burak, N.T. The influence of the Thomson effect on the performance of a thermoelectric power generator. *Solid-State Electron.* **1964**, *7*, 465–471. [CrossRef]

14. Min, G.; Rowe, D.M.; Kontostavlakis, K. Thermoelectric figure-of-merit under large temperature differences. *J. Phys. D: Appl. Phys.* **2004**, *37*, 1301. [CrossRef]

15. Chen, J.; Yan, Z.; Wu, L. The influence of Thomson effect on the maximum power output and maximum efficiency of a thermoelectric generator. *J. Appl. Phys.* **1996**, *79*, 8823–8828. [CrossRef]

16. Fraisse, G.; Ramousse, J.; Sgorlon, D.; Goupil, C. Comparison of different modeling approaches for thermoelectric elements. *Energy Convers. Manag.* **2013**, *65*, 351–356. [CrossRef]

17. Sandoz-Rosado, E.J.; Weinstein, S.J.; Stevens, R.J. On the Thomson effect in thermoelectric power devices. *Int. J. Therm. Sci.* **2013**, *66*, 1–7. [CrossRef]

18. Ponnusamy, P.; de Boor, J.; Müller, E. Using the constant properties model for accurate performance estimation of thermoelectric generator elements. *Appl. Energy* **2020**, *262*, 114587. [CrossRef]

19. Zhang, T. Effects of temperature-dependent material properties on temperature variation in a thermoelement. *J. Electron. Mater.* **2015**, *44*, 3612–3620. [CrossRef]

20. Wu, H.; Zhao, L.-D.; Zheng, F.; Wu, D.; Pei, Y.; Tong, X.; Kanatzidis, M.G.; He, J. Broad temperature plateau for thermoelectric figure of merit ZT> 2 in phase-separated PbTe 0.7 S 0.3. *Nat. Commun.* **2014**, *5*, 1–9. [CrossRef]

21. Sankhla, A.; Patil, A.; Kamila, H.; Yasseri, M.; Farahi, N.; Mueller, E.; de Boor, J. Mechanical Alloying of Optimized Mg2 (Si, Sn) Solid Solutions: Understanding Phase Evolution and Tuning Synthesis Parameters for Thermoelectric Applications. *ACS Appl. Energy Mater.* **2018**, *1*, 531–542. [CrossRef]

22. Kamila, H.; Sahu, P.; Sankhla, A.; Yasseri, M.; Pham, H.-N.; Dasgupta, T.; Mueller, E.; de Boor, J. Analyzing transport properties of p-type Mg_2Si–Mg_2Sn solid solutions: Optimization of thermoelectric performance and insight into the electronic band structure. *J. Mater. Chem. A* **2019**, *7*, 1045–1054. [CrossRef]

23. Kim, H.S.; Kikuchi, K.; Itoh, T.; Iida, T.; Taya, M. Design of segmented thermoelectric generator based on cost-effective and light-weight thermoelectric alloys. *Mater. Sci. Eng. B* **2014**, *185*, 45–52. [CrossRef]

24. Zhao, L.-D.; Lo, S.-H.; Zhang, Y.; Sun, H.; Tan, G.; Uher, C.; Wolverton, C.; Dravid, V.P.; Kanatzidis, M.G. Ultralow thermal conductivity and high thermoelectric figure of merit in SnSe crystals. *Nature* **2014**, *508*, 373. [CrossRef]

25. Seifert, W.; Ueltzen, M.; Müller, E. One-dimensional modelling of thermoelectric cooling. *Phys. Status Solidi A* **2002**, *194*, 277–290. [CrossRef]

26. Lamba, R.; Kaushik, S. Thermodynamic analysis of thermoelectric generator including influence of Thomson effect and leg geometry configuration. *J. Energy Convers. Manag.* **2017**, *144*, 388–398. [CrossRef]

27. Garrido, J.; Casanovas, A.; Manzanares, J.A. Thomson Power in the Model of Constant Transport Coefficients for Thermoelectric Elements. *J. Electron. Mater.* **2019**, *48*, 5821–5826. [CrossRef]

4

Variational Autoencoder Reconstruction of Complex Many-Body Physics

Ilia A. Luchnikov [1,2]**, Alexander Ryzhov** [1]**, Pieter-Jan Stas** [3]**, Sergey N. Filippov** [2,4,5] **and Henni Ouerdane** [1,*]

[1] Center for Energy Science and Technology, Skolkovo Institute of Science and Technology, 3 Nobel Street, Skolkovo, Moscow Region 121205, Russia; Ilia.Luchnikov@skoltech.ru (I.A.L.); a.ryzhov@skoltech.ru (A.R.)

[2] Moscow Institute of Physics and Technology, Institutskii Per. 9, Dolgoprudny, Moscow Region 141700, Russia; sergey.filippov@phystech.edu

[3] Department of Applied Physics, Stanford University 348 Via Pueblo Mall, Stanford, CA 94305, USA; pjstas@stanford.edu

[4] Valiev Institute of Physics and Technology of Russian Academy of Sciences, Nakhimovskii Pr. 34, Moscow 117218, Russia

[5] Steklov Mathematical Institute of Russian Academy of Sciences, Gubkina St. 8, Moscow 119991, Russia

* Correspondence: h.ouerdane@skoltech.ru

Abstract: Thermodynamics is a theory of principles that permits a basic description of the macroscopic properties of a rich variety of complex systems from traditional ones, such as crystalline solids, gases, liquids, and thermal machines, to more intricate systems such as living organisms and black holes to name a few. Physical quantities of interest, or equilibrium state variables, are linked together in equations of state to give information on the studied system, including phase transitions, as energy in the forms of work and heat, and/or matter are exchanged with its environment, thus generating entropy. A more accurate description requires different frameworks, namely, statistical mechanics and quantum physics to explore in depth the microscopic properties of physical systems and relate them to their macroscopic properties. These frameworks also allow to go beyond equilibrium situations. Given the notably increasing complexity of mathematical models to study realistic systems, and their coupling to their environment that constrains their dynamics, both analytical approaches and numerical methods that build on these models show limitations in scope or applicability. On the other hand, machine learning, i.e., data-driven, methods prove to be increasingly efficient for the study of complex quantum systems. Deep neural networks, in particular, have been successfully applied to many-body quantum dynamics simulations and to quantum matter phase characterization. In the present work, we show how to use a variational autoencoder (VAE)—a state-of-the-art tool in the field of deep learning for the simulation of probability distributions of complex systems. More precisely, we transform a quantum mechanical problem of many-body state reconstruction into a statistical problem, suitable for VAE, by using informationally complete positive operator-valued measure. We show, with the paradigmatic quantum Ising model in a transverse magnetic field, that the ground-state physics, such as, e.g., magnetization and other mean values of observables, of a whole class of quantum many-body systems can be reconstructed by using VAE learning of tomographic data for different parameters of the Hamiltonian, and even if the system undergoes a quantum phase transition. We also discuss challenges related to our approach as entropy calculations pose particular difficulties.

Keywords: complex systems thermodynamics; machine learning; quantum phase transition; Ising model; variational autoencoder

1. Introduction

The development of the dynamical theory of heat or classical equilibrium thermodynamics as we know it was possible only with empirical data collection, processing, and analysis, which led, through a phenomenological approach, to the definition of two fundamental physical concepts, the actual pillars of the theory: energy and entropy [1]. It is with these two concepts that the laws (or principles) of thermodynamics could be stated and the absolute temperature be given a first proper definition. Though energy remains as fully enigmatic as entropy from the ontological viewpoint, the latter concept is not completely understood from the physical viewpoint. This of course did not preclude the success of equilibrium thermodynamics as evidenced not only by the development of thermal sciences and engineering, but also because of its cognate fields that owe it, at least partly or as an indirect consequence, their birth, from quantum physics to information theory.

Early attempts to refine and give thermodynamics solid grounds started with the development of the kinetic theory of gases and of statistical physics, which in turn permitted studies of irreversible processes with the development of nonequilibrium thermodynamics [2–6] and later on finite-time thermodynamics [7–9], thus establishing closer ties between the concrete notion of irreversibility and the more abstract entropy, notably with Boltzmann's statistical definition [10] and Gibbs' ensemble theory [11]. Notwithstanding conceptual difficulties inherent to the foundations of statistical physics, such as, e.g., irreversibility and the ergodic hypothesis [12,13], entropy acquired a meaningful statistical character and the scope of its definitions could be extended beyond thermodynamics, thus paving the way to information theory, as information content became a physical quantity per se, i.e., something that can be measured [14]. Additionally, although quantum physics developed independently from thermodynamics, it extended the scope of statistical physics with the introduction of quantum statistics, led to the definition of the von Neumann entropy [15], and also introduced new problems related to small, i.e., mesoscopic and nanoscopic systems [16,17], down to nuclear matter [18], where the concepts of thermodynamic limit and ensuing standard definitions of thermodynamic quantities may be put at odds.

Quantum physics problems that overlap with thermodynamics are typically classified into different categories: ground state characterization [19], thermal state characterization at finite temperature [20], the so-called eigenstate thermalization hypothesis [21–25], calculation of the dynamics of either closed or open systems [26,27], state reconstruction from tomographic data [28], and quantum system control, which, given the complexity for its implementation, requires the development of new methods [29]. Among the rich variety of methods applicable to such problems, including, e.g., mean-field approach [30], slave particle approach [31], dynamical mean-field theory [32], nonperturbative methods based on functional integrals [33], we believe two large families of techniques are of particular interest for numerical studies of many-body systems when strong correlations must be accounted for: One is based on the quantum Monte Carlo (QMC) framework [34], which is powerful to overcome the curse of dimensionality by using the stochastic estimation of high-dimensional integrals; the other family encompasses methods that search solutions in the parametric set of functions, also called ansatz. The most used ansatzes are based on different tensor network architectures [35,36] as tensor network-based methods show state-of-the-art performance for the characterization of one-dimensional strongly correlated quantum systems. One can solve either the ground-state problem by using the variational matrix product state (MPS) ground state search [37] or a dynamical problem using a time-evolving block decimation (TEBD) algorithm [38]. Quantum criticality of one-dimensional systems also can be studied by using a more advanced architecture called multiscale entanglement renormalization ansatz (MERA) [39]. The application of tensor networks is not restricted to one-dimensional systems, and one can describe an open quantum dynamics [40], characterize the numerical complexity of an open quantum dynamics [41,42], perform tomography of non-Markovian quantum processes by using tensor networks [43,44], analyze properties of two dimensional quantum lattices by using projected entangled pair states (PEPS) [45], or solve classical statistical physics problems [46,47].

The cross-fertilization of quantum physics and thermodynamics has benefited much from the powerful quantum formalism and computational techniques; however, as thermodynamic concepts evolved from intuitive/phenomenological definitions to classical-mechanics constructs, extended with quantum physics and formalism when needed, thermodynamics, in spite of its undeniable theoretical and practical successes, never managed to fully mature into a genuine fundamental theory that firmly rests on strong basic postulates. On one hand, this led a growing number of physicists to consider thermodynamics as incomplete, and on the other, to think quantum theory as the underlying framework from which equilibrium and nonequilibrium thermodynamics emerge. Quantum thermodynamics [48,49] is a fairly recent field of play, where new ideas are tested while revisiting old problems related to cycles, engines, refrigerators, and entropy production, to name a few [50,51]. Further, quantum technology is a burgeoning field at the interface of physics and engineering, which seeks to develop devices able to harness quantum effects for computing and secure communication purposes [52,53]. The wide scale development of such a kind of systems, which irreversibly interact with an infinite environment, rests on the ability to properly simulate the open quantum dynamics of their many-body properties and analyze coherence and dissipation at the quantum level.

How fast quantum thermodynamics will progress is difficult to anticipate as there exist numerous unsolved problems, especially those related to the proper characterization of the physical processes, e.g., what qualifies as heat or work on ultrashort time and length scales, where averages become irrelevant is unclear, and how the laws of thermodynamics may be systematically adapted still may be debated. To mitigate risks of slow progress, one may resort to approaches that do not rely on models of systems, but rather on data, the idea being to gain actual knowledge and understanding from data irrespective of how complex the studied system is. Machine learning (ML) provides perfectly suited tools for that purpose [54]. ML has a rather long history that can be dated back with the works of Bayes (1763) on prior knowledge that can be used to calculate the probability of an event as formulated by Laplace (1812). Much later (1913), Markov chains were proposed as a tool to describe sequences of events, each being characterized by a probability of occurrence that depends on the actuality of the previous event only. The main milestone is in 1950, with Turing's machine that can learn [55], shortly followed in 1951 by the first neural network machine [56]. Thanks to the huge increase in computational power over the last two decades, ML is now used for a wide variety of problems [54], and quantum machine learning now shows extraordinary potential for faster and more efficient than ever treatment of complex quantum systems problems [57], one major challenge still residing in the development of the hardware capable to harness and transform this potentiality into actual tool.

With the recent success in the field of deep learning, tools other than those based on tensor networks work as well as an ansatz. Restricted Boltzmann machine has been successfully applied as an ansatz to a ground state search, dynamics calculation, and quantum tomography [58–60], as well as convolution neural network to the two-dimensional frustrated $J_1 - J_2$ model [61]. The deep autoregressive model was applied very efficiently and elegantly to a ground state search of many-body quantum system and to classical statistical physics as well [62,63]. It was also recently shown how ML can establish and classify with high accuracy the chaotic or regular behavior of quantum billiards models and XXZ spin chains [64]. Thus, it can be useful to transfer deep architectures from the field of deep learning to the area of many-body quantum systems. A variational autoencoder (VAE) was used for sampling from probability distributions of quantum states in [65]; in the present work, we show that state-of-the-art generative architecture called conditional VAE can be applied to describe the whole family of the ground states of a quantum many-body system. For that purpose, using quantum tomography (albeit in an approximate fashion as discussed below) and reconstruction tools developed in [66], we consider the paradigmatic Ising model in a transverse-field as an illustration of the usefulness and efficiency of our approach. The use of VAE in such a problem is justified by the simplicity of VAE training, as well as its expressibility [67].

The article is organized as follows. In Section 2, we give a brief recap of the physics of the Ising model in a transverse field. In Section 3, we develop our generative model in the framework of the tensor network. Section 4 is devoted to the variational autoencoder architecture. The results are shown and discussed in Section 5. The article ends with concluding remarks, followed a by a short series of appendices.

2. Transverse-Field Ising Model

Among the rich variety of condensed matter systems, magnetic materials are a source of many fruitful problems, whose studies and solutions inspired discussions and new models beyond their immediate scope. The Kondo effect (existence of a minimum of electrical resistivity at low temperature in metals due to the presence of magnetic impurities) is one such problem [68,69], as it provides an excellent basis for studies of quantum criticality and absolute zero-temperature phase transitions [70,71] and, also, on a more fundamental level, a concrete example of asymptotic freedom [69]. Assuming infinite on-site repulsion, the single-impurity Anderson model [68,72] was used to establish a correspondence between Hamiltonian language and path integral for the development of nonperturbative methods in quantum field theory [73,74]. One other important model is that of the Heisenberg Hamiltonian, defined for the study of ferromagnetic materials, and which, assuming a crystal subjected to an external magnetic field $\boldsymbol{B}$, reads [75] as

$$H = -\sum_{\langle i,j \rangle} J_{ij} \hat{S}^i \hat{S}^j - \boldsymbol{h} \cdot \sum_j \hat{S}^j \tag{1}$$

where, for ease of notations, we introduced $\boldsymbol{h} = g\mu_B \boldsymbol{B}$, with g being the Landé factor and $\mu_B = e\hbar/2m_e$ being the Bohr magneton (e: elementary electric charge, and m_e: electron mass); J_{ij} is a parameter that characterizes the nearest-neighbors exchange interaction between electron spins on the crystal sites i and j (the quantum spins $\hat{S}^i$ and $\hat{S}^j$ are vector operators whose components are proportional to the Pauli matrices). For simplicity, one may consider $J_{ij} \equiv J$ constant. If $J > 0$, then the system is ferromagnetic and if $J < 0$ the system is antiferromagnetic. Hereafter, we fix the electron's magnetic moment $g\mu_B = 1$.

Although Equation (1) has a fairly simple form, the exact calculation of the partition function is

$$Z = \operatorname{Tr} e^{-\beta H} \tag{2}$$

where $\beta = 1/k_B T$ is the inverse thermal energy, which is possible on the analytical level with the mean-field approximation that simplifies the Hamiltonian (1), and also for one-dimensional systems, one difficulty of the Heisenberg Hamiltonian being that the three components of a spin vector operator do not commute. That said, Heisenberg's Hamiltonian is very useful to, e.g., study spin frustration [76], entanglement entropy [77], and also serve as a test case for density-matrix renormalization group algorithms [78]. Under zero field, Heisenberg's Hamiltonian is also a simplified form of the Hubbard model at half-filling, thus including ferromagnetism in the scope of strongly correlated systems studies.

A particular, but very important, approximation of Heisenberg's Hamiltonian, whose significance lies in physics, especially for the study of critical phenomena, cannot be underestimated: the so-called Ising model. In its initial formulation [79], Ising spins are N classical variables, which may take ± 1 as values and form a one-dimensional (1D) system characterized by free or periodic boundary conditions. The classical partition function Z may be calculated analytically for the 1D Ising model, and quantities such as the average total magnetization are obtained directly [80]:

$$M = \frac{1}{\beta} \frac{\partial \ln Z}{\partial h} \tag{3}$$

In the present work, we consider a 1D quantum spin chain whose Hilbert space is given by $\mathcal{H} = \otimes_i^N \mathbb{C}^2$. The system is described by the transverse-field Ising (TFI) Hamiltonian [81]:

$$H = -J \sum_{\langle i,j \rangle} \sigma_z^i \sigma_z^{i+1} + h_x \sum_{i=1}^{N} \sigma_x^i. \tag{4}$$

where σ_α^i ($\alpha \equiv x, z$) is the Pauli matrix for the α-component of the i-th spin in the chain, and h_x is the magnetic field applied in the transverse direction x. In this case, the spins are no longer the classical Ising ones and the two terms that compose the Hamiltonian H do not commute, therefore requiring a full quantum approach. An example of a real-world system that may be studied as a quantum Ising chain is cobalt niobate ($CoNb_2O_6$); in this case, the spins that undergo the phase transition as the transverse field varies are those of the Co^{2+} ions [82]. The spin states are denoted $|+\rangle_i$ and $|-\rangle_i$ at ion site i. There are two possible ground states: when all N spins are in the state $|+\rangle$ or in the state $|-\rangle$, i.e., when they are all aligned, which defines the ferromagnetic phase.

The phase transition from the ferromagnetic phase to the paramagnetic phase that we speak of now is of a quantum nature, and not of a thermal nature, as here it is driven only by the external magnetic field. More precisely, when the transverse field h_x is applied with sufficient strength, the spins align along the x direction, and the spin state at site i is given as the superposition $(|+\rangle_i + |-\rangle_i) / \sqrt{2}$, which is nothing else but the eigenstate of the x-component of the spin. Therefore, in this particular case, there is no need to raise the temperature of the system initially in the ferromagnetic phase beyond the Curie temperature to make it a paramagnet: the many-body system remains in its ground state, but its properties have changed. Further, note that unlike for the ferromagnetic phase, the quantum paramagnetic phase has spin-inversion symmetry. An insightful discussion on quantum criticality can be found in Reference [83].

Now, we briefly comment on the quantity $\beta = 1/k_B T$ in the context of quantum phase transitions, which, strictly speaking, can only occur at temperature $T = 0$ K. In fact, close to the absolute zero, where $\beta \to \infty$, their signatures can be observed as quantum fluctuations dominate thermal fluctuations in the criticality region, where the quantum critical point lies. The imaginary time formalism [84], where $\exp(-\beta H)$ is interpreted as an evolution operator, and the partition function Z as a path integral, provides a way to map a quantum problem onto a classical one with the introduction of the imaginary time β resulting from a Wick rotation in the complex plane, thus yielding one extra dimension to the model. In classical thermodynamics, to observe a phase transition in a system requires that its size (i.e., the number of constituents N) tends to infinity so that the order parameter is non-analytic at the transition point; so, for the quantum transition, the thermodynamic limit entails the limit $\beta \to \infty$ also: the 1D TFI model is mapped onto an equivalent 2D classical Ising model [85]. The imaginary time formalism permits implementation of classical Monte Carlo simulations to study quantum systems. Further discussion, including the sign problem for the quantum spin-1/2 system, is available in Reference [4].

We have chosen the transverse-field Ising model as an illustrative case for our study for several reasons. First, as this system is 1-dimensional, we can apply an MPS variational ground state solver [37], and therefore obtain the ground state solution in MPS representation. We can then perform fast and exact sampling for generation of large data sets for the training of the VAE. Next, this model can be solved analytically, which allows us to adequately benchmark our results. Finally, this model shows a nontrivial behavior around the quantum phase transition point at $h_x = 1$, and thus constitutes an interesting example to apply a VAE.

3. Generative Model as a Quantum State

Many-body quantum physics is rich in high-dimensional problems. Often, however, with increasing dimensionality, these become extremely difficult or impossible to solve. One solving method is through the reformulation of the quantum mechanical problem as a statistical problem, when possible. This way, machine learning can be used to effectively solve such a problem, as machine learning is a tool for the solving of high-dimensional statistical problems [86]. Probabilistic interpretation allows for using powerful sampling-based methods that work efficiently with high dimensional data.

An example of the reformulation of a quantum problem as a statistical problem is with informationally complete (IC) positive-operator valued measures (POVMs) [87]. POVMs describe the most general measurements of a quantum system. Each particular POVM is defined by a set of positive semidefinite operators M^α, with the normalization condition $\sum_\alpha M^\alpha = \mathbb{1}$, where $\mathbb{1}$ is the identity operator. The fact that the POVM is informationally complete means that using measurement outcomes one can reconstruct the state of a system with arbitrary accuracy.

The probability of measurement outcome for a quantum system with the density operator ρ is governed by Born's rule: $P[\alpha] = \text{Tr}(\varrho M^\alpha)$, where $\{M^\alpha\}$ is a particular POVM and α is an outcome result. In other words, any density matrix can be mapped on a mass function, although not all mass functions can be mapped on a density matrix [88,89]. Some mass functions lead to non-positive semidefinite "density matrices", which is not physically allowed. As such, quantum theory is a constrained version of probability theory. For a many-body system, these constraints can be very complicated, and direct consideration of quantum theory as a constrained probability theory is not fruitful. However, if one can access the samples of the IC POVM induced mass function, which is by definition physically allowed, this mass function can be reconstructed using generative modeling [66,67]. Samples can be obtained either by performing generalized measurements over the quantum system or by in silico simulation.

In the present work, we simulate measurements of the ground state of a spin chain with the TFI Hamiltonian, Equation (4). As a local (one spin) IC POVM, we use the so-called symmetric IC POVM for qubits (tetrahedral) POVM [90]:

$$M^\alpha_{\text{tetra}} = \frac{1}{4}\left(\mathbb{1} + s^\alpha \sigma\right), \ \alpha \in (0,1,2,3), \ \sigma = (\sigma_x, \sigma_y, \sigma_z),$$

$$s^0 = (0,0,1), \ s^1 = \left(\frac{2\sqrt{2}}{3}, 0, -\frac{1}{3}\right), \ s^2 = \left(-\frac{\sqrt{2}}{3}, \sqrt{\frac{2}{3}}, -\frac{1}{3}\right), \ s^3 = \left(-\frac{\sqrt{2}}{3}, -\sqrt{\frac{2}{3}}, -\frac{1}{3}\right). \quad (5)$$

Note that the many-spin generalization of local IC POVM can easily be obtained by considering the tensor product of local ones:

$$M^{\alpha_1,\ldots,\alpha_N}_{\text{tetra}} = M^{\alpha_1}_{\text{tetra}} \otimes M^{\alpha_2}_{\text{tetra}} \otimes \cdots \otimes M^{\alpha_N}_{\text{tetra}}. \quad (6)$$

To simulate measurements outcome under the IC POVM described above, we implement the following numerical scheme: First, we run a variational MPS ground state solver to obtain the ground state of the TFI model in the MPS form:

$$\Omega_{i_1, i_2, \ldots, i_N} = \sum_{\beta_1, \beta_2, \ldots, \beta_{N-1}} A^1_{i_1 \beta_1} A^2_{\beta_1 i_2 \beta_2} \cdots A^N_{\beta_N} {}_{1 i_N} \quad (7)$$

where we use the tensor notation instead of the bra-ket notation for further simplicity, and we obtain the MPS representation of IC POVM induced mass function:

$$P[\alpha_1, \alpha_2, \ldots, \alpha_N] = \sum_{\delta_1, \delta_2, \ldots, \delta_{N-1}} \pi_{\alpha_1 \delta_1} \pi_{\delta_1 \alpha_2 \delta_2} \cdots \pi_{\delta_{N-1} \alpha_N},$$

$$\pi_{\delta_{n-1} \alpha_n \delta_n} = \pi \underbrace{{}_{\beta_{n-1} \beta'_{n-1}}}_{\text{multi-index } \delta_{n-1}} \alpha_n \underbrace{{}_{\beta_n \beta'_n}}_{\text{multi-index } \delta_n} = [M_{\text{tetra}}]^{\alpha_n}_{ij} A^n_{\beta_{n-1} j \beta_n} [A^n]^*_{\beta'_{n-1} i \beta'_n} \quad (8)$$

whose diagrammatic representation [35] is shown in Figure 1. Next, we produce a set of samples of size M: $\{\alpha^i_1, \alpha^i_2, \ldots, \alpha^i_N\}^M_{i=1}$ from the given probability. The sampling can be efficiently implemented as shown in Appendix B. We call this set of samples (outcome measurements) a data set, which may then be used to train a generative model $p[\alpha_1, \alpha_2, \ldots, \alpha_N | \theta]$ to emulate the true mass function $P[\alpha_1, \alpha_2, \ldots, \alpha_N]$. Here, θ is the set of parameters of the generative model, which is trained by maximizing the logarithmic likelihood $\mathcal{L}(\theta) = \sum_{i=1}^{M} \log p[\alpha_1^i, \alpha_2^i, \ldots, \alpha_N^i | \theta]$ with respect to the parameters θ [91]. The trained generative model fully characterizes a quantum state. The density matrix is obtained by applying an inverse transformation to the mass function [92]:

$$\varrho = \sum_{\alpha_1, \alpha_2, \ldots, \alpha_N} p[\alpha_1, \alpha_2, \ldots, \alpha_N | \theta] [M_{\text{tetra}}^{\alpha_1}]^{-1} \otimes [M_{\text{tetra}}^{\alpha_2}]^{-1} \otimes \cdots \otimes [M_{\text{tetra}}^{\alpha_N}]^{-1},$$

$$[M_{\text{tetra}}^{\alpha}]^{-1} = \sum_{\alpha'} T_{\alpha\alpha'}^{-1} M_{\text{tetra}}^{\alpha'}, \tag{9}$$

$$T_{\alpha\alpha'} = \text{Tr}\left(M_{\text{tetra}}^{\alpha} M_{\text{tetra}}^{\alpha'} \right),$$

the diagrammatic representation of which is given in Figure 2. Note that the summation included in the density matrix representation is numerically intractable, but we can estimate it using samplings from the generative model.

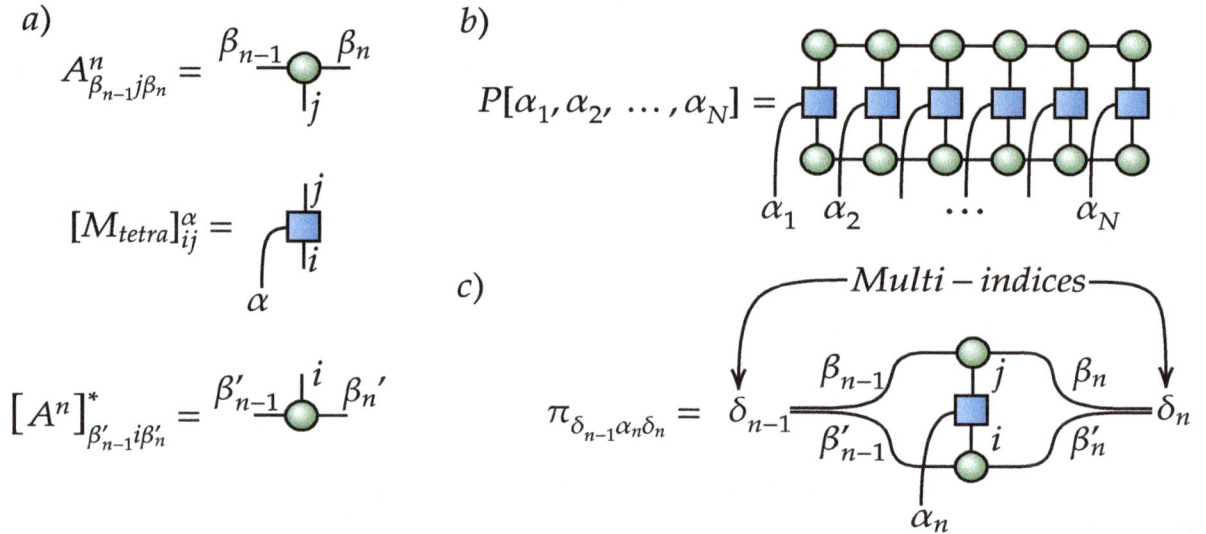

Figure 1. Tensor diagrams for (**a**) building blocks, (**b**) matrix product state (MPS) representation of measurement outcome probability, and (**c**) its subtensor.

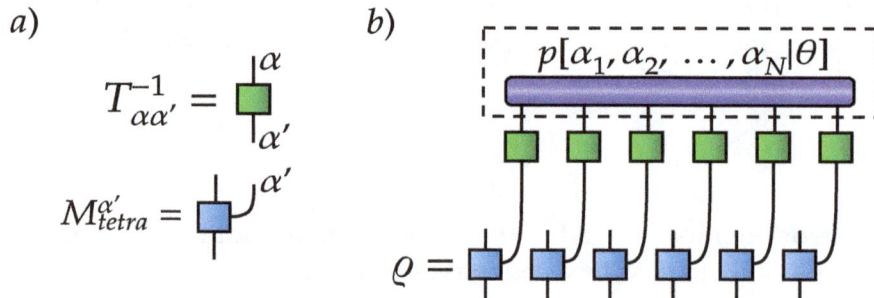

Figure 2. Tensor diagrams for (**a**) building blocks and (**b**) inverse transformation from a mass function to a density matrix.

Our goal is to use a generative model as an effective representation of quantum states to calculate the mean values of observables such as, e.g., two-point and higher-order correlation functions. An explicit expression of the two-point correlation function obtained by sampling from the trained generative model is shown in Figure 3. To obtain the ground state of the TFI model, we use a variational MPS ground state search, and we pick the bond dimension of MPS equal to 25 and perform 5 DMRG sweeps to get an approximate ground state in the MPS form. We use the variational MPS solver provided by the mpnum toolbox [93].

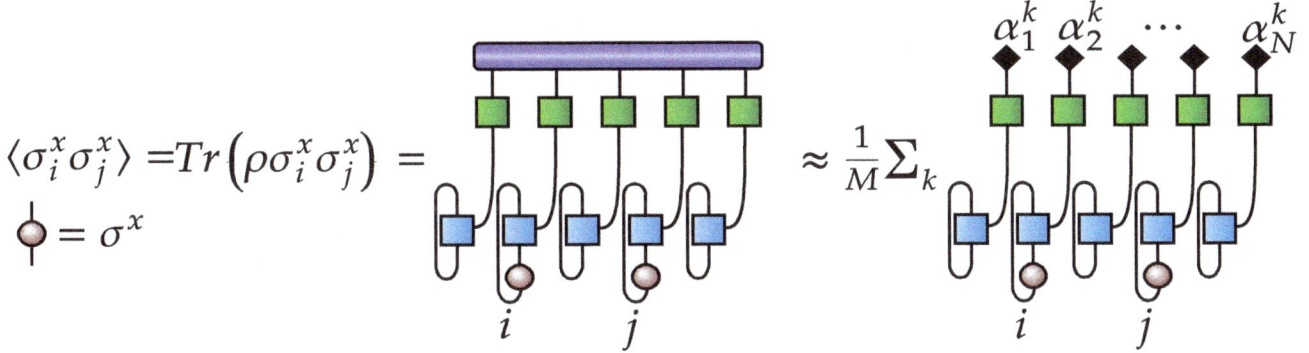

Figure 3. Tensor diagrams representing calculation of two-point correlation function.

4. Variational Autoencoder Architecture

In our work, we use a conditional VAE [94] to represent quantum states. A conditional VAE is a generative model expressed by the following probability distribution,

$$p[x|\theta, h] = \int p[x|z, \theta, h] p[z] dz, \tag{10}$$

where x is the data we want to simulate; θ represents the VAE parameters, which can be tuned to get the desired probability distribution over x; h is the condition; and z is a vector of latent variables. In our case, x is the quantum measurement outcome in one-hot notation. A collection of measurement outcomes is a matrix of size $N \times 4$, where N is the number of particles in the chain and 4 is the number of possible outcomes of the tetrahedral IC POVM, which is either [1000], [0100], [0010], or [0001]. h is the external magnetic field. The probability distribution $p[x|z, \theta, h]$ can thus be written as

$$p[x|z, \theta, h] = \prod_{i=1}^{N} \prod_{j=1}^{4} \pi_{ij}(z, h, \theta)^{x_{ij}}, \tag{11}$$

where $\pi_{ij}(z, h, \theta)$ is the neural network in our architecture, and, more precisely, π_{ij} is the probability of the jth outcome of the POVM for the ith spin with $\sum_{j=1}^{N} \pi_{ij} = 1$ and $\pi_{ij} \geq 0$. The quantity $p[z]$ is the prior distribution over latent variables, which is simply given by $\mathcal{N}(0, I) = \frac{1}{\sqrt{2\pi}^N} \exp\left\{-\frac{1}{2} z^T z\right\}$, with I being the identical covariance matrix. We take the number of latent variables equal to the number of spins, N. Essentially, we want to optimize our VAE so that its probability matches the probability of the quantum measurement outcomes as closely as possible. This can be done using the well-known maximum likelihood estimation:

$$\theta_{MLE} = \underset{\theta}{\operatorname{argmax}} \sum_{i=1}^{M} \log(p[x_i|\theta, h]), \tag{12}$$

where $\{x_i\}_{i=1}^{M}$ is the data set of outcome measurements. We cannot simply maximize this function using, for example, a gradient descent method, due to the presence of hidden variables in the structure of this function. However, we can overcome this problem by using the Evidence Lower Bound (ELBO) [95] and the reparametrization trick shown in [96]. The detailed description of the procedure is given in the Appendix A.

Once trained, the VAE is a simple and efficient way to produce new samples from its probability distribution. It can be done in three steps. First, we produce a sample from the prior distribution $p[z] = \mathcal{N}(0, I)$. Next, we feed this sample and the external magnetic field value into the neural network decoder $\pi_{ij}(z, \theta, h)$, which returns the matrix of probabilities. Finally, we sample from the matrix of probability $\pi_{ij}(z, \theta, h)$ to generate "fake" outcome measurements. A visual representation of the sampling method is shown in Figure 4.

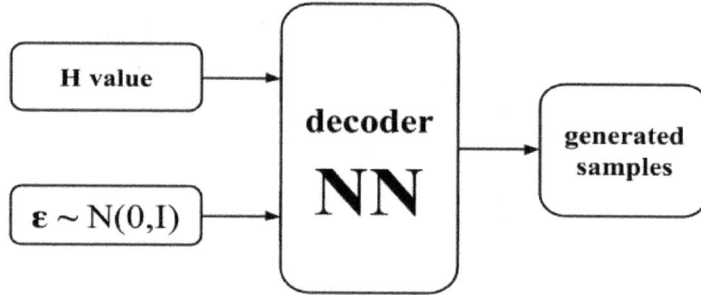

Figure 4. Sampling scheme with the trained variational autoencoder (VAE).

In many problems, gradients of observables with respect to different model parameters yield quantities of interest. For example, one may consider the magnetic differential susceptibility tensor $\chi_{ij} = \partial \mu_i / \partial h_j$. It can be done efficiently by using backpropagation through the VAE architecture but, as samples from the VAE are discrete, a straightforward backpropagation is impossible. In recent papers [97–99], a method called the Gumbel-softmax was introduced to overcome this difficulty through continuous relaxation. The spirit, and therefore the physical meaning of the method, may be understood with a short discussion of the so-called simulated annealing technique, which is often used to solve discrete optimization problems. Broadly speaking, the simulated annealing rests on the introduction of a parameter that acts as an artificial "temperature", which varies continuously to modify the state of the system in search of a global optimum. Starting from a given state, for some values of the temperature, if the system mostly explores the neighboring states, moving among them and possibly in the vicinity of the "better" ones, i.e., with lower energy, it may get and remain close to a local optimum, or local energy minimum in the thermodynamic language; however, to avoid remaining in a locally optimal region, "bad" moves leading to worse (i.e., higher energy) states are useful to explore the temperature space more completely improving the chance to find a global optimum or at least to be near it. To each move an energy variation, ΔE, is associated; it is the continuous character of the fictitious temperature that makes the discrete problem continuous as the probability $\exp(-\Delta E)/k_B T$ of acceptance of a state is continuous. Although this approach has been known for a long time [100], it remains topical and under active development [101,102]. The method of continuous relaxation we use also exploits such an artificial temperature to make discrete samples continuous.

The Gumbel-softmax trick, consists of three steps:

1. We calculate the matrix of log probabilities, taking element-wise logarithm of decoder network output:
$$\log \Pi = \begin{bmatrix} \log \pi_{11} & \log \pi_{12} \ldots \log \pi_{1N} \\ \log \pi_{21} & \log \pi_{22} \ldots \log \pi_{2N} \\ \log \pi_{31} & \log \pi_{32} \ldots \log \pi_{3N} \\ \log \pi_{41} & \log \pi_{42} \ldots \log \pi_{4N} \end{bmatrix},$$

2. We generate a matrix of samples from the standard Gumbel distribution G and sum it up element-wise with the matrix of log probabilities $\log \Pi$: $Z = \log \Pi + G$,

3. Finally, we take the softmax function of the result from the previous step: $x_{\text{soft}}^{\text{fake}}(T) = \text{softmax}(Z/T)$, where T is a temperature of softmax. The softmax functions is defined by the expression $\text{softmax}(x_{ij}) = \dfrac{\exp(x_{ij})}{\Sigma_i \exp(x_{ij})}$.

The quantity $x_{soft}^{fake}(T)$ has a number of remarkable properties: first, it becomes an exact one-hot sample when $T \to 0$; second, we can backpropagate through soft samples for any T> 0. The method is validated in the next section.

Before we proceed to the presentation and discussion of our results, and to better see the added value of the VAE, it is instructive to compare MPS and VAE (NN) in terms of expressibility, i.e., "estimation of MPS states via incomplete local measurements" vs "VAE reconstruction". As the state of the system is assumed to be unknown, and some measurement outcomes are only known for different magnetic fields, these outcomes are too few for exact tomography. Further, it is known that for a given bond dimension d, the entangled entropy cannot be larger than $\log(d)$; in other words, the bond dimension of MPS places an upper bound on the entangled entropy. Thus, the MPS representation describes well only quantum states with low entangled entropy, i.e., quantum states which satisfy the area law [103,104]. The situation with neural network quantum states (NQS) is different: there is no such a restriction for NQS. Moreover, the existence of NQS with volume-law entanglement [105] shows a promising development of new, and possibly powerful, NN-based approaches to representing many-body quantum systems.

5. Results

Here, we show that the VAE trained on a set of preliminary measurements is capable to describe the physics of the whole family of TFI models. We validate our results by comparing VAE-based calculations with numerically exact calculations performed by variational MPS algorithm [35]. Additionally, to assess the capabilities of the VAE, we consider a spin chain with 32 spins. We calculate the MPS representation of the ground state and extract information from it by performing measurements over the state. The external field in the x-direction is varied from 0 to 2 with a step of 0.1. The VAE is trained on a data set (TFI measurement outcomes) consisting of 10.5 million samples in total: 21 external fields h_x with 500,000 samples per field.

To evaluate the VAE performance, we simply compare directly the numerically exact correlation functions with those reconstructed with our VAE. Those of $n = 1, \ldots, 32$, $\langle \sigma_z^1 \sigma_z^n \rangle$, and $\langle \sigma_x^1 \sigma_x^n \rangle$ are shown in Figures 5 and 6, respectively, and we compare the numerically exact and the VAE-based average magnetizations along x, given by $\langle \sigma_x^n \rangle$ for each position of the spin along the chain, in Figure 7. We see that the VAE captures well the physics of the one- and two-point correlation functions. Figure 8 shows the total magnetizations, μ_x and μ_z, in the x and z directions, respectively, with $\mu_i = \frac{1}{N} \sum_{j=1}^{N} \langle \sigma_i^j \rangle$, and we see that the VAE is a tool well-suited for the description of the quantum phase transition and also finite-size effects: whereas for the infinite TFI chain, i.e., in the thermodynamic limit, the phase transition is observed at $h_x = 1$, and the finite size of the system yields a shift of the critical point at $h_x \approx 0.9$. Also note that in the $T \to 0$ limit, the magnetization M defined in Equation (3) coincides exactly with the magnetization μ defined above.

A backpropagation algorithm combined with the Gumbel-softmax trick may be used to evaluate the derivative of an output over an input. We use this approach to calculate some elements of a magnetic differential susceptibility tensor $\chi_{ij} = \partial \mu_i / \partial h_j$, in particular, χ_{xx} and χ_{zx} shown in Figure 9. The backpropagation-based magnetic differential susceptibility agrees well with the numerically calculated one (central differences). The main advantage of the backpropagation-based calculation is its numerical efficiency. The VAE may thus be trained with an arbitrary set of external parameters, i.e., not only h_x, but also h_y and h_z, and yield the full differential susceptibility tensor.

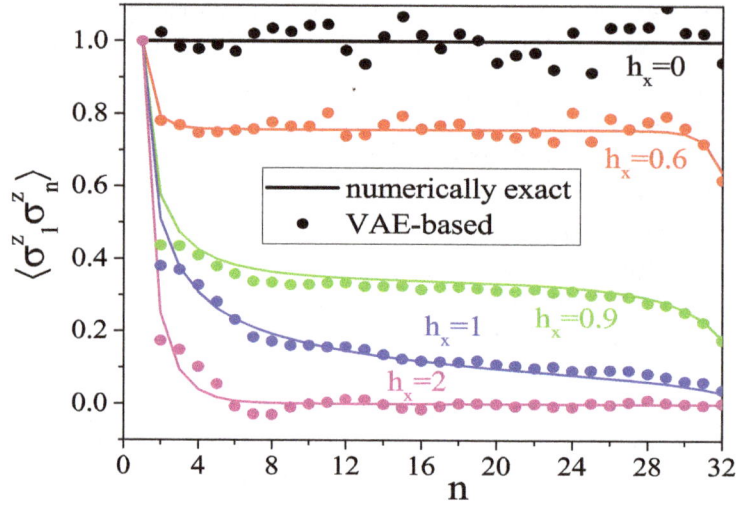

Figure 5. Two-point correlation function $\langle \sigma_1^z \sigma_n^z \rangle$ for different values of external magnetic field h_x.

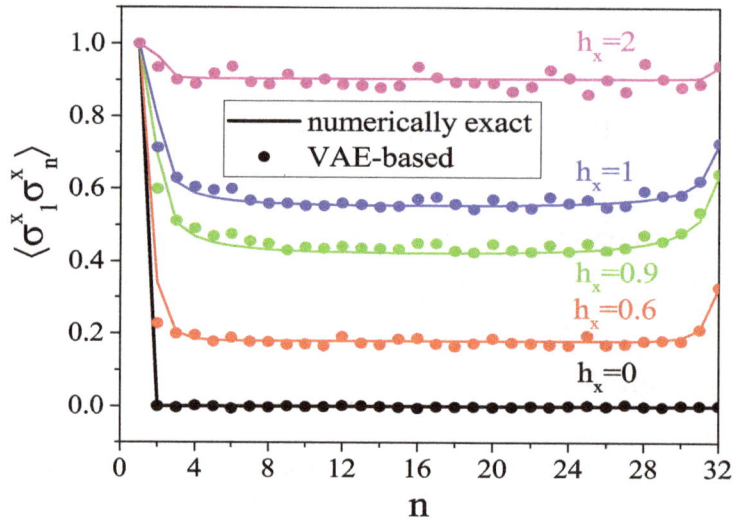

Figure 6. Two-point correlation function $\langle \sigma_1^x \sigma_n^x \rangle$ for different values of external magnetic field h_x.

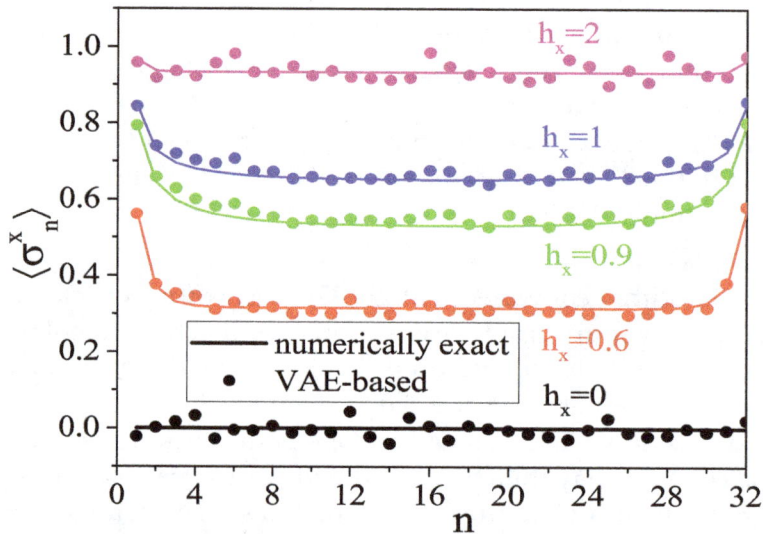

Figure 7. Average magnetization per site along x for different values of external magnetic field h_x.

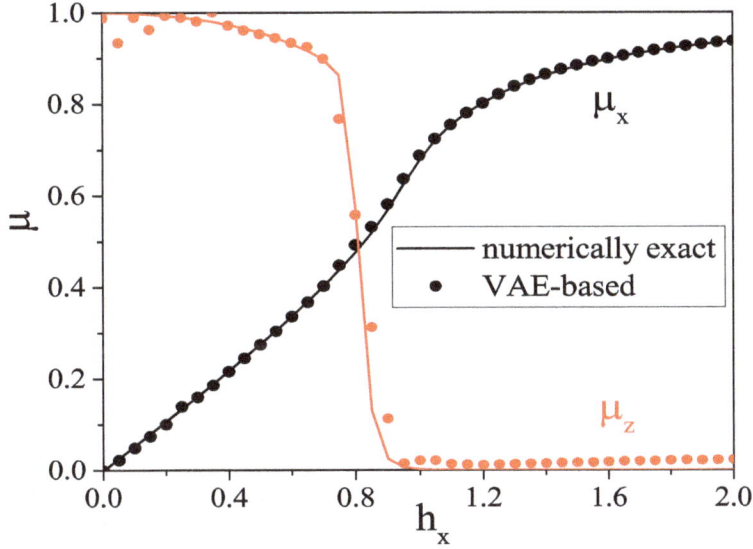

Figure 8. Total magnetization along x and z axes for different values of external magnetic field h_x. The location of the critical region is slightly shifted towards smaller values of h_x due to the finite size of the chain.

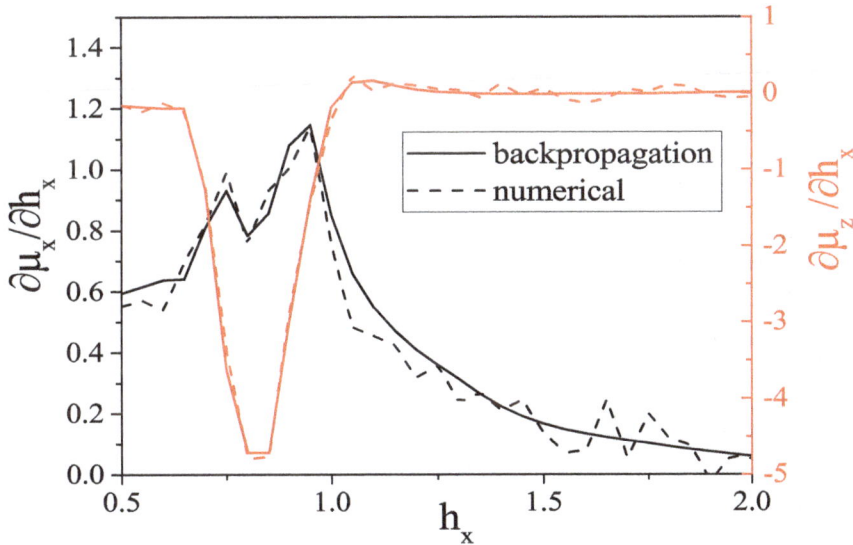

Figure 9. Backpropagation-based and numerical-based (central differences) values of χ_{xx} and χ_{zx} for different values of external magnetic field h_x. Both derivatives slightly fluctuate due to VAE error.

At this stage, we could conclude that the VAE is capable to describe the physics of one- and two-point correlation functions, and therefore the TFI physics. However, notwithstanding the ability of the VAE to yield correlation functions that fit well numerically-exact correlation functions, this is not yet a full proof that it represents quantum states well. To address this point, we consider a small spin chain (five spins with TFI Hamiltonian and an external magnetic field $h_x = 0.9$) for which we calculate both the exact mass function and that estimated from VAE samples. Figure 10 shows that the VAE result again fits the numerically exact mass function with high accuracy. Further, we calculate the Bhattacharyya coefficient [106]: $\text{BC}(p_{\text{vae}}, p_{\text{exact}}) = \sum_\alpha p_{\text{exact}}[\alpha] \sqrt{\frac{p_{\text{vae}}[\alpha]}{p_{\text{exact}}[\alpha]}}$ as a function of the external magnetic field h_x. Results reported in Figure 11 show that $\text{BC}(p_{\text{vae}}, p_{\text{exact}}) > 0.99$ over the whole h_x range, which thus proves that the VAE represents a quantum state well, at least for small spin chains.

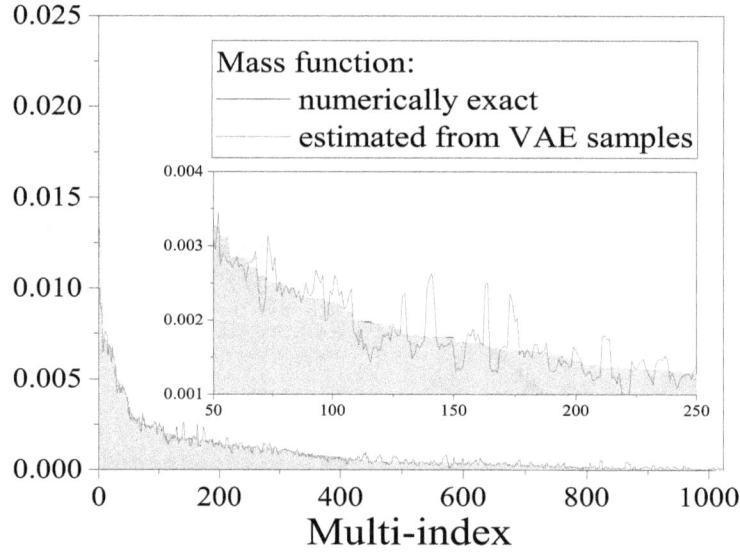

Figure 10. Comparison of two positive-operator valued measure (POVM)-induced mass functions ($P[\alpha] = \mathrm{Tr}(\rho M^{\alpha})$) for a chain of size 5: numerically exact mass function and reconstructed from VAE samples mass function. A sequence of indices α has been transformed into a single multi-index. Indices have been ordered to put numerically exact probability in descending order. A good agreement between the mass functions is observed.

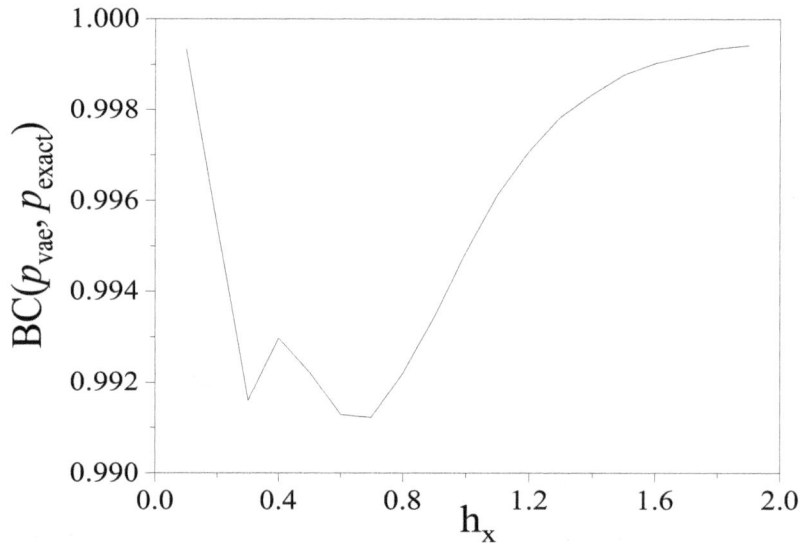

Figure 11. Dependence of the classical fidelity on the external magnetic field. A high predictive accuracy is demonstrated for the whole set of fields.

The structure of the entanglement is an another interesting subject that we would like to validate. The essence of entanglement between two parts of the chain, which is split into n left spins and $N - n$ right spins, can be described by the Réniy entropy of the left part of this chain: $S_{\alpha} = \frac{1}{1-\alpha} \log \mathrm{Tr}\rho_{n}^{\alpha}$, where ρ_{n} is the density matrix of the first n spins in the chain. We estimate the Rényi entropy of order 2: $S_2 = -\log(\mathrm{Tr}\rho^2)$, as it can be efficiently calculated from the matrix product representation of the density matrix and from the VAE samples. However, as sample-based estimation of the entangled entropy has a variance that grows exponentially with the number of spins, we consider a small spin chain of size 10. A direct comparison between the numerically exact and the VAE-based entangled entropies is shown for different values of n in Figure 12. For this particular case, the VAE clearly overestimates the entangled entropy. This undesirable effect is indeed observed for all sizes of spin chains, and even for the spin chain of size 5, for which we have an excellent agreement between the numerically exact mass function and the VAE-based result. The entropy S_2 is sensitive to small errors in the mass function, but it also appears that the primary method of state reconstruction used in the present work has the following shortcomings.

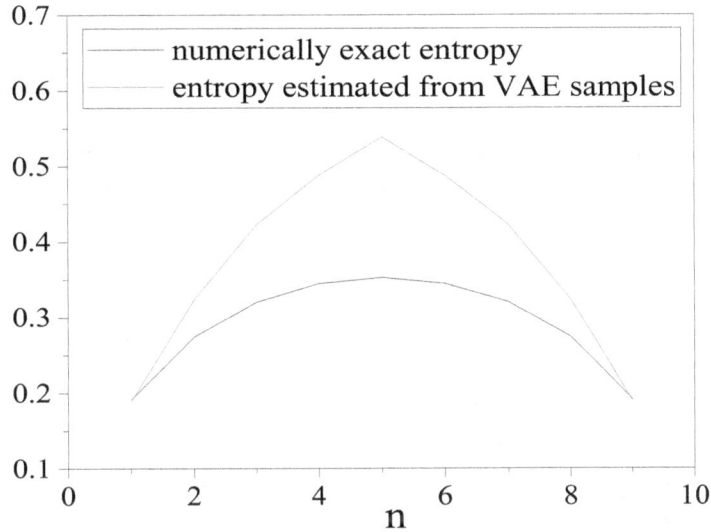

Figure 12. Comparison of the numerically exact Rényi entropy and that reconstructed from the VAE samples for different values of n.

1. If one reconstructs a pure state, the VAE smooths the spectrum of the density matrix and approximates the pure state by a slightly mixed state, as illustrated with a simple example in Figure 13.
2. The VAE does not account the positivity constraints, which yields negative eigenvalues for the density matrix. These negative eigenvalues even appear in the spectrum of the reduced density matrix, as shown in Figure 13.

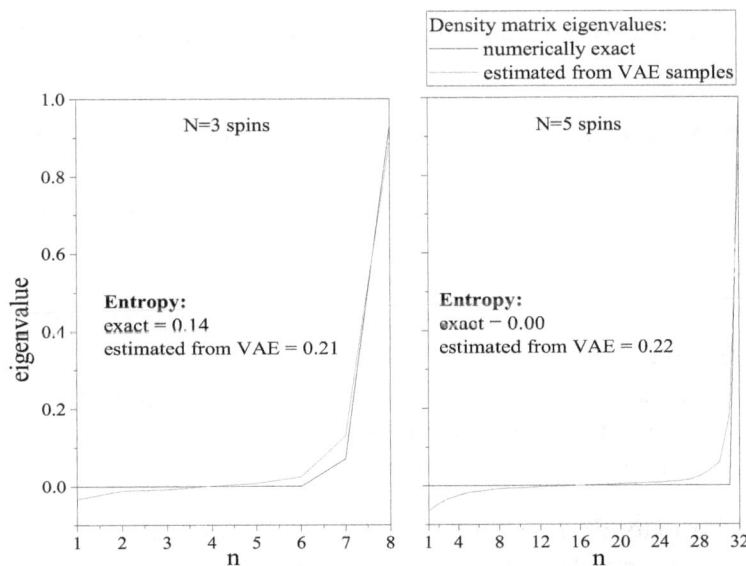

Figure 13. Comparison of numerically exact spectra of density matrices and VAE-estimated spectra. The ground state spectra of the spin chain of size 5 with an external magnetic field $h = 0.9$ is shown on the right panel, and the spectra of the reduced density matrix (last 3 spins) are shown on the left panel.

These drawbacks hinder a robust description of the entanglement structure. In addition to the mismatch between the Rényi entropies (S_2), the entropy of a reduced density matrix can be larger than the entropy of the whole density matrix, which is erroneous. This particular issue, now identified, may be resolved by introduction of a particular regularization term into the VAE loss. This is the object of future work.

Finally, it is also instructive to comment on the memory costs of the use of either MPS or VAE, which is somehow a tricky question, as it is unclear for any NN-based architecture what numbers of layers and neurons per layer are needed because there is no criterion for NN, whereas for the MPS and tensor networks, there is one. Thus, a direct comparison of NN architectures and tensor networks (MPS, etc.) is certainly a difficult task, and in our opinion, likely an impossible one. At this stage, we may say the following. For a given spin chain of size N and maximal entangled entropy between subchains $S = -\mathrm{Tr}\rho \log \rho$, the MPS requires to store approximately $2N \exp(2S)$ complex numbers; this follows from the fact that one then considers N subtensors of size $\exp(S) \times 2 \times \exp(S)$, where $\exp(S)$ is the typical (approximate) size of bond dimension. For a VAE, although it seems that there are no entropic restrictions, the proper quantitative characterization of the "neural network" complexity of a quantum state still is an open question (for tensor networks, it is the entangled entropy). A VAE contains two neural networks: encoder and decoder. To store a feed-forward neural network, one has to store $\sum_i l_{i-1} \times l_i + l_i$ real numbers, with l_i being the number of neurons in the layer number i. In general, one may conclude that the MPS is preferable for low entangled states, and the VAE is preferable for highly entangled states.

6. Conclusions

The thermodynamic study of complex many-body quantum systems still requires the development of new methods, including those that may stem from machine learning. The quantum Ising model, which is of particular importance for practical purposes [107,108], provides a rich framework to test these new methods that are also useful to obtain deeper physical insight into its nonequilibrium dynamics properties such as, e.g., quantum fluctuations propagation [109]. In the present work, we studied the ability of a VAE to reconstruct the physics of quantum many-body systems, using the transverse-field Ising model as a nontrivial example. We used the IC POVM to map the quantum problem onto a probabilistic domain and vice versa. We trained the VAE on a set of samples from the transformed quantum problem, and our numerical experiments show the following results.

- For a large system (32 spins), the VAE's reliability is verified by comparing one- and two-point correlation functions.
- For small system (five spins), the VAE's reliability is verified by direct comparison of mass functions.
- The VAE can capture a quantum phase transition.
- The response functions (magnetic differential susceptibility tensor) can be obtained using backpropagation through VAE.
- Despite the very good agreement between the VAE-based mass function and the true mass function, the VAE shows limited performance with the determination of the entangled entropy. This is point is the object of further development.

Our method can be extended to any other thermodynamic system by introduction of the temperature as an external parameter, thereby considering also thermal phase transitions. As one can calculate different thermodynamic quantities by applying backpropagation through VAE, a worthwhile and highly complex system to study would be water under its difference phases, so as to test recent new ideas and models [110,111].

Our code for our numerical experiments is available on the GitHub repository website [112].

Author Contributions: Conceptualization, I.A.L., S.N.F., and H.O.; methodology, I.A.L. and A.R.; software, I.A.L., A.R., and P.J.S.; validation, all authors; writing-original draft preparation, I.A.L., A.R., P.J.S., and H.O; writing-review and editing, S.N.F. and H.O.

Acknowledgments: The authors thank Stepan Vintskevich for fruitful discussions. The authors also thank Google Colaboratory for providing access to GPU for the acceleration of computations.

Abbreviations

The following abbreviations are used in this manuscript.

VAE	Variational Autoencoder
MPS	Matrix product state
TFI	Transverse-field Ising
IC	Informationally incomplete
POVM	Positive-operator valued measure
ELBO	Evidence lower bound
NN	Neural network
KL	Kullback–Leibler
DMRG	Density matrix renormalization group

Appendix A. VAE: Training and Implementation Details

When training our VAE, we find the arg maximum of the logarithmic likelihood $\mathcal{L}(\theta)$ w.r.t. its parameters θ:

$$\theta_{\mathrm{MLE}} = \underset{\theta}{\mathrm{argmax}}\mathcal{L}(\theta) = \underset{\theta}{\mathrm{argmax}}\log(p[x|\theta,h]), \tag{A1}$$

Equation (A1) cannot directly be evaluated, because of hidden variables in the structure of $p[x|\theta,h]$. We can, however, simplify this problem by introducing a distribution over hidden variables z. Remember that the probability distribution can be described as $p[x|\theta,h] = \int p[x|z,\theta,h]p[z]dz$, so that the expression for the log likelihood becomes

$$\mathcal{L}(\theta) = \log\left(\int p[x|z,\theta,h]p[z]dz\right). \tag{A2}$$

We can then use a mathematical trick that might seem counterintuitive at first glance, but ultimately becomes quite powerful. We multiply the function inside the integral by $\frac{q[z|x,\tilde{\theta},h]}{q[z|x,\tilde{\theta},h]} = 1$, where $q[z|x,\tilde{\theta},h]$ is some arbitrary distribution that can be adjusted with $\tilde{\theta}$, so that

$$\mathcal{L}(\theta) = \log\left(\int p[x|z,\theta,h]p[z]dz\right) = \log\left(\int \frac{q[z|x,\theta,h]}{q[z|x,\tilde{\theta},h]}p[x|z,\theta,h]p[z]dz\right)$$
$$= \log\left(\mathbb{E}_{q[z|x,\tilde{\theta},h]}p[x|z,\theta,h]\frac{p[z]}{q[z|x,\tilde{\theta},h]}\right) \tag{A3}$$

where the quantity $\mathbb{E}_{f[x]}$ denotes the expectation value w.r.t some distribution $f[x]$. We can then use Jensen's inequality to show that

$$\log\left(\mathbb{E}_{q[z|x,\tilde{\theta},h]}p[x|z,\theta,h]\frac{p[z]}{q[z|x,\tilde{\theta},h]}\right) \geq \mathbb{E}_{q[z|x,\tilde{\theta},h]}\log\left(p[x|z,\theta,h]\frac{p[z]}{q[z|x,\tilde{\theta},h]}\right). \tag{A4}$$

where the rhs of this inequality is the lower bound of the log likelihood, as it will always be greater than or equal to the lower bound, and equality can always be achieved by a proper choice of q if it is in a complex enough family.

Maximizing the lower bound is equivalent to maximizing the log likelihood. We can decompose this lower bound term into two terms:

$$\mathcal{L}(\theta) \geq \text{ELBO}(\theta, \tilde{\theta}) = \mathbb{E}_{q[z|x,\tilde{\theta},h]} \log \left(p[x|z,\theta,h] \right) - \int q[z|x,\tilde{\theta},h] \log \frac{q[z|x,\tilde{\theta},h]}{p[z]} dz \qquad (A5)$$

Note that the second term is equivalent to the Kullback–Leibler divergence $KL(q[z|x,\tilde{\theta},h] \,||\, p[z])$. In our case, we picked the particular distribution forms that reflect the structure of our problem:

$$p[x|z,\theta,h] = \prod_{i=1}^{N} \prod_{j=1}^{4} \pi_{ij}(z,\theta,h)^{x_{ij}},$$

$$q[z|x,\tilde{\theta},h] = \mathcal{N}(\mu_i(x,\tilde{\theta},h), \text{Diag}(\sigma_i^2(x,\tilde{\theta},h))), \qquad (A6)$$

$$P[z] = \mathcal{N}(0,I)$$

where μ_i and σ_i are given by the encoder neural network, and π_{ij} is given by the decoder neural network, with $\sum_{j=1}^{4} \pi_{ij} = 1$ and $\pi_{ij} \geq 0$, which can be achieved by applying the softmax funtion to the output of the neural network. Now, we can use the reparametrization trick to change the variable in the integral $z = \sigma_j(x,\tilde{\theta},h)\varepsilon + \mu_j(x,\tilde{\theta},h)$, where $\varepsilon_j \sim \mathcal{N}(0,I)$, to simplify this expression to

$$\text{ELBO}(\theta,\tilde{\theta}) = \sum_{i=1}^{N}\sum_{j=1}^{4} x_{ij} \left\langle \log\left(\pi_{ij}(\sigma_i(x,\tilde{\theta},h)\varepsilon + \mu_i(x,\tilde{\theta},h), \theta, h) \right) \right\rangle_{\varepsilon_j \sim \mathcal{N}(0,I)}$$
$$- \sum_{i=1}^{N} \left(\log \sigma_i(x,\tilde{\theta},h) - \frac{\sigma_i^2(x,\tilde{\theta},h) + \mu_i^2(x,\tilde{\theta},h) - 1}{2} \right). \qquad (A7)$$

The first term is the cross-entropy, which pushes the probability distribution to be as close as possible to the data. The second term is the regularizer, which forces the latent variable z not to diverge too much from the normal distribution $\mathcal{N}(0,I)$, so that the VAE can be used to generate new data once it is trained. Note that both x_{ij} and σ_i must be positive. Instead of adding a constraint to the VAE, which would be difficult to do, we train the VAE for the variables $\Pi = \log \pi$ and $\xi = 2\log\sigma$. Equation (A7) then becomes

$$\text{ELBO}(\theta,\tilde{\theta}) = \sum_{i=1}^{N}\sum_{j=1}^{4} x_{ij} \left\langle \Pi_{ij}(e^{\xi_i(x,\tilde{\theta},h)/2}\varepsilon + \mu_i(x,\tilde{\theta},h), \theta, h) \right\rangle_{\varepsilon_j \sim \mathcal{N}(0,I)}$$
$$- \frac{1}{2}\sum_{i=1}^{N} \left(\xi_i(x,\tilde{\theta},h) - e^{\xi_i(x,\tilde{\theta},h)} - \mu_i^2(x,\tilde{\theta},h) + 1 \right). \qquad (A8)$$

Now, $\text{ELBO}(\theta,\tilde{\theta})$ can be effectively optimized using gradient descent methods, averaging over ε can be done by sampling. Generalizing to a data set of size M: $\{x^k\}_{k=1}^{M}$ can be easily done and is shown by

$$\text{ELBO}(\theta,\tilde{\theta}) = \sum_{k=1}^{M}\sum_{i=1}^{N}\sum_{j=1}^{4} x_{ij}^k \left\langle \Pi_{ij}(e^{\xi_i(x^k,\tilde{\theta},h)/2}\varepsilon + \mu_i(x^k,\tilde{\theta},h), \theta, h) \right\rangle_{\varepsilon_j \sim \mathcal{N}(0,I)}$$
$$- \frac{1}{2}\sum_{k=1}^{M}\sum_{i=1}^{N} \left(\xi_i(x^k,\tilde{\theta},h) - e^{\xi_i(x^k,\tilde{\theta},h)} - \mu_i^2(x^k,\tilde{\theta},h) + 1 \right). \qquad (A9)$$

A visual representation of the VAE architecture is shown in Figure A1.

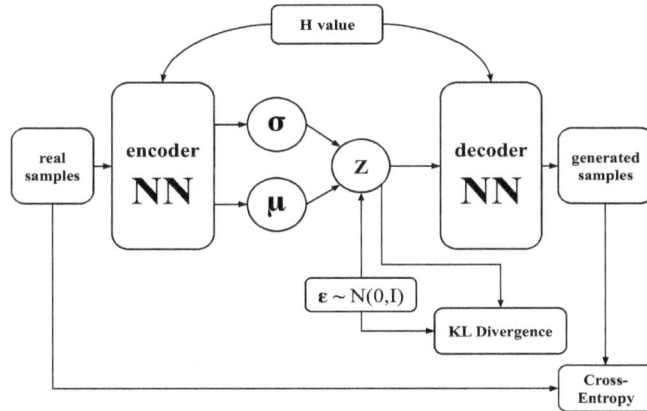

Figure A1. Architecture of the variational autoencoder.

To solve the optimization problem, we use Adam optimizer [113] with standard parameters ($\text{lr} = 0.001, \beta_1 = 0.9, \beta_2 = 0.999$). For the encoder and decoder, we use fully-connected neural networks with two hidden layers and 256 neurons on each. We train the VAE using batches of size 100,000 samples and for 750 epochs.

Appendix B. Sampling from POVM-Induced Mass Function

The mass function induced by POVM $P[\alpha_1, \alpha_2, \ldots, \alpha_N]$ has a form of matrix product state. Thus, one can easily calculate any marginal mass function because a summation over any α can be done locally. Any conditional mass functions can be also calculated by using marginal mass functions. Thus, one can calculate chain decomposition of the whole mass function:

$$P[\alpha_1, \alpha_2, \ldots, \alpha_N] = P[\alpha_N]P[\alpha_{N-1}|\alpha_N]P[\alpha_{N-2}|\alpha_{N-1}, \alpha_N] \ldots P[\alpha_1|\alpha_2, \ldots, \alpha_N] \qquad (A10)$$

With this decomposition, one can produce a sample $\tilde{\alpha}_N$ from $P[\alpha_N]$ first, then a sample $\tilde{\alpha}_{N-1}$ from $P[\alpha_{N-1}|\tilde{\alpha}_N]$, and continue up to the end of the chain. The obtained set $\{\tilde{\alpha}_1, \tilde{\alpha}_2, \ldots, \tilde{\alpha}_N\}$ is a valid sample from the mass function.

References

1. Muller, I. *A History of Thermodynamics*; Springer: Berlin/Heidelberg, Germany, 2007.
2. Onsager, L. Reciprocal Relations in Irreversible Processes. I. *Phys. Rev.* **1931**, *37*, 405–426. [CrossRef]
3. De Groot, S.R. *Thermodynamics of Irreversible Processes*; Interscience: New York, NY, USA, 1958.
4. Le Bellac, M.; Mortessagne, F.; Batrouni, G.G. *Equilibrium and Non-Equilibrium Statistical Thermodynamics*; Cambridge University Press: Cambridge, UK, 2006.
5. Apertet, Y.; Ouerdane, H.; Goupil, C.; Lecoeur, P. Revisiting Feynman's ratchet with thermoelectric transport theory. *Phys. Rev. E* **2014**, *90*, 012113. [CrossRef] [PubMed]
6. Goupil, C.; Ouerdane, H.; Herbert, E.; D'Angelo, Y.; Lecoeur, P. Closed-loop approach to thermodynamics. *Phys. Rev. E* **2016**, *94*, 032136. [CrossRef]
7. Andresen, B. Current trends in finite-time thermodynamics. *Angew. Chem.-Int. Edit.* **2011**, *50*, 2690–2704. [CrossRef] [PubMed]
8. Ouerdane, H.; Apertet, Y.; Goupil, C.; Lecoeur, P. Continuity and boundary conditions in thermodynamics: From Carnot's efficiency to efficiencies at maximum power. *Eur. Phys. J. Spec. Top.* **2015**, *224*, 839–864. [CrossRef]
9. Apertet, Y.; Ouerdane, H.; Goupil, C.; Lecoeur, P. True nature of the Curzon-Ahlborn efficiency. *Phys. Rev. E* **2017**, *96*, 022119. [CrossRef]
10. Boltzmann, L. Uber die beziehung dem zweiten Haubtsatze der mechanischen Warmetheorie und der Wahrscheinlichkeitsrechnung respektive den Satzen uber das Warmegleichgewicht. *Wiener Berichte* **1877**, *76*, 373–435.

11. Gibbs, J.W. *Elementary Principles in Statistical Mechanics*; Charles Scribner's Sons: New York, NY, USA, 1902.

12. Penrose, O. Foundations of statistical mechanics. *Rep. Prog. Phys.* **1979**, *42*, 1937–2006. [CrossRef]

13. Goldstein, S.; Lebowitz, J.L.; Zanghì, N. Gibbs and Boltzmann entropy in classical and quantum mechanics. *arXiv* **2019**, arXiv:1903.11870. Available online: https://arxiv.org/abs/1903.11870 (accessed on 6 November 2019).

14. Shannon, C.E. A mathematical theory of communication. *Bell Labs Tech. J.* **1948**, *27*, 379–423. [CrossRef]

15. von Neumann, J. *Mathematical Foundations of Quantum Mechanics. New Edition*; Princeton University Press: Princeton, NJ, USA, 2018.

16. Datta, S. *Electronic Transport in Mesoscopic Systems*; Cambridge University Press: Cambridge, UK, 1995.

17. Heikillä, T.T. *The Physics of Nanoelectronics*; Oxford University Press: Oxford, UK, 2013.

18. Chomaz, P.; Colonna, M.; Randrup, J. Nuclear spinodal fragmentation. *Phys. Rep.* **2004**, *389*, 263–440. [CrossRef]

19. Bressanini, D.; Morosi, G.; Mella, M. Robust wave function optimization procedures in quantum Monte Carlo methods. *J. Chem. Phys.* **2002**, *116*, 5345–5350. [CrossRef]

20. Feiguin, A.E.; White, S.R. Finite-temperature density matrix renormalization using an enlarged Hilbert space. *Phys. Rev. B* **2005**, *72*, 220401. [CrossRef]

21. Deutsch, J.M. Quantum statistical mechanics in a closed system. *Phys. Rev. A* **1991**, 43, 2046–2049. [CrossRef]

22. Srednicki, M. Chaos and quantum thermalization. *Phys. Rev. E* **1994**, *50*, 888–901. [CrossRef]

23. Rigol, M.; Dunjko, V.; Olshanii, M. Thermalization and its mechanism for generic isolated quantum systems. *Nature* **2008**, *452*, 854–858. [CrossRef]

24. Dymarsky, A.; Lashkari, N.; Liu, H. Subsystem eigenstate thermalization hypothesis *Phys. Rev. E* **2018**, *97*, 012140. [CrossRef] [PubMed]

25. Dymarsky, A. Mechanism of macroscopic equilibration of isolated quantum systems *Phys. Rev. B* **2019**, *99*, 224302. [CrossRef]

26. Carleo, G.; Becca, F.; Schiró, M.; Fabrizio, M. Localization and glassy dynamics of many-body quantum systems. *Sci. Rep.* **2012**, *2*, 243. [CrossRef] [PubMed]

27. Chen, L.; Gelin, M.; Zhao, Y. Dynamics of the spin-boson model: A comparison of the multiple Davydov D_1, $D_{1.5}$, D_2 Ansätze. *Chem. Phys.* **2018**, *515*, 108–118. [CrossRef]

28. Lanyon, B.; Maier, C.; Holzäpfel, M.; Baumgratz, T.; Hempel, C.; Jurcevic, P.; Dhand, I.; Buyskikh, A.; Daley, A.; Cramer, M.; et al. Efficient tomography of a quantum many-body system. *Nat. Phys.* **2017**, *13*, 1158. [CrossRef]

29. Liao, H.J.; Liu, J.G.; Wang, L.; Xiang, T. Differentiable programming tensor networks. *Phys. Rev. X* **2019**, *9*, 031041. [CrossRef]

30. Fetter, A.L.; Walecka, J.D. *Quantum Theory of Many-Particle Systems*; Dover: New York, NY, USA, 2003.

31. Frésard, R.; Kroha, J.; Wölfle, P., The pseudoparticle approach to strongly correlated electron systems. In *Strongly Correlated Systems*; Avella, A., Mancini, F., Eds.; Springer: Berlin/Heidelberg, Germany, 2011; Volume 171.

32. Georges, A.; Kotliar, G.; Krauth, W.; Rozenberg, M.J. Dynamical mean-field theory of strongly correlated fermion systems and the limit of infinite dimensions. *Rev. Mod. Phys.* **1996**, *68*, 13–125. [CrossRef]

33. Negele, J.W.; Orland, H. *Quantum Many-Particle Systems*; Perseus Books: New York, NY, USA, 1998.

34. Foulkes, W.; Mitas, L.; Needs, R.; Rajagopal, G. Quantum Monte Carlo simulations of solids. *Rev. Mod. Phys.* **2001**, *73*, 33. [CrossRef]

35. Orús, R. A practical introduction to tensor networks: Matrix product states and projected entangled pair states. *Ann. Phys.* **2014**, *349*, 117–158. [CrossRef]

36. Orús, R. Tensor networks for complex quantum systems. *Nat. Rev. Phys.* **2019**, *1*, 538–550. [CrossRef]

37. Schollwöck, U. The density-matrix renormalization group in the age of matrix product states. *Ann. Phys.* **2011**, *326*, 96–192. [CrossRef]

38. Vidal, G. Efficient classical simulation of slightly entangled quantum computations. *Phys. Rev. Lett.* **2003**, *91*, 147902. [CrossRef]

39. Evenbly, G.; Vidal, G. Quantum criticality with the multiscale entanglement renormalization ansatz. In *Strongly Correlated Systems*; Springer: Berlin/Heidelberg, Germany, 2013; pp. 99–130.

40. Pollock, F.A.; Rodríguez-Rosario, C.; Frauenheim, T.; Paternostro, M.; Modi, K. Non-Markovian quantum processes: Complete framework and efficient characterization. *Phys. Rev. A* **2018**, *97*, 012127. [CrossRef]

41. Luchnikov, I.; Vintskevich, S.; Ouerdane, H.; Filippov, S. Simulation complexity of open quantum dynamics: Connection with tensor networks. *Phys. Rev. Lett.* **2019**, *122*, 160401. [CrossRef]

42. Taranto, P.; Pollock, F.A.; Modi, K. Memory strength and recoverability of non-Markovian quantum stochastic processes. *arXiv* **2019**, arXiv:1907.12583. Available online: https://arxiv.org/abs/1907.12583 (accessed on 6 November 2019).

43. Milz, S.; Pollock, F.A.; Modi, K. Reconstructing non-Markovian quantum dynamics with limited control. *Phys. Rev. A* **2018**, *98*, 012108. [CrossRef]

44. Luchnikov, I.A.; Vintskevich, S.V.; Grigoriev, D.A.; Filippov, S.N. Machine learning of Markovian embedding for non-Markovian quantum dynamics. *arXiv* **2019**, arXiv:1902.07019. Available online: https://arxiv.org/abs/1902.07019 (accessed on 6 November 2019).

45. Verstraete, F.; Murg, V.; Cirac, J.I. Matrix product states, projected entangled pair states, and variational renormalization group methods for quantum spin systems. *Adv. Phys.* **2008**, *57*, 143–224. [CrossRef]

46. Levin, M.; Nave, C.P. Tensor renormalization group approach to two-dimensional classical lattice models. *Phys. Rev. Lett.* **2007**, *99*, 120601. [CrossRef]

47. Evenbly, G.; Vidal, G. Tensor network renormalization. *Phys. Rev. Lett.* **2015**, *115*, 180405. [CrossRef] [PubMed]

48. Gemmer, J.; Michel, M. *Quantum Thermodynamics*; Springer: Berlin/Heidelberg, Germany, 2009.

49. Kosloff, R. Quantum thermodynamics and open-systems modeling. *J. Phys. Chem.* **2019**, *150*, 204105. [CrossRef]

50. Allahverdyan, A.E.; Johal, R.S.; Mahler, G. Work extremum principle: Structure and function of quantum heat engines. *Phys. Rev. E* **2008**, *77*, 041118. [CrossRef]

51. Thomas, G.; Johal, R.S. Coupled quantum Otto cycle. *Phys. Rev. E* **2011**, *83*, 031135. [CrossRef]

52. Makhlin, Y.; Schön, G; Shnirman, A. Quantum-state engineering with Josephson-junction devices. *Rev. Mod. Phys.* **2001**, *73*, 357–400. [CrossRef]

53. Navez, P.; Sowa, A.; Zagoskin, A. Entangling continuous variables with a qubit array. *Phys. Rev. B* **2019**, *100*, 144506. [CrossRef]

54. Bishop, C.M. *Pattern Recognition and Machine Learning*; Springer: Berlin/Heidelberg, Germany, 2006.

55. Turing, M.A. Computing machinery and intelligence. *Mind* **1950**, *59*, 433–460. [CrossRef]

56. Crevier, D. *AI: The Tumultuous Search for Artificial Intelligence*; BasicBooks: New York, NY, USA, 1993.

57. Biamonte, J.; Wittek, P.; Pancotti, N.; Rebentrost, P.; Wiebe, N.; Lloyd, S. Quantum machine learning. *Nature* **2017**, *549*, 195–202. [CrossRef] [PubMed]

58. Carleo, G.; Troyer, M. Solving the quantum many-body problem with artificial neural networks. *Science* **2017**, *355*, 602–606. [CrossRef] [PubMed]

59. Torlai, G.; Mazzola, G.; Carrasquilla, J.; Troyer, M.; Melko, R.; Carleo, G. Neural-network quantum state tomography. *Nat. Phys.* **2018**, *14*, 447. [CrossRef]

60. Tiunov, E.S.; Tiunova, V.V.; Ulanov, A.E.; Lvovsky, A.I.; Fedorov, A.K. Experimental quantum homodyne tomography via machine learning. *arXiv* **2019**, arXiv:1907.06589. Available online: https://arxiv.org/abs/1907.06589 (accessed on 6 November 2019).

61. Choo, K.; Neupert, T.; Carleo, G. Study of the two-dimensional frustrated J1-J2 model with neural network quantum states. *Phys. Rev. B* **2019**, *100*, 124125. [CrossRef]

62. Sharir, O.; Levine, Y.; Wies, N.; Carleo, G.; Shashua, A. Deep autoregressive models for the efficient variational simulation of many-body quantum systems. *arXiv* **2019**, arXiv:1902.04057. Available online: https://arxiv.org/abs/1902.04057 (accessed on 6 November 2019).

63. Wu, D.; Wang, L.; Zhang, P. Solving statistical mechanics using variational autoregressive networks. *Phys. Rev. Lett.* **2019**, *122*, 080602. [CrossRef]

64. Kharkov, Y.A.; Sotskov, V.E.; Karazeev, A.A.; Kiktenko, E.O.; Fedorov, A.K. Revealing quantum chaos with machine learning. *arXiv* **2019**, arXiv:1902.09216. Available online: https://arxiv.org/abs/1902.09216 (accessed on 6 November 2019).

65. Rocchetto, A.; Grant, E.; Strelchuk, S.; Carleo, G.; Severini, S. Learning hard quantum distributions with variational autoencoders. *npj Quantum Inf.* **2018**, *4*, 28. [CrossRef]

66. Carrasquilla, J.; Torlai, G.; Melko, R.G.; Aolita, L. Reconstructing quantum states with generative models. *Nat. Mach. Intell.* **2019**, *1*, 155. [CrossRef]

67.   Generative Models for Physicists. Lecture note. Available online: http://wangleiphy.github.io/lectures/PILtutorial.pdf (accessed on 7 November 2019).

68.   Hewson, A.C. *The Kondo Problem to Heavy Fermions*; Cambridge University Press: Cambridge, UK, 1993.

69.   Coleman, P. Heavy fermions: Electrons at the edge of magnetism. In *Handbook of Magnetism and Advanced Magnetic Materials*; Kronmúller, H., Parkin, S., Eds.; John Wiley & Sons: Chichester, UK, 2007.

70.   Sachdev, S. *Quantum Phase Transitions*; Cambridge University Press: Cambridge, UK, 2000.

71.   Coleman, P.; Schofield, A. Quantum criticality. *Nature* **2000**, *433*, 226–229. [CrossRef] [PubMed]

72.   Anderson, P.W. Localized Magnetic States in Metals. *Phys. Rev.* **1961**, *124*, 41–53. [CrossRef]

73.   Frésard, R.; Ouerdane, H.; Kopp, T. Slave bosons in radial gauge: a bridge between path integral and Hamiltonian language. *Nucl. Phys. B* **2007**, *785*, 286–306. [CrossRef]

74.   Frésard, R.; Ouerdane, H.; Kopp, T. Barnes slave-boson approach to the two-site single-impurity Anderson model with non-local interaction. *EPL* **2008**, *82*, 31001. [CrossRef]

75.   Diu, B.; Guthmann, C.; Lederer, D.; Roulet, B. *Physique Statistique*; Éditions Hermann: Paris, France, 1996.

76.   Mila, F., Frustrated spin systems. In *Many-Body Physics: From Kondo to Hubbard*; Pavarini, E., Koch, E., Coleman, P., Eds.; Verlag des Forschungszentrum Jülich: Kreis Düren, Rheinland, 2015.

77.   Refael, G.; Moore, J.E. Entanglement Entropy of Random Quantum Critical Points in One Dimension. *Phys. Rev. Lett.* **2004**, *93*. [CrossRef]

78.   Schollwóck, U. The density-matrix renormalization group. *Rev. Mod. Phys.* **2005**, *77*, 259–315. [CrossRef]

79.   Ising, E. Beitrag zur Theorie des Ferromagnetismus. *Z. Phys.* **1925**, *31*, 253–258. [CrossRef]

80.   Kramers, H.A.; Wannier, G.H. Statistics of the two-dimensional ferromagnet. Part I. *Phys. Rev.* **1941**, *60*, 252–262. [CrossRef]

81.   Ovchinnikov, A.A.; Dmitriev, D.V.; Krivnov, V.Y.; Cheranovskii, V.O. Antiferromagnetic Ising chain in a mixed transverse and longitudinal magnetic field. *Phys. Rev. B* **2003**, *68*, 214406. [CrossRef]

82.   Coldea, R.; Tennant, D.A.; Wheeler, E.M.; Wawrzynska, E.; Prabhakaran, D.; Telling, M.; Habicht, K.; Smeibidl, P.; K, K. Quantum criticality in an Ising chain: Experimental evidence for emergent E_8 symmetry. *Science* **2010**, *327*, 177–180. [CrossRef] [PubMed]

83.   Sachdev, S.; Keimer, B. Quantum criticality. *Phys. Today* **2011**, *64*, 29–35. [CrossRef]

84.   Matsubara, T. A new approach to quantum statistical mechanics. *Prog. Theor. Exp.* **1955**, *14*, 351–378. [CrossRef]

85.   Kogut, J.B. An introduction to lattice gauge theory and spin systems. *Rev. Mod. Phys.* **1979**, *51*, 659–713. [CrossRef]

86.   Krizhevsky, A.; Sutskever, I.; Hinton, G.E. Imagenet classification with deep convolutional neural networks. In Proceedings of the NIPS: Advances in Neural Information Processing Systems 25, Stateline, NV, USA, 3–8 December 2012; pp. 1097–1105.

87.   Holevo, A.S. *Probabilistic and Statistical Aspects of Quantum Theory*; Springer: Berlin/Heidelberg, Germany, 2011; Volume 1.

88.   Filippov, S.N.; Man'ko, V.I. Inverse spin-s portrait and representation of qudit states by single probability vectors. *J. Russ. Laser Res.* **2010**, *31*, 32–54. [CrossRef]

89.   Appleby, M.; Fuchs, C.A.; Stacey, B.C.; Zhu, H. Introducing the Qplex: A novel arena for quantum theory. *Eur. Phys. J. D* **2017**, *71*, 197. [CrossRef]

90.   Caves, C.M. Symmetric informationally complete POVMs - UNM Information Physics Group internal report (1999). Available online: http://info.phys.unm.edu/~caves/reports/infopovm.pdf (accessed on 7 November 2019).

91.   Myung, I.J. Tutorial on maximum likelihood estimation. *J. Math. Psychol.* **2003**, *47*, 90–100. [CrossRef]

92.   Filippov, S.N.; Man'ko, V.I. Symmetric informationally complete positive operator valued measure and probability representation of quantum mechanics. *J. Russ. Laser Res.* **2010**, *31*, 211–231. [CrossRef]

93.   mpnum: A Matrix Product Representation Library for Python. Available online: https://mpnum.readthedocs.io/en/latest/ (accessed on 7 November 2019).

94.   Sohn, K.; Lee, H.; Yan, X. Learning structured output representation using deep conditional generative models. In Proceedings of the NIPS: Advances in Neural Information Processing Systems 28, Montreal, QC, Canada, 7–12 December 2015; pp. 3483–3491.

95.   Kingma, D.P.; Welling, M. Auto-encoding variational Bayes. *arXiv* **2013**, arXiv:1312.6114. Available online: https://arxiv.org/abs/1312.6114 (accessed on 6 November 2019).

96. Rezende, D.J.; Mohamed, S.; Wierstra, D. Stochastic backpropagation and approximate inference in deep generative models. In Proceedings of the 31st International Conference on Machine Learning (ICML), Beijing, China, 21–26 June 2014; Volume 32.

97. Jang, E.; Gu, S.; Poole, B. Categorical reparameterization with Gumbel-softmax. *arXiv* **2016**, arXiv:1611.01144. Available online: https://arxiv.org/abs/1611.01144 (accessed on 6 November 2019).

98. Kusner, M.J.; Hernández-Lobato, J.M. Gans for sequences of discrete elements with the Gumbel-softmax distribution. *arXiv* **2016**, arXiv:1611.04051. Available online: https://arxiv.org/abs/1611.04051 (accessed on 6 November 2019).

99. Maddison, C.J.; Mnih, A.; Teh, Y.W. The concrete distribution: A continuous relaxation of discrete random variables. *arXiv* **2016**, arXiv:1611.00712. Available online: https://arxiv.org/abs/1611.00712 (accessed on 6 November 2019).

100. Metropolis, N.; Rosenbluth, A.W.; Rosenbluth, M.N.; Teller, A.H.; Teller, E. Equation of State Calculations by Fast Computing Machines. *J. Chem. Phys.* **1953**, *21*, 1087–1092. [CrossRef]

101. Das, A.; Chakrabarti, B.K. Colloquium: Quantum annealing and analog quantum computation. *Rev. Mod. Phys.* **2008**, *80*, 1061–1081. [CrossRef]

102. Yavorsky, A.; Markovich, L.A.; Polyakov, E.A.; Rubtsov, A.N. Highly parallel algorithm for the Ising ground state searching problem. *arXiv* **2019**, arXiv:1907.05124. Available online: https://arxiv.org/abs/1907.05124 (accessed on 6 November 2019).

103. Verstraete, F.; Cirac, J.I. Matrix product states represent ground states faithfully. *Phys. Rev. B* **2006**, *73*, 094423. [CrossRef]

104. Eisert, J.; Cramer, M.; Plenio, M.B. Colloquium: Area laws for the entanglement entropy. *Rev. Mod. Phys.* **2010**, *82*, 277–306. [CrossRef]

105. Deng, D.-L.; Li, X.; Das Sarma, S. Quantum entanglement in neural network states. *Phys. Rev. X* **2017**, *7*, 021021. [CrossRef]

106. Bhattacharyya, A. On a measure of divergence between two statistical populations defined by their probability distributions. *Bull. Calcutta Math. Soc.* **1943**, *35*, 99–109.

107. Boixo, S.; Ronnow, T.F.; Isakov, S.V.; Wang, Z.; Wecker, D.; Lidar, D.A.; Martinis, J.M.; Troyer, M. Evidence for quantum annealing with more than one hundred qubits. *Nat. Phys.* **2014**, *10*, 218–224. [CrossRef]

108. Denchev, V.S.; Boixo, S.; Isakov, S.V.; Ding, N.; Babbush, R.; Smelyanskiy, V.; Martinis, J.; Neven, H. What is the computational value of finite-range tunneling? *Phys. Rev X* **2016**, *6*, 031015. [CrossRef]

109. Navez, P.; Tsironis, G.P.; Zagoskin, A.M. Propagation of fluctuations in the quantum Ising model. *Phys. Rev. B* **2017**, *95*, 064304. [CrossRef]

110. Volkov, A.A.; Artemov, V.G.; Pronin, A.V. A radically new suggestion about the electrodynamics of water: Can the pH index and the Debye relaxation be of a common origin? *EPL* **2014**, *106*, 46004. [CrossRef]

111. Artemov, V.G. A unified mechanism for ice and water electrical conductivity from direct current to terahertz. *Phys. Chem. Chem. Phys.* **2019**, *21*, 8067–8072. [CrossRef] [PubMed]

112. Github Repository with Code. Available online: https://github.com/LuchnikovI/Representation-of-quantum-many-body-states-via-VAE (accessed on 7 November 2019).

113. Kingma, D.P.; Ba, J. Adam: A method for stochastic optimization. *arXiv* **2014**, arXiv:1412.6980. Available online: https://arxiv.org/abs/1412.6980 (accessed on 6 November 2019).

5

The Ohm Law as an Alternative for the Entropy Origin Nonlinearities in Conductivity of Dilute Colloidal Polyelectrolytes

Ioulia Chikina [1], **Valeri Shikin** [2] and **Andrey Varlamov** [3,*]

[1] LIONS, NIMBE, CEA, CNRS, Universitè Paris-Saclay, CEA Saclay, 91191 Gif-sur-Yvette, France
[2] ISSP, RAS, Chernogolovka, 142432 Moscow, Russia
[3] CNR-SPIN, c/o DICII-Universitá di Roma Tor Vergata, Via del Politecnico, 1, 00133 Roma, Italy
[*] Correspondence: andrey.varlamov@spin.cnr.it

Abstract: We discuss the peculiarities of the Ohm law in dilute polyelectrolytes containing a relatively low concentration $n_\odot$ of multiply charged colloidal particles. It is demonstrated that in these conditions, the effective conductivity of polyelectrolyte is the linear function of $n_\odot$. This happens due to the change of the electric field in the polyelectrolyte under the effect of colloidal particle polarization. The proposed theory explains the recent experimental findings and presents the alternative to mean spherical approximation which predicts the nonlinear I–V characteristics of dilute colloidal polyelectrolytes due to entropy changes.

Keywords: polyelectrolytes; Ohm law; colloids

1. Introduction

Polyelectrolytes are polymers whose repeating units contain a group of electrolytes. These groups dissociate in aqueous solutions, making the polymers charged. Polyelectrolyte properties resemble those of both electrolytes and polymers, and, like salts, their solutions are electrically conductive. The incorporation of the nano- and micro-meter-sized charged colloidal particles can dramatically change the electrical and heat transport properties of such systems. For instance, the authors of Ref. [1] study the electrical transport in charged colloidal suspensions of iron oxide nanoparticles (maghemite) dispersed in an aqueous medium, while in Ref. [2], the thermal and electrical transport is investigated in ionically stabilized magnetic nanoparticles dispersed in aqueous potassium ferro/ferricyanide electrolytes. Both groups report the unusual effect of multiply charged colloidal particles on conductivity of the dilute polyelectrolytes. It turns out that the latter grows linearly with an increase of colloidal particle concentration.

This finding seems to be non-trivial from the point of view of the percolation theory (see, for example, [3]). Indeed, in accordance with the latter, the conductivity of a mixture between dielectric (in our case water molecules) and conducting (colloidal particle with counter-ions coat) components remains minute until the fraction of the conducting phase approaches the percolation threshold, and only in the vicinity of the latter, the conductivity growths smoothly have a value of dielectric component that is similar to to that of a metallic one.

Before discussing this contradiction, let us make an excursus into the physics of semiconductors. In the theory of semiconductors [3], the regions of weak and strong doping (i.e., introduction of charged impurities or structural defects with the purpose of changing the electrical properties of a semiconductor) are distinguished. In the low doping regime, the impurity concentration $n_\odot$ is so small that the distances between them significantly exceed the Debye length λ_0 and the bare radius of the colloidal particle R_0, i.e.

$$n_\odot \left(\lambda_0 + R_0\right)^3 \ll 1, \quad R_0 \leq \lambda_0, \tag{1}$$

and the intrinsic charge carriers of semiconductor completely screen the electric fields produced by the charged impurities (see Figure 1). In the strong doping regime, when criterion (1) is violated, the fields produced by the dopants are screened only partially and their interaction becomes significant.

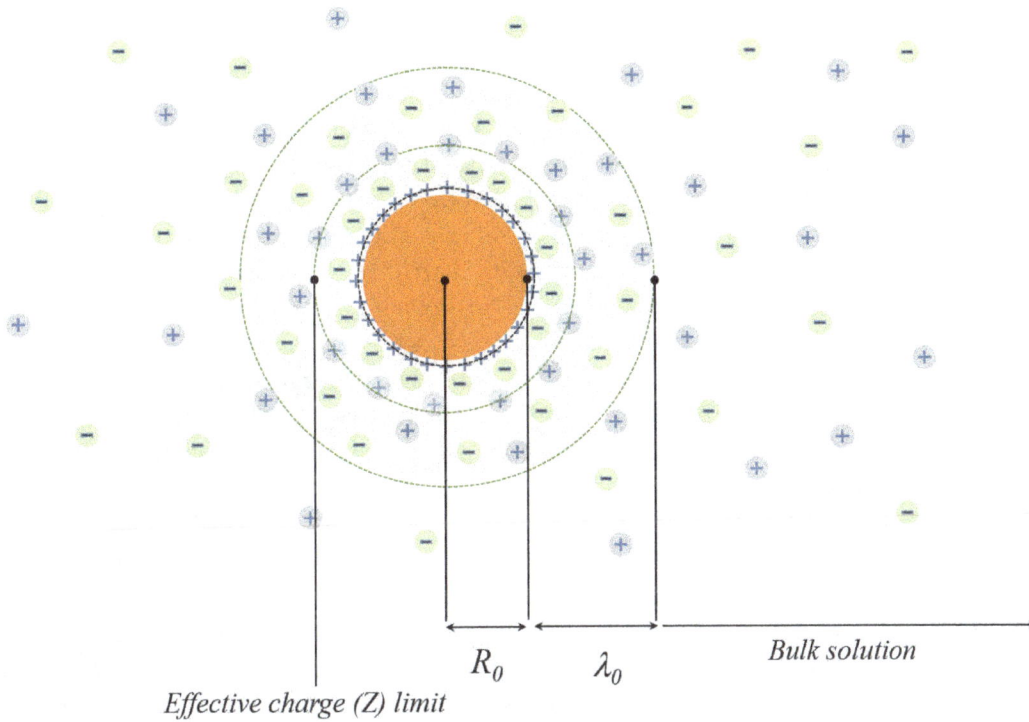

Figure 1. The schematic presentation of the multiply charged colloidal particle surrounded by the cloud of counter-ions.

Returning to the case of the dilute colloidal polyelectrolyte, one can map its properties to the ones of the weak doped semiconductor and identify $n_\odot$ with the concentration of the colloidal particles, while λ_0 should be related to their characteristic size. The latter is determined by the known concentration n_0 of the counterions of the electrolyte hosting charged colloidal particles.

The criterion (1) is in a reasonable agreement to the common concepts of the physics of dilute polyelectrolytes developed in the 40s of the last century by Derjaguin, Landau, Verwey, and Overbeek [4,5] and known as DLVO formalism. Namely, if the colloidal particles are neutral, they are not stationary in dilute solution and coagulating due to van der Waals forces acts between them. In order to prevent such coagulation processes, one should immerse individual colloidal particles in the electrolyte specific for each sort of them. The latter are called stabilizing electrolytes.

Being immersed (or synthesized within) in an electrolyte solution, the nanoparticles acquire surface ions (e.g., hydroxyl groups, citrate, etc. [6–8]) resulting in a very large structural charge eZ ($|Z| \gg 10$). Its sign can be both positive or negative, depending on the surface group type. The latter, in return, attracts counterions from the surrounding solvent creating an electrostatic shielding coat of the size λ_0 with an effective charge $-eZ$. In these conditions, nano-particles approaching between them to the distances $r \leq \lambda_0$ begin to repel each other without flocculation [4,5,9]. The region of an essential interaction between them in terms of the criterion (1) corresponds to the condition

$$n_\odot^c \left(\lambda_0 + R_0\right)^3 \sim 1. \tag{2}$$

In Ref. [1,2], the massive multiply charged colloidal particles are surrounded by the clouds of counter-ions screening their positive charge. Such formations, according to Ref. [3], should not affect the conductivity of the dilute polyelectrolyte until the shells of the neighbor charged complexes do not overlap among themselves (see Equation (2)). The results of both Ref. [1] and [2] demonstrate the opposite: the conductivity of dilute colloidal polyelectrolyte grows linearly with increase of concentration already in the range $n_\odot \ll n_\odot^c$, where there is not yet place for percolation effects.

This contradiction can be removed by noticing that the presence of the multiply charged colloidal particles has an effect not only on the value of conductivity of a solution but also on the local value of the electric field:

$$j(n_\odot) = \sigma(n_\odot)E(n_\odot). \tag{3}$$

It is important to note that the factors in Equation (3) are affected by the presence of the multiply charged colloidal particles in different ways. While the conductivity of the electrolyte at low concentrations of multiply charged colloidal particles ($n_\odot \leq n_\odot^c$) remains almost unchanged, their effect on the local electric field in this range of concentrations is essential. This happens due to polarization of the colloidal particles by an external electric field which, in accordance with the Le Chatelier's principle, results in the decrease of the effective value of the field. Consequently, the growth of conductivity [1,2] as a function of concentration $n_\odot$ is observed in experiments. When the concentration of multiply charged colloidal particles reaches the percolation threshold ($n_\odot = n_\odot^c$), the role of the factors in Equation (3) is reversed. Here, the subsystem of colloidal particles forms clusters and cannot be considered more as the gas of polarized highly conducting particles. Yet, in this range of concentrations, the new channel of percolation charge transfer is opened and the total conductivity of the electrolyte growth further increase by $n_\odot$.

The state-of-the-art in transport phenomena in polyelectrolytes was recently reviewed in Ref. [1]. Focusing mainly on the results of the microscopic approach (so called mean spherical approximation theory (MSA)) [10–12], the authors discuss mobility, diffusion coefficient, and the effective charge space distribution of the colloidal particles as the function of their concentration. Yet, in Ref. [1], there is not any information concerning the effect of clusters of polarization on the charge transfer process in such complex systems. This aspect of the problem is the subject of our work.

2. Effective Electric Field in Bulk of Colloidal Polyelectrolyte

The colloidal polyelectrolyte is a weakly conducting liquid with the small but finite fraction of relatively highly (due to $Z \gg 1$) conducting inclusions, i.e., colloidal particles. The collective polarization of these inclusions occurs when the external electric field E_0 is applied. This phenomenon is analogous to polarization of neutral atoms in gas. The only difference is that the neutral atoms reside in vacuum, while the charged conducting clusters of colloidal polyelectrolyte are immersed in a less, but still conducting, medium. Hence, our goal is to account for this peculiarity and find the effective field which governs the charge transport in such a complex system.

2.1. Electric Field in Absence of Current

The space distribution of the effective electric field of the colloidal particle is determined by the Poisson equation (see [3,9])

$$\Delta\varphi = \frac{4\pi}{\epsilon}\rho(r), \quad \rho(r) = |e|[n_+(r) - n_-(r)], \tag{4}$$

where ϵ is the dielectric permittivity of stabilizing electrolyte.

The concentrations of the screening counterions $n_\pm(r)$ is determined self-consistently via the value of local electrostatic potential

$$n_\pm(r) = n_0 \exp\left[e_\pm\varphi(r)/T\right], \tag{5}$$

$n_0 = n_0^+ = n_0^-$ is the counterions bare concentration, occuring due to the complete dissociation of the electrolyte which stabilizes the gas of colloidal particles.

In assumption $e\varphi(r) < T$ the Poisson equation can be linearized and takes form

$$\Delta\varphi = \varphi/\lambda_0^2, \quad \lambda_0^{-2} = \frac{8\pi e^2}{\epsilon T}n_0. \tag{6}$$

This equation should be solved accounting for the boundary conditions

$$r\varphi(r)|_{r\to R_0} \to Z|e|, \quad \varphi(r)|_{r\to\infty} \to 0, \tag{7}$$

what results in the standard screened Coulomb potential:

$$\varphi(r) = Ze\frac{\exp(-\frac{r}{\lambda_0})}{r}. \tag{8}$$

The values Z, R_0 and n_0 of the electrolyte, which stabilizes the colloidal solution can be determined by independent experiments (for example, by measurements of the electrophoretic forces, osmotic pressure, etc. [1]).

One should remember that even strongly diluted polyelectrolytes can undergo the transition to the state of a Wigner crystal in the case of strongly charged colloidal particles ($Z \gg 1$). For description of this, observed experimentally [13–15], phenomenon the authors of [16] assumed that the interaction between two colloidal particles has the same form of Yukawa potential (8), yet with the renormalized effective charge $Z^* \ll Z$, explicitly depending on the colloidal particles density $n_\odot$. The value of Z^* is determined in the Wigner-Seitz model from the new boundary condition

$$\frac{\partial\varphi}{\partial r}\Big|_{r\to n_\odot^{-1/3}} = 0$$

replacing that ones, valid for the isolated charged particle in the screening media (see Equation (7)). For some range of the colloidal particles densities $n_\odot$ the conditions $Z \gg 1$ and $Z^* \ll 1$ can be satisfied simultaneously. The former characterizes the properties of the multiply charged colloidal particles, while the latter is determined by the strength of their interaction and $n_\odot$. In the range of densities $n_\odot$ satisfying Equation (1), the effect of the effective charge Z^* on the Ohmic transport is negligible.

2.2. Electric Field in Presence of Current

When a stationary current flows through the polyelectrolyte, an internal electric field $\vec{E}$ appears in it. In the approximation of a very diluted solution, one can start considerations from the effect of presence of the isolated colloidal particle on a flowing current. Namely, one should find the perturbation of the internal electric field which would provide the homogeneity of the transport current far from the colloidal particle. A corresponding problem recalls that one of classic hydrodynamics: calculus of the associated mass of the particle moving in the ideal liquid [17].

We choose the center of spherical coordinates coinciding with the colloidal particle and direct the $z-$axis along the electric field $\vec{E}_0$. We assume that the conductivity of the electrolyte in the absence of colloidal particles is σ_0. The highly charged colloidal particle we will model as the conducting solid sphere of the radius $R \simeq (R_0 + \lambda_0)$ (see Figure 1) with conductivity $\sigma_\odot > \sigma_0$. Analysis of the charge transport in multi-phase systems (see [18]) is based on the requirements

$$div\vec{j} = 0, \quad \vec{j} = \sigma\vec{E}. \tag{9}$$

When the medium conductivity is invariable in space the constancy of the current, this automatically means the homogeneity of the electric field. The situation changes when the system is inhomogeneous and $\sigma \neq const$. The continuity Equation (9) in this case should be solved with the boundary conditions accounting for the current flow through the boundaries between domains of diverse conductivity. According to Ref. [18,19], the tangential components of electric field intensity at the boundary must be continuous, while the normal ones provide the continuity of the charge transfer. Applying these rules to our simple model of the highly charged colloidal particle in the less conductive medium, one can write

$$j_n^0 = J_n^\odot, \quad \text{or} \quad \sigma_0 E_0 = \sigma_\odot E_\odot. \tag{10}$$

Solution of the system of Equations (9) and (10) for the electrostatic potential in the vicinity of the colloidal particle ($r \geq R$) acquires the form:

$$\varphi(r,\theta) = -E_0 r \cos\theta + \left(\frac{\gamma-1}{\gamma+2}\right) E_0 \frac{R^3}{r^2} \cos\theta, \tag{11}$$

with $\gamma = \sigma_\odot / \sigma_0$. In the limit $\gamma \to 1$ the electric field remains unperturbed, $\vec{E} = -\nabla\varphi \to \vec{E}_0$. In the opposite case, $\gamma \gg 1$, the dipole perturbation takes the form corresponding to the case of metallic inclusion of the radius R in the weakly conducting environment (Ref. [18]):

$$\varphi(r,\theta) = -E_0 r \cos\theta \left(1 - \frac{R^3}{r^3}\right). \tag{12}$$

One can see that in accordance with the intuitive expectations, the presence of an isolated colloidal particle in an electrolyte leads to the appearance of the local perturbation of the electric field of the dipole type $\nabla\varphi \propto r^{-3}$ with the value of the dipole moment of one colloidal particle

$$p_\odot = \left(\frac{\gamma-1}{\gamma+2}\right) R^3 E_0. \tag{13}$$

Returning to the initial problem of the rarefied gas of colloidal particles of concentration $n_\odot$ in the electrolyte media, one can introduce the effective dielectric permittivity $\epsilon_\odot$. It can be related to the dipole moment (13) by means of the Clausius–Mossotti relation (see Ref. [18]) and in terms of the material parameters of the problem which is read as:

$$\epsilon_\odot = 1 + 4\pi \left(\frac{\gamma-1}{\gamma+2}\right) R^3 n_\odot. \tag{14}$$

One can try to make the model of colloidal particles more realistic assuming that the latter has the structure of a thick-walled sphere; a "nut" with the conducting shell and the insulating core of the bare radius R_0. This intricacy leads to the change in the expression for the corresponding dipole momentum: instead of Equation (13) it takes the form (see Ref. [18])

$$\tilde{p}_\odot = \frac{(2\gamma+1)(\gamma-1)}{(2\gamma+1)(\gamma+2) - 2(\gamma-1)^2 R_0^3/R^3} \left(R^3 - R_0^3\right) E_0. \tag{15}$$

This formula contains two geometrical parameters: R and R_0. The latter should be determined from some independent measurements. The difference $R - R_0$ can be identified with the Debye length λ_0 or to consider it as the fitting parameter.

3. Ohmic Transport in a Weak Colloidal Polyelectrolyte

Equation (14) demonstrates that growth of the nano-particle concentration $n_\odot$ leads to increase of the dielectric constant $\epsilon_\odot$, which, in its turn, results in the decrease of the effective electric field in an electrolyte. The latter, in conditions of the fixed transport current, is perceived as the growth of conductivity with an increase of the colloidal particles concentration:

$$\sigma(n_\odot) = j\epsilon_\odot / E_0 = \sigma_0 \left[1 + 4\pi n_\odot \frac{p_\odot(E_0)}{E_0}\right]. \tag{16}$$

This expression can be already used for the experimental data processing.

3.1. Approximation of the Conducting Spheres

Substituting the dipole moment taken in the approximation of Equation (13) in Equation (16) one finds

$$\frac{\Delta\sigma_{CP}(n_\odot)}{\sigma_0} = \frac{\sigma(n_\odot) - \sigma_0}{\sigma_0} = 4\pi n_\odot \left(\frac{\gamma - 1}{\gamma + 2}\right) R^3, \tag{17}$$

where $\Delta\sigma_{CP}$ is the excess conductivity due to the presence of colloidal particles. The left-hand-side of this equation can be extracted from the data presented in Figure 2. Indeed, in the interval of the nanoparticles concentrations $0 \leq \phi \leq 0.6\%$ the behavior of conductivity $\sigma(n_\odot)$ is almost linear and $\sigma(n_\odot)/\sigma_0 - 1 = 0.7$. In turn, the concentration $\varphi = 0.6\%$ corresponds to $n_\odot^{(1)} = 5.45 \times 10^{15}$ cm^{-3}.

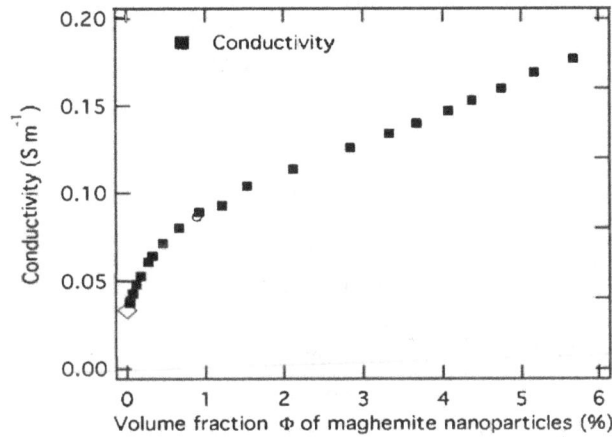

Figure 2. Experimental values of electrical conductivity of water based polyelectrolyte solution as a function of colloidal concentrations. Measurements were performed in pH = 3.1 solutions containing maghemite nanoparticles with an average diameter of 12 nm. More detailed information on the colloidal solution preparation methods and the nature of other ions is found in Ref. [1,2].

For further estimations, it will be crucial that Equation (17) is sensitive to the value of γ only when it is not very large. When $\gamma \gg 1$ (we will justify this limit below) the combination $(\gamma - 1)/(\gamma + 2) \to 1$ and it ceases to influence the evaluations based on Equation (17). This allows us to find this limit

$$R_{exp}^{(1)} = 2.17 \times 10^{-6} cm,$$

$$n_\odot^{(1)} \left[R_{exp}^{(1)}\right]^3 = 0.055 \ll 1. \tag{18}$$

One can see that these values, together with the nanoparticle concentration $n_\odot^{(1)}$, confirm the validity of the assumed above approximation (1). The plausible reasons for the discovered considerable difference between $R_{exp}^{(1)}$ and the value of bare radius $R_0^{(1)} = 6 \times 10^{-7}$ cm given in Ref. [1] will be discussed below.

The above found conductivity correction $\Delta\sigma_{CP}(n_\odot) \propto n_\odot R^3$ (see Equation (17)) caused by presence of nanoparticles in electrolyte can be confidently distinguished from the standard Onsager–Debye conductivity (σ_{OD}) of the diluted 1:1 electrolyte [20–22]. Indeed, first of all, the concentration dependencies of these conductivities are different: $\Delta\sigma_{CP}(n_\odot) \propto n_\odot$ while $\sigma_{OD}(n_\odot) \propto \sqrt{n_\odot}$.

Let us focus on the unusual dependence of the excess conductivity (17) of the nanoparticle size: $\Delta\sigma_{CP}$ growths with increase of R. Usually, this dependence is supposed to be opposite (the larger radius of the sphere in Stokes viscous law, the lower its mobility, and hence, the conductivity).

One can analyze the available experimental data on the conductivity of the stabilized diluted colloidal solution [1,2] in the conditions described by Equation (2). In accordance with Equation (17), the excess conductivities for different sizes of nanoparticles in assumption of the same concentration should scale as $[R_0^{(1)}/R_0^{(2)}]^3$. Taking the value $R_0^{(1)} = 6\ nm$ from [1] and $R_0^{(2)} = 3.8\ nm$ from [2] one finds that the ratio

$$\frac{\Delta\sigma_{CP}^{(1)}}{\sigma_0^{(1)}} / \frac{\Delta\sigma_{CP}^{(2)}}{\sigma_0^{(2)}} = \left(\frac{6}{3.8}\right)^3 \approx 4 \tag{19}$$

Experimental data for this value give even more striking difference:

$$\frac{\Delta\sigma_{CP}^{(1)}}{\sigma_0^{(1)}} / \frac{\Delta\sigma_{CP}^{(2)}}{\sigma_0^{(2)}} = \frac{0.7}{0.06} \approx 11.7. \tag{20}$$

3.2. Approximation of the Conducting Thick-Walled Spheres

Here, it is necessary to note that the value $R_{exp}^{(1)}$ obtained in the simple approximation of Equations (13) and (16) and the measured in Ref. [1] bare radius of the colloidal particle R_0 form a relatively small numerical parameter, $[R_0/R_{exp}^{(1)}]^3 \simeq 0.02$. It makes sense to improve the experimental data proceeding replacing the value $p_\odot$ in Equation (16) by the two parametric expressions (15). Tending $\gamma \to \infty$ in it one finds

$$\frac{\sigma(n_\odot) - \sigma_0}{\sigma_0} = 4\pi n_\odot [R_{exp}^{(1)}]^3 \left[1 - \frac{3}{\gamma} - \frac{9}{2\gamma}\frac{R_0^3}{\left([R_{exp}^{(1)}]^3 - R_0^3\right)}\right] \tag{21}$$

From this expression, it is clear that the approximation (17) is valid when $\gamma \gg 1$.

The parameter γ requires special discussion. In the DLVO colloidal model, it is assumed that some bare core exists which is able to cause the van der Waals forces between colloidal particles in dilute, non-stabilizing solutions. The conducting properties of this core is not so essential. For example, one can suppose this bare core of the radius R_0 to be a semiconductor possessing its intrinsic charge carriers which are confined in its volume. If the solvent possesses the stabilizing properties its own mobile charge carriers, counterions have the same properties as the intrinsic charge carriers of the bare core. The requirement of electrochemical potential constancy leads to the charge exchange between the bare core and the solvent. Such exchange results in the formation of the Debye shell (see Equations (4)–(8)), where the concentration of counterions considerably exceeds that in the solvent bulk. We assumed above that the value of corresponding conductivity $\sigma(n_\odot)$ considerably exceeds σ_0 of the electrolyte conductivity in absence of the nanoparticles. This assumption ($\gamma \gg 1$) breaks when the average value of electrochemical potential in the Debye shell $e\phi_\odot$ exceeds the temperature. The authors of Ref. [23] state that in these conditions the Debye shell of the DLVO colloid can crystallize due to Coulomb forces and the latter becomes an insulator with $\sigma(n_\odot) \leq \sigma_0$.

4. Conclusions

The main result of this work consists of the proposition of an alternative scenario explaining the linear growth of the polyelectrolyte conductivity versus the concentration of colloidal particles observed in Ref. [1,2] in the conditions of the validity of Equation (1). It drastically differs from the existing ideas of the transport in electrolytes resulting in the empirical Kohlrausch's law (see [22,24])

$$\Delta\sigma \sim \sqrt{n_\odot}. \tag{22}$$

The speculations justifying Equation (22) were firstly proposed in early papers such as Ref. [20,21] and the recent efforts to improve this mechanism were undertaken in Ref. [25].

The fact of the observation of the Ohmic transport in strong electrolytes (Ref. [1,2]) denies the applicability of Kohlrausch's law in the interval of a very low concentration of the colloidal particles. Conversely, the mechanism proposed above, based on the analogy to the percolation mechanism of conductivity occurring in doped semiconductors, allows to get an excellent agreement in the observed linear dependence. Moreover, it also provides very reasonable values of the microscopic parameters of the problem.

One can believe that the validity of Kohlrausch's law is restored in the domain of higher concentrations and the crossover point between the two regimes (16) and (22) is determined by the condition (2), as is shown in Figure 2. One can find the pro-arguments for this statement in the experimental curve shown in Figure 2 of Ref. [1], where the regimes are changed in the vicinity of the concentration $n_{\odot}^{(1)} = 5.45 \times 10^{15} \ cm^{-3}$.

The question that arises is why such linear growth below the percolation threshold was never reported in measurements performed on semiconductors. The answer probably consists of the overwhelming supremacy of the colloidal particle dipole momentum in comparison to that of the dopant in semiconductors.

It would be interesting to compare the values of effective charge Z extracted from the experiments on conductivity of [2] and the review article [1]. Unfortunately, this is not easy to do because of the analysis of the data for different Z results in very different values of R_0. It is why one cannot judge the influence of the effective charge Z on the bare radius of the colloidal particle R_0.

The relative insensibility of the polyelectrolyte conductivity on the value of parameter Z is not extended on the Seebeck coefficient. The measurements of [2] demonstrate the existence in its kinetics of the two different phases; the initial and steady ones. The authors dealt with two types of colloids; one is almost electroneutral ($Z \geq 1$) and the other is supposed to have $Z \gg 1$.

Author Contributions: I.C.: Methodology, Formal analysis, Writing—review and editing, Visualization, Funding acquisition. V.S.: Methodology, Formal analysis, Writing—review and editing, Visualization. A.V.: Methodology, Formal analysis, Writing—review and editing, Visualization, Funding acquisition. All authors have read and agreed to the published version of the manuscript.

Acknowledgments: The authors acknowledge multiple and useful discussions with Sawako Nakamae.

References

1. Lucas, I.T.; Durand-Vidal, S.; Bernard, O.; Dahirel, V.; Dubois, E.; Dufrêche, J.F.; Gourdin-Bertin, S.; Jardat, M.; Meriguet, G.; Roger, G. Influence of the volume fraction on the electrokinetic properties of maghemite nanoparticles in suspension. *Mol. Phys.: Int. J. Interface Chem. Phys.* **2014**, *112*, 1463–1471. [CrossRef]

2. Salez, T.J.; Huang, B.; Rietjens, M.; Bonetti, M.; Wiertel-Gasquet, C.; Roger, M.; Filomeno, C.L.; Dubois, E.; Perzynski, R.; Nakamae, S. Can charged colloidal particles increase the thermoelectric energy conversion efficiency? *Phys. Chem. Chem. Phys.* **2017**, *19*, 9409–9416; doi:10.1039/C7CP01023K. [CrossRef] [PubMed]

3. Shklovski, B.I.; Efros, A.L. *Electronic Properties of Doped Semiconductors*, 1st ed.; Springer-Verlag: Berlin, Germany, 1984; pp. 94–107.

4. Derjaguin, B.V.; Landau, L.D. Theory of the stability of strongly charged lyophobic sols and of the adhesion of strongly charged particles in solutions of electrolytes. *Acta Phys. Chem. URSS* **1941**, *14*, 633–662. [CrossRef]

5. Verwey, E.; Overbeek, J. *Theory of the Stability of Lyophobic Colloids*, 1948 ed.; Elsevier: Amsterdam, The Netherlands, 1948; pp. 131–136.

6. Riedl, J.C.; Akhavan Kazemi, M.A.; Cousin, F.; Dubois, E.; Fantini, S.; Lois, S.; Perzynski, R.; Peyre, V. Colloidal dispersions of oxide nanoparticles in ionic liquids : Elucidating the key parameters. *Nanoscale Adv.* **2020**. [CrossRef]

7. Bacri, J.C.; Perzynski, R.; Salin, D.; Cabuil, V.; Massart, R. Ionic ferrofluids: A crossing of chemistry and physics. *J. Magn. Magn. Mater.* **1990**, *85*, 27–32. [CrossRef]

8. Dubois, E.; Cabuil, V.; Boué, F.; Perzynski, R. Structural analogy between aqueous and oily magnetic fluids. *J. Chem. Phys.* **1999**, *111*, 7147–7160. [CrossRef]

9. Landau, L.D.; Lifshitz, E.M. Statistical Physics. In *Course of Theoretical Physics*, 3rd ed.; Elsevier: Amsterdam, The Netherlands, 2011; Volume 5, pp. 276–278.

10. Rotenberg, B.; Dufrêche, J.F.; Turq, P. Frequency-dependent dielectric permittivity of salt-free charged lamellar systems. *J. Chem. Phys. B* **2005**, *123*, 154902–154903. [CrossRef] [PubMed]

11. Durand-Vidal, S.; Jardat, M.; Dahirel, V.; Bernard, O.; Perrigaud, K.; Turq, P. Determining the radius and the

apparent charge of a micelle from electrical conductivity measurements by using a transport theory: Explicit equations for practical use. *J. Chem. Phys. B* **2006**, *110*, 15542–15547. [CrossRef] [PubMed]

12. Jardat, M.; Dahirel, V.; Durand-Vidal, S; Lucas, I.; Bernard, O.; Turq, P. Effective charges of micellar species obtained from Brownian dynamics simulations and from an analytical transport theory. *Mol. Phys.* **2004**, *104*, 3667–3674. [CrossRef]

13. Heltner, P.; Papir, Y.; Krieger, I. Diffraction of light by nonaqueous ordered suspensions. *J. Phys. Chem.* **1971**, *75*, 1881–1886. [CrossRef]

14. Kose, A.; Ozake, T.; Takano, K.; Kobayschi, Y.; Hachisu, S. Direct observation of ordered latex suspension by metallurgical microscope. *J. Colloid Interface Sci.* **1973**, *44*, 330–338. [CrossRef]

15. Williams, R.; Crandall, R. The structure of crystallized suspensions of polystyrene spheres. *Phys. Lett. A* **1974**, *48*, 225–226. [CrossRef]

16. Alexander, S.; Chaikin, P.; Grant, P.; Morales, G.; Pincus, P. Charge renormalization, osmotic pressure, and bulk modulus of colloidal crystals: Theory. *J. Chem. Phys.* **1984**, *80*, 5776–5781. [CrossRef]

17. Landau, L.D.; Lifshitz, E.M. *Course of Theoretical Physics: Vol. 6, Fluid Mechanics*, 2nd ed.; Elsevier: Amsterdam, The Netherlands, 2013; p. 46.

18. Jackson, J.D. *Classical Electrodynamics*, 3rd ed.; John Wiley & Sons, Inc.: New York, NY, USA, 1999; p. 114.

19. Dykhne, A.M. Conductivity of a Two-Dimensional Two-Phase System. *Sov. JETP* **1971**, *32*, 63–65.

20. Onsager, L. On the theory of electrolytes. *Physica Z* **1927**, *28*, 277–298.

21. Debye, P.; Huckel, E. The theory of the electrolyte II-The border law for electrical conductivity. *Physica Z* **1923**, *24*, 305–325.

22. Lifshitz, E.M.; Pitaevskii, L.P. *Physical Kinetics: Course of Theoretical Physics—Volume 10*, 1st ed.; Butterworth-Heinenann Ltd.: London, UK, 2002; p. 125.

23. Grosberg, A.; Nguyen, T.; Shklovskii, B. The physics of charge inversion in chemical and biological systems. *Rev. Mod. Phys.* **2002**, *74*, 329–345. [CrossRef]

24. Robinson, R.; Stokes, R. *Electrolyte Solutions*, 1959 ed.; Butterworths Scientific Publications: London, UK, 1959; p. 119.

25. Lizana, L.; Grossberg, A. Exact expressions for the mobility and electrophoretic mobility of a weakly charged sphere in a simple electrolyte. *Europhys. Lett.* **2013**, *104*, 68004–68009. [CrossRef]

6

Entropy of Conduction Electrons from Transport Experiments

Nicolás Pérez [1,*], **Constantin Wolf** [1,2], **Alexander Kunzmann** [1,2], **Jens Freudenberger** [1,3], **Maria Krautz** [4], **Bruno Weise** [4], **Kornelius Nielsch** [1,2,5] and **Gabi Schierning** [1]

[1] Institute for Metallic Materials, IFW-Dresden, 01069 Dresden, Germany; c.wolf@ifw-dresden.de (C.W.); a.kunzmann@ifw-dresden.de (A.K.); j.freudenberger@ifw-dresden.de (J.F.); k.nielsch@ifw-dresden.de (K.N.); g.schierning@ifw-dresden.de (G.S.)
[2] Institute of Materials Science, TU Dresden, 01062 Dresden, Germany
[3] Institute of Materials Science, TU Bergakademie Freiberg, 09599 Freiberg, Germany
[4] Institute for Complex Materials, IFW-Dresden, 01069 Dresden, Germany; m.krautz@ifw-dresden.de (M.K.); b.weise@ifw-dresden.de (B.W.)
[5] Institute of Applied Physics, TU Dresden, 01062 Dresden, Germany
* Correspondence: n.perez.rodriguez@ifw-dresden.de

Abstract: The entropy of conduction electrons was evaluated utilizing the thermodynamic definition of the Seebeck coefficient as a tool. This analysis was applied to two different kinds of scientific questions that can—if at all—be only partially addressed by other methods. These are the field-dependence of meta-magnetic phase transitions and the electronic structure in strongly disordered materials, such as alloys. We showed that the electronic entropy change in meta-magnetic transitions is not constant with the applied magnetic field, as is usually assumed. Furthermore, we traced the evolution of the electronic entropy with respect to the chemical composition of an alloy series. Insights about the strength and kind of interactions appearing in the exemplary materials can be identified in the experiments.

Keywords: electronic entropy; Seebeck coefficient; transport; LaFeSi; FeRh; CuNi

1. Introduction

Entropy provides information about the degrees of freedom or ordering of a statistical collectivity, i.e., it is macroscopically seen and treated as an entity. This order directly correlates with changes in the density of states of the respective statistical collectivity. For electrons in crystalline solids, this information is usually extracted from band structure theory assumptions. It is valid in the case that the sometimes quite stringent assumptions of the theoretical model are met. Experimental systems inherently deviate from the ideal solid state model. Due to this, the density of states calculated theoretically is sometimes not enough to describe the electronic properties in real systems. Typical cases where changes in the electronic density of states occur are charge order/disorder phenomena, such as the formation of charge density waves phases, superconducting phases, Fermi liquid systems, or other correlated electron systems. Further systems that are challenging to describe by theoretical solid state considerations are disordered solids, such as alloys, amorphous materials, materials with complex elementary cells, or materials containing a high number of defects induced, for instance, by the fabrication technology.

A usual approach to evaluate the total electronic entropy S_E of a crystalline solid from experimental data is to analyse the low temperature specific heat capacity, c_p, measurements under the assumption of a free electron gas [1]. Here the Sommerfeld coefficient is the relevant value, experimentally obtained by fitting the low-temperature c_p data. While this is currently the most widely applied method for an

S_E characterization of crystalline solids, there are some intrinsic drawbacks to this method. These come on the one hand from the assumption of a free electron gas and on the other hand from the fact that the relevant materials properties can only be inspected at low temperature [1]. Both rule out the investigation S_E changes at phase transition, especially those occurring at temperatures above 20 K, and such that induce electronic ordering phenomena.

Within this article, we discuss a recently suggested method for the S_E characterization [2] that overcomes some of the limitations of the low temperature c_p analysis, providing a tool for investigating such mentioned electronic systems by a direct experimental approach. We herein utilize the thermodynamic description of the Seebeck coefficient, α, originally described by Onsager [3,4], and later referred to by Ioffe [5] in order to describe the S_E of solids. The inherent advantage of the thermodynamic interpretation of α is that it is not bound to any model, provided the statistical description of the system is significant.

The idea to measure the S_E through the measurement of macroscopic electronic properties like the Seebeck of Thomson effect has been discussed in literature [4–8], and dates, in principle, back to Thomson (Lord Kelvin) who interpreted that the Thomson effect could be seen as the specific heat of electrons, whereas the Seebeck coefficient would be the electronic entropy (divided by the charge of the electrons) [9]. Rockwood [9] pointed out that the measurement of thermoelectric transport properties necessarily only addresses the electrons that participate in the transport. He therefore specified the term "electronic transport entropy" to distinguish from a "static electronic entropy". Furthermore, thermoelectric transport measurement could never be done under truly reversible conditions since the sample needs to be exposed to a temperature gradient and is therefore not under isothermal conditions. Still, he came to the conclusion that the measurement of the thermoelectric coefficients would most likely provide the only practical and generally valid method by which partial molar entropies of electrons could be obtained. Peterson and Shastry construed the Seebeck coefficient as particle number derivative of the entropy at constant volume and constant temperature [8]. Despite this given theoretical framework, examples in which Seebeck coefficient measurements were used to quantitatively deduce S_E are rare and recent but still prove the broad applicability. Our group showed that S_E of a magneto-caloric phase transition could be obtained by thermoelectric transport characterization [2]. Small entities of particles like quantum dots can likewise be characterized [10]. At high temperatures, molten semiconductors and metals were similarly studied [11]. Within this paper, we will discuss the broad applicability of this method. For the following discussion, we refer to the description of the electronic entropy per particle, S_N, as derived within a recent review, providing an applied view on the thermodynamic interpretation of α [12]:

$$S_N = \alpha \cdot e \tag{1}$$

where e is the charge of the particle.

In simple metals, a formal expression of α can be derived from band structure arguments as in the case of the Mott formula [13]. Often, a single parabolic band model is assumed. Herein, the relation between α and the density of states becomes evident, thus establishing a direct connection between α and S_E. While the general thermodynamic interpretation of α does not rely on any kind of model, the Mott formula already contains simplifications and assumptions. From the description of the quantity S_N as introduced in Equation (1), it is suggested that there exists an absolute value of S_N since α is a quantity that also has an experimentally accessible defined zero-level rather than a relative one where only changes in the quantity can be considered. The case of $\alpha = 0$ occurs, for example, (i) in the superconducting state of matter, where electrons all condense at the state of lowest energy possible and therefore per definition a situation of zero entropy [14] and (ii) in the compensated case that electrons and holes exactly transport the same amount of heat, i.e., intrinsic semiconductors have zero Seebeck coefficients [15]. The latter is an often-seen zero crossing of an n-type conductivity mechanism to a p-type conductivity mechanism. Then, the measured $\alpha = 0$ corresponds to the overall observable α of the material. Naturally, the contributions of the individual bands contain electronic entropy contributions with $S_{E, \text{individual subband}} \neq 0$. The full evaluation of S_E from α requires a correct description of the collectivity of electrons in the system. This is the point in the complete line of argumentation where assumptions and simplifications necessarily enter the picture. In order to experimentally obtain the entropy of the entity of electrons that participate in the transport, referred to as electronic entropy, S_E, the number of electrons contributing to the Seebeck voltage needs to be known. In principle, any experimental procedure to obtain the charge carrier density, n, could be used. Herein, it is, as, for instance, suggested in [2,11]:

$$S_E = n \cdot S_N = n \cdot \alpha \cdot e \tag{2}$$

In this work, we measure the ordinary Hall coefficient R_H to obtain n, using the relation $R_H = 1/(n \cdot e)$. By doing so, we introduce the strong assumption of a parabolic single-band transport model that is inherent to any Hall measurement. Combining both quantities, we can give a measure of S_E:

$$S_E = \alpha/R_H \tag{3}$$

We present examples that highlight the relevance of the entropy interpretation of α and provide insight into the electronic properties: (1) magneto-structural phase transitions of an intermetallic Ni-doped iron rhodium phase, $Fe_{0.96}Ni_{0.02}Rh_{1.02}$ (FeRh) [16], and an intermetallic lanthanum iron silicon phase, $LaFe_{11.2}Si_{1.8}$ (LaFeSi) [17–21]; (2) alloying in the copper–nickel (CuNi) solid solution series.

2. Materials and Methods

All samples characterized within this work were obtained by arc melting, and followed by specific temperature treatments to ensure a homogenous microstructure. Details about the fabrication and structural characterization of the samples can be found in [22] and in [23] for $LaFe_{11.2}Si_{1.8}$ (LaFeSi). The samples investigated in the present paper stem from the same batches as the indicated references. In the case of the CuNi alloy series, the processing followed a combination of homogenization (973 K, 5 h) with quenching in H_2O, hot rolling (1173 K) and recrystallization (973 K, 1 h).

The transport characterization was performed depending on the temperature range using physical property measurement systems of the Quantum Design DynaCool series and the Versalab series using the thermal transport option for α and the electrical transport option for the Hall characterization in standard Hall bar geometry [24]. For the CuNi alloy series, a Linseis LSR 3 device was used to measure the near-room temperature α (315 K) and electrical conductivity, σ.

The microstructure of the samples was routinely investigated by scanning electron microscopy and X-ray diffraction.

3. Results and Discussion

As briefly discussed above, the entity of carriers needs to be known for the statistical interpretation of α. Following Equation (2), we utilize n obtained from a Hall-effect measurement. Herein, one has to be aware of the fact that this evaluation method may be affected by multi-channel transport, induced by multiple bands. However, given a minimal set of regularities, we can compare a homogenous series of samples or one sample under different experimental conditions consistently.

3.1. Magneto-Structural Phase Transition

The first example is related to meta-magnetic phase transitions in two magneto-caloric materials, namely Ni-doped FeRh and LaFeSi. They represent examples for a system that can be described with a band magnetism model (FeRh) [25] and a system with a component of localized ionic magnetism (LaFeSi) [26]. General information on the total entropy change in the phase transition of FeRh can be found in Ref. [2] and references therein, as well as a discussion of S_E of this phase transition derived by transport measurements. Additionally, LaFeSi is a well-studied material with respect to magnetic and lattice entropy [26–28]. Due to soft phonon states close to transition, the lattice entropy change is large [29], but a combined contribution of lattice entropy and S_E was suggested [27]. Details on the transport properties of LaFeSi are given in literature with respect to α [28,29] and the anomalous Hall effect [30].

The impact of the applied magnetic field on the transport of the mobile charge carriers shows a clear distinct signature in both materials, which we will discuss in the following. Both magnetic systems behave differently, as best seen in α. In the case of FeRh (Figure 1a), it can be seen that the temperature of the phase transition depends on the magnetic field. This is a striking difference of the S_E evaluation by transport experiments and calorimetric measurements that—for intrinsic reasons—do not allow this difference to be unveiled. The α far from the phase transition is independent of the strength of the magnetic field, as emphasized in the inset in Figure 1a that shows an enlarged view of the data in the main panel. In contrast, in the case of LaFeSi (Figure 1c), the α far from the phase transition shows a clear difference in the value depending on the magnetic field. Interestingly, the magnitude of α increases as a magnetic field is applied. The inset to Figure 1c shows the measured Hall coefficient, and the black lines indicate the levels used for the entropy evaluation as was similarly done in [2]. We get a value corresponding to the ΔS_E at the phase transition, as depicted in Figure 1b,d. In both cases, we see ΔS_E of a comparable magnitude around 4 J K^{-1} kg^{-1}. Moreover, the absolute values of the obtained S_E are also comparable. Furthermore, in both cases, an increase of ΔS_E is observed when a magnetic field is applied. However, the apparent origin of the increase in ΔS_E for both materials is different. In the case of FeRh, the first order meta-magnetic transition shifts to lower temperatures as the field is applied (Figure 1a,b). Accordingly, α follows a monotonic trend until the phase transition occurs. In the case of the LaFeSi, the amount of Si ($x = 1.8$) is on the threshold for changing the transition type to the second order [28]. Therefore, the transition temperature does not shift significantly, and only a slight broadening is observed. In this case, it is the change of the over-all entropy level with the applied magnetic field (Figure 1c,d) that causes the increase in ΔS_E. In the case of LaFeSi, this could be an indication of the interaction between itinerant electrons and localized moments, causing the increase of S_E with magnetic field. There is no such interaction in the FeRh case, as magnetism resides to a dominant part within the conduction electrons. Besides minor numerical corrections to the presented results (compare discussion Ref. [2]), it is clear that this method of analysis provides an insight to the interactions relevant to the conduction electrons that go beyond what typical calorimetric experiments can offer.

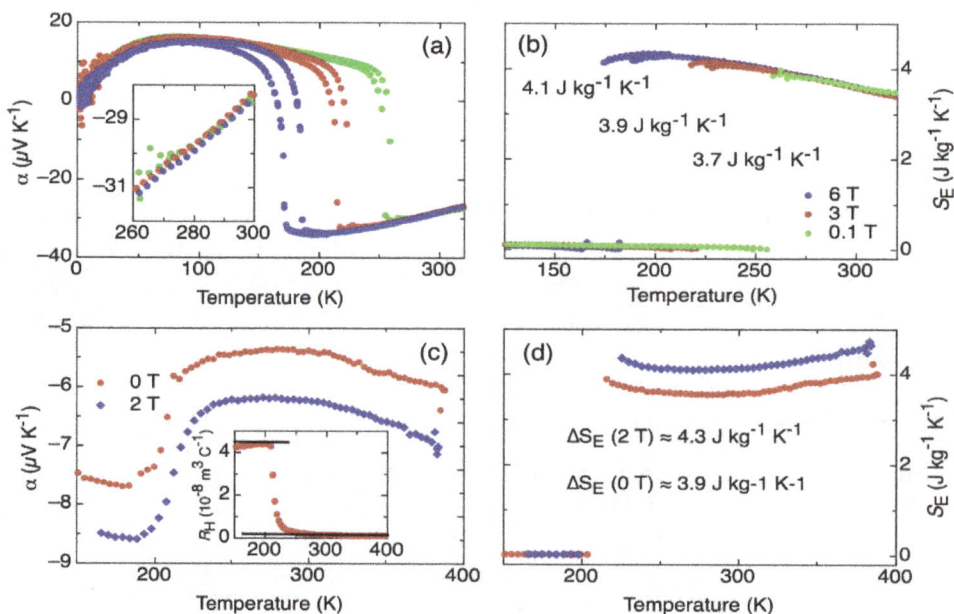

Figure 1. Seebeck coefficient and entropy evaluation in Ni-doped FeRh (**a**,**b**) and LaFeSi (**c**,**d**). Inset to (**a**): enlarged view of the high temperature region. Inset to (**c**): Measured Hall coefficient of LaFeSi.

3.2. Alloying

The differential evaluation of a systematic series of homogenized CuNi alloys with respect to their $|\alpha|$, σ, n, and S_E at room temperature is shown in Figure 2. Herein, it is the specific situation of alloys that they typically cannot be accurately calculated or predicted by usual band structure models. However, the full alloy series is experimentally accessible. There are no structural phase transitions reported, and, also, all investigated samples were homogenous with respect to their microstructure and composition by scanning electron microscopy and X-ray diffraction. The dependence of σ on the Ni content (Figure 2a) presents two minima at around 30 at.%-Ni and at around 70 at.%-Ni, which are better seen in the inset to Figure 2a, where the data of the main panel are presented in logarithmic vertical scale.

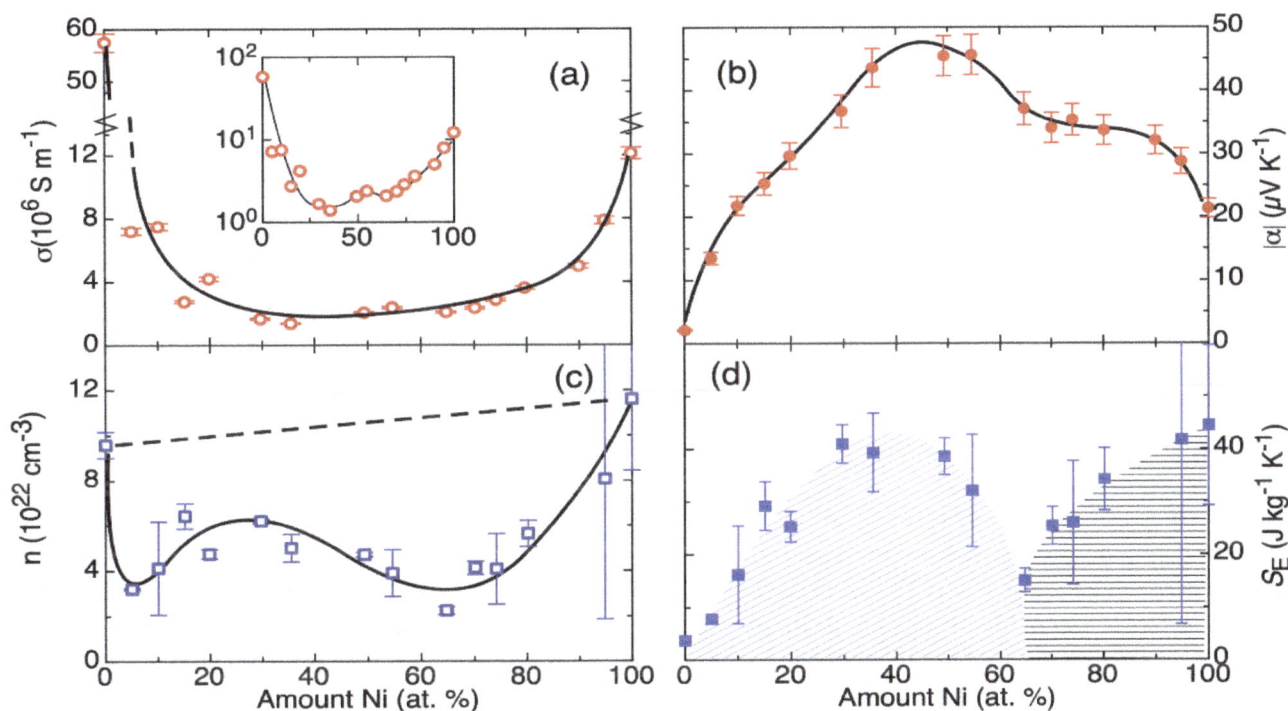

Figure 2. Thermoelectric and transport properties across alloy system Cu–Ni at room temperature, alloy composition was obtained with Energy-Dispersive X-Ray spectroscopy: (**a**) electrical conductivity, (**b**) the Seebeck coefficient in absolute values, (**c**) the carrier concentration derived from the Hall coefficient, (**d**) calculated electronic entropy. Lines and shades are guides to the eye.

In the trend of $|\alpha|$ (Figure 2b), a broad maximum can be seen slightly below to the equiatomic composition, close to the composition of the highest chemical disorder, a similar situation to that of other entropic parameters of such alloys [31], but a shoulder at a composition of about 70 at.%-Ni is also evident. This observation of high $|\alpha|$ for a high chemical disorder reflects the general finding that high configurational entropy is a prerequisite for the observation of large $|\alpha|$ [32]. Because of the close relationship between large $|\alpha|$ and high configurational entropy, it was recently suggested to even use configurational entropy as a gene-like performance indicator for the computational search of new thermoelectric materials [33].

The parameters σ and $|\alpha|$ follow inverse trends with respect to one another. Additionally, these trends match the description of α under the Mott formula [13]. Consequently, the investigated alloy series represents a good electronic model system. There is no clear trend in the data of n. (Figure 2c) The pure metals Cu and Ni have the highest n. Different effects superimpose to a more sophisticated dependence of n on the alloy composition: (i) the effect of change in the average lattice parameter by the alloying [31] should create a gradual increase in n as the amount of Nickel increases; (ii) additionally, with the addition of Ni (Ni: $3d^8 4s^2$; 2 electrons per Ni atom) into the Cu matrix (Cu: $3d^{10} 4s^1$; 1 electron per Cu atom) more charge will also be added [34]. A linear increase is schematically depicted by the dashed line in Figure 2c. The overall result of these measurements is a clear minimum at approximately 65 at.%-Ni. This already indicates that additional degrees of complexity add to this simplified picture.

The combination of $|\alpha|$ and n to extract the S_E allows us to gain additional information compared to the individual transport coefficients. Figure 2d shows a curve in S_E with maximum at approximately 30% of Ni and an additional clear minimum at approximately 65% of Ni. Coming from the Cu-side of the phase diagram, the increase in S_E points out an increase in the available states for the transport electrons, which may be intuitively understood: the disorder in the non-periodic electrostatic potential leads to an increase in the entropy of the transport electrons. This increase in S_E reaches a maximum close to the point where the maximum chemical disorder is expected, following the trend of $|\alpha|$. Coming from the Ni-side of the phase diagram, $|\alpha|$ increases and n decreases. The $|\alpha|$, similar to the Cu-side of the phase diagram, shows higher values because of a higher degree of chemical disorder in the system. But the $|\alpha|$ does not follow a monotonic trend; instead, it has a plateau. This, combined with the reduction of n in the same composition region, results in a sharp minimum of S_E. This minimum exactly coincides with the onset of ferromagnetism in the alloy series. Hence, the entropy evaluation provides an insight on how the magnetic ordering mechanism in this alloy affects the localization of charges, possibly due to interactions between d- and s-orbitals. While there is no one-to-one correspondence between the experiment and the microscopic origin, it still provides a meaningful measure of the intensity of correlations in the electronic transport system, which are not easily accessible by usual ab-initio methods.

4. Conclusions

In conclusion, this proposed method provides a good instrument for the characterization of electronic interactions or correlations in the material, although the absolute values of S_E or ΔS_E obtained may, in some cases, need to be corrected (further discussed in [1]). In the case of magnetocaloric materials, the effect of the magnetic field on the electronic entropy change can be traced. In the case of alloys, the effect of the atomic disorder can also be traced on the free electrons. In order to gain deeper insight on the physics of disordered systems or systems with concurring interactions, the goal of future research might be to develop the statistical methods under the point of view of thermodynamics that would allow us to describe the statistical collectivity of electrons. In this way, we could transform the qualitative results of our experiments into quantitative predictions.

Author Contributions: Conceptualization: N.P. and G.S.; methodology: C.W., B.W., M.K., A.K., J.F., and N.P.; resources and supervision: K.N.; writing—original draft preparation, N.P., C.W., G.S.; writing—review and editing, N.P., C.W., G.S. All authors have read and agreed to the published version of the manuscript.

Acknowledgments: The authors want to thank D. Seifert and R. Uhlemann in IFW Dresden for technical support. Further, G.S. and N.P. want to thank Anja Waske and Sebastian Fähler in IFW Dresden for fruitful discussions. M.K. and B.W. greatfully acknowledge financial support from the Germany Federal Ministry for Economic Affairs and Energy under Project Number 03ET1374B.

References

1.   Ashcroft, N.W.; Mermin, N.D. *Solid State Physics*; Brooks/Cole Thomson Learning: South Melbourne, Australia, 2012.

2.   Pérez, N.; Chirkova, A.; Skokov, K.P.; Woodcock, T.G.; Gutfleisch, O.; Baranov, N.V.; Nielsch, K.; Schierning, G. Electronic entropy change in Ni-doped FeRh. *Mater. Today Phys.* **2019**, *9*, 100129. [CrossRef]

3.   Onsager, L. Reciprocal Relations in Irreversible Processes. I. *Phys. Rev.* **1931**, *37*, 405–426. [CrossRef]

4.   Onsager, L. Reciprocal Relations in Irreversible Processes. II. *Phys. Rev.* **1931**, *38*, 2265–2279. [CrossRef]

5. Ioffe, A.F.; Stil'Bans, L.S.; Iordanishvili, E.K.; Stavitskaya, T.S.; Gelbtuch, A.; Vineyard, G. Semiconductor Thermoelements and Thermoelectric Cooling. *Phys. Today* **1959**, *12*, 42. [CrossRef]

6. Rockwood, A.L. Partial Molar Entropy and Partial Molar Heat Capacity of Electrons in Metals and Superconductors. *J. Mod. Phys.* **2016**, *7*, 199–218. [CrossRef]

7. Rockwood, A.L. Partial molar entropy of electrons in a jellium model: Implications for thermodynamics of ions in solution and electrons in metals. *Electrochim. Acta* **2013**, *112*, 706–711. [CrossRef]

8. Peterson, M.; Shastry, B.S. Kelvin formula for thermopower. *Phys. Rev. B* **2010**, *82*, 195105. [CrossRef]

9. Rockwood, A.L. Relationship of thermoelectricity to electronic entropy. *Phys. Rev. A* **1984**, *30*, 2843–2844. [CrossRef]

10. Kleeorin, Y.; Thierschmann, H.; Buhmann, H.; Georges, A.; Molenkamp, L.W.; Meir, Y. How to measure the entropy of a mesoscopic system via thermoelectric transport. *Nat. Commun.* **2019**, *10*, 5081. [CrossRef]

11. Rinzler, C.C.; Allanore, A. Connecting electronic entropy to empirically accessible electronic properties in high temperature systems. *Philos. Mag.* **2016**, *96*, 3041–3053. [CrossRef]

12. Goupil, C.; Seifert, W.; Zabrocki, K.; Müller, E.; Snyder, G.J.; Müller, E. Thermodynamics of Thermoelectric Phenomena and Applications. *Entropy* **2011**, *13*, 1481–1517. [CrossRef]

13. Cutler, M.; Mott, N.F. Observation of Anderson Localization in an Electron Gas. *Phys. Rev.* **1969**, *181*, 1336–1340. [CrossRef]

14. Roberts, R.B. The absolute scale of thermoelectricity. *Philos. Mag.* **1977**, *36*, 91–107. [CrossRef]

15. da Rosa, A. Thermoelectricity. In *Fundamentals of Renewable Energy Processes*; Elsevier: Amsterdam, The Netherlands, 2013; pp. 149–212.

16. Nikitin, S.; Myalikgulyev, G.; Tishin, A.; Annaorazov, M.; Asatryan, K.; Tyurin, A. The magnetocaloric effect in Fe49Rh51 compound. *Phys. Lett. A* **1990**, *148*, 363–366. [CrossRef]

17. Fukamichi, K.; Fujita, A.; Fujieda, S. Large magnetocaloric effects and thermal transport properties of La(FeSi)13 and their hydrides. *J. Alloys Compd.* **2006**, *408–412*, 307–312. [CrossRef]

18. Fujieda, S.; Hasegawa, Y.; Fujita, A.; Fukamichi, K. Thermal transport properties of magnetic refrigerants La(FexSi1−x)13 and their hydrides, and Gd5Si2Ge2 and MnAs. *J. Appl. Phys.* **2004**, *95*, 2429–2431. [CrossRef]

19. Shen, B.G.; Sun, J.R.; Hu, F.X.; Zhang, H.W.; Cheng, Z.H. Recent Progress in Exploring Magnetocaloric Materials. *Adv. Mater.* **2009**, *21*, 4545–4564. [CrossRef]

20. Hu, F.-X.; Shen, B.-G.; Sun, J.; Cheng, Z.; Rao, G.-H.; Zhang, X.-X. Influence of negative lattice expansion and metamagnetic transition on magnetic entropy change in the compound LaFe$_{11.4}$Si$_{1.6}$. *Appl. Phys. Lett.* **2001**, *78*, 3675–3677. [CrossRef]

21. Hu, F.-X.; Ilyn, M.; Tishin, A.M.; Sun, J.R.; Wang, G.J.; Chen, Y.F.; Wang, F.; Cheng, Z.H.; Shen, B.G. Direct measurements of magnetocaloric effect in the first-order system LaFe11.7Si1.3. *J. Appl. Phys.* **2003**, *93*, 5503–5506. [CrossRef]

22. Baranov, N.; Barabanova, E. Electrical resistivity and magnetic phase transitions in modified FeRh compounds. *J. Alloys Compd.* **1995**, *219*, 139–148. [CrossRef]

23. Glushko, O.; Funk, A.; Maier-Kiener, V.; Kraker, P.; Krautz, M.; Eckert, J.; Waske, A. Mechanical properties of the magnetocaloric intermetallic LaFe11.2Si1.8 alloy at different length scales. *Acta Mater.* **2019**, *165*, 40–50. [CrossRef]

24. Haeusler, J. Die Geometriefunktion vierelektrodiger Hallgeneratoren. *Electr. Eng.* **1968**, *52*, 11–19. [CrossRef]

25. Lu, W.; Nam, N.T.; Suzuki, T. First-order magnetic phase transition in FeRh–Pt thin films. *J. Appl. Phys.* **2009**, *105*, 07A904. [CrossRef]

26. Gercsi, Z.; Fuller, N.; Sandeman, K.G.; Fujita, A. Electronic structure, metamagnetism and thermopower of LaSiFe12and interstitially doped LaSiFe12. *J. Phys. D Appl. Phys.* **2017**, *51*, 034003. [CrossRef]

27. Jia, L.; Liu, G.J.; Sun, J.R.; Zhang, H.W.; Hu, F.-X.; Dong, C.; Rao, G.; Shen, B.G. Entropy changes associated with the first-order magnetic transition in LaFe13−xSix. *J. Appl. Phys.* **2006**, *100*, 123904. [CrossRef]

28. Hannemann, U.; Lyubina, J.; Ryan, M.P.; Alford, N.M.; Cohen, L.F. Thermopower of LaFe 13−x Si x alloys. *EPL* **2012**, *100*, 57009. [CrossRef]

29. Gruner, M.; Keune, W.; Landers, J.; Salamon, S.; Krautz, M.; Zhao, J.; Hu, M.Y.; Toellner, T.; Alp, E.E.; Gutfleisch, O.; et al. Moment-Volume Coupling in La(Fe1−x Si x)13. *Phys. Status Solidi (B)* **2017**, *255*, 1700465. [CrossRef]

30. Landers, J.; Salamon, S.; Keune, W.; Gruner, M.; Krautz, M.; Zhao, J.; Hu, M.Y.; Toellner, T.S.; Alp, E.E.; Gutfleisch, O.; et al. Determining the vibrational entropy change in the giant magnetocaloric material LaFe11.6Si1.4 by nuclear resonant inelastic x-ray scattering. *Phys. Rev. B* **2018**, *98*, 024417. [CrossRef]

31. Madelung, O. *Cr-Cs–Cu-Zr*; Springer: Berlin/Heidelberg, Germany, 1994.

32. Gruen, D.M.; Bruno, P.; Xie, M. Configurational, electronic entropies and the thermoelectric properties of nanocarbon ensembles. *Appl. Phys. Lett.* **2008**, *92*, 143118. [CrossRef]

33. Liu, R.; Chen, H.; Zhao, K.; Qin, Y.; Jiang, B.; Zhang, T.; Sha, G.; Shi, X.; Uher, C.; Zhang, W.; et al. Entropy as a Gene-Like Performance Indicator Promoting Thermoelectric Materials. *Adv. Mater.* **2017**, *29*, 1702712. [CrossRef]

34. Hurd, C.M. *The Hall Effect in Metals and Alloys*; Springer Science and Business Media LLC: Berlin, Germany, 1972.

7

Power Conversion and its Efficiency in Thermoelectric Materials

Armin Feldhoff

Institute of Physical Chemistry and Electrochemistry, Leibniz University Hannover, Callinstraße 3A, D-30167 Hannover, Germany; armin.feldhoff@pci.uni-hannover.de

Abstract: The basic principles of thermoelectrics rely on the coupling of entropy and electric charge. However, the long-standing dispute of energetics versus entropy has long paralysed the field. Herein, it is shown that treating entropy and electric charge in a symmetric manner enables a simple transport equation to be obtained and the power conversion and its efficiency to be deduced for a single thermoelectric material apart from a device. The material's performance in both generator mode (thermo-electric) and entropy pump mode (electro-thermal) are discussed on a single voltage-electrical current curve, which is presented in a generalized manner by relating it to the electrically open-circuit voltage and the electrically closed-circuited electrical current. The electrical and thermal power in entropy pump mode are related to the maximum electrical power in generator mode, which depends on the material's power factor. Particular working points on the material's voltage-electrical current curve are deduced, namely, the electrical open circuit, electrical short circuit, maximum electrical power, maximum power conversion efficiency, and entropy conductivity inversion. Optimizing a thermoelectric material for different working points is discussed with respect to its figure-of-merit zT and power factor. The importance of the results to state-of-the-art and emerging materials is emphasized.

Keywords: thermoelectrics; power conversion; efficiency; voltage-electrical current curve; working point; entropy pump mode; generator mode; power factor; figure of merit; Altenkirch-Ioffe model

---

1. Introduction

1.1. Controversial Points of View

Entropy is a central quantity in thermoelectrics, but seldom has it been addressed as such. The basic physical quantity that is known today as entropy is widely considered to be a derived quantity according to the approaches by Clausius [1–3] and Boltzmann [4–6] to quantify its value in certain situations. Both the perception of entropy as a derived quantity and the underestimation of its role in thermal processes are seen as residual outcomes of the Ostwald-Boltzmann battle, which is worth recalling and constitutes another chapter in the tragicomical history of thermodynamics [7]. In the frame of this work, entropy is considered to be a basic quantity. The benefits of this controversial point of view are made obvious on the example of thermoelectric materials.

1.2. Implications of Natural Philosophy

Clausius intended to borrow terms for important quantities from the ancient languages, so that they may be adopted unchanged in all modern languages. He proposed to call the quantity S, which had been introduced by him, the entropy of the body, from the Greek word τροπη (tropy), transformation [1–3]. Intentionally, he formed the word entropy to be as similar as possible to the word energy. In his opinion, the two quantities to be denoted by these words are so nearly allied in their physical meanings that a certain similarity in designation is desirable [1–3].

The importance of entropy was underlined by Gibbs in the very first words of his treatise on thermodynamics: "The comprehension of the laws which govern any material system is greatly facilitated by considering the energy and entropy of the system in the various states of which it is capable" [8,9]. However, the "Energeticist" [10] school in Germany, which rejected atomism and other matter theories, postulated energy as the primary substance in nature, and considered entropy as a superfluous derived concept [11–13]. The protagonist was Ostwald, cofounder of physical chemistry and its Nestor in Germany, and behind it was the natural philosophy of Mach [6,14,15]. Soon, the "Energeticist" school attracted much critical attention not only by the British pioneers [16] but also from a younger generation of German physicists [11]. The young Sommerfeld witnessed a memorable debate at the 1895 Assembly of the German Society of Scientists and Physicians in Lübeck, in which Boltzmann "like a bull defeated the torero [Helm as substitute to Ostwald] despite all his art of fencing [14]." In a follow-up critique, Boltzmann [17,18] condemned Ostwald's "Energetics" not only for perceived mathematical and physical error, but also for its false promise of easy rewards [11]. However, Ostwald never admitted that he had been defeated, and the object of the dispute has been kept alive to the present day [19,20]. Even though the personalities have changed over time, the battle has been newly inflamed in the controversy regarding the Karlsruhe Physics Course [21], which resulted in removing the entropy-treating educational course from German schools [22].

Today, the dissipation or "degradation" of energy is often treated without clear reference to entropy [19,20]. Preference is given to thermal energy ("heat") or enthalpy. Textbooks on classical thermodynamics take the approach of Clausius to quantify entropy in equilibrium conditions as the definition of entropy, which then is perceived as an energy-derived quantity. The success of Boltzmann's principle (called so by Einstein [6]) to quantify entropy in partitioned systems in equilibrium [23] renders it often to be a statistics-derived quantity [24]. However, the special cases considered herein do show only certain aspects of entropy, which should be considered in a wider context. By not considering entropy as a central basic quantity, clearness is lost, and uncertainty even creeps over authors who endeavor for accuracy and clarity when it comes to the description of thermal phenomena.

1.3. Evolution of Thermodynamics

The field of thermodynamics has evolved from the aim of understanding the thermodynamical engine (i.e., the steam engine) [11], which by principle operates under non-equilibrium conditions. However, for several reasons, thermodynamics has been limited to equilibrium conditions for a long time. For its suggestion to use entropy under non-equilibrium conditions, Planck's PhD thesis [25] was heavily criticized [19,20]. Planck was likely then intimidated and did not deepen this approach to entropy [19,20]. Alternately, the elegance and success of Gibbs' treatise on using equilibrium conditions did pave the way for thermodynamics under equilibrium conditions.

It took several decades until Callen [26,27] and de Groot [28] independently formulated a theory to describe thermodynamic systems in non-equilibrium conditions. This theory was helpful for quantitatively describing thermoelectric phenomena. However, the primary focus was the entropy production in irreversible processes and, thus, the excess entropy. No attention was given to entropy itself and its ability, which in older terms could be mentioned as the motive power of entropy, to drive a steam engine [29–31] or thermoelectric generator [32–34].

1.4. Modern Thermodynamics

Consistent with Falk [35], Fuchs [32], and Strunk [23,31], the author holds the view that entropy should be considered as a fundamental quantity. The characteristics of a fundamental quantity unfold from its relations with other fundamental quantities. Concise theories have been developed by Fuchs [32], Job & Rüffler [36,37], and Strunk [23,31,38].

In context of the development of physical concepts, it is worth noting that the basic physical quantity that is known today as entropy, was named quantity of heat by Joseph Black (1728–1799) [39–41] and calorique by Sadi Carnot (1796–1832) [29,30,40]. Indeed, calorique is the French word for quantity of heat. In his 1911 Presidential address to the Physical Society of London, Hugh Longbourne Callendar [29] outlined Carnot's calorique (i.e., entropy) as a quantity, that "any schoolboy could understand". Moreover, Callendar underlined that Carnot's calorique reappeared as a triple integral in Kelvin's 1852 paper, as the thermodynamic function of Rankine and as equivalence-value of a transformation in the 1854 paper of Clausius, and as entropy in the 1865 paper of Clausius [2] along with an abstract redefinition. No one at that time appears to have realized that entropy was merely calorique under another name. Callendar closed his remarks with the advice to distinguish a quantity of heat from a quantity of thermal energy.

Traditionally, thermal energy is called "heat". Concordant with Callendar [29] and Fuchs [32], in the author's opinion, heat is not energy, and entropy is the true measure of a quantity of heat as opposed to a quantity of thermal energy. Thus, the use this term for thermal energy should be avoided [42]. For clarity, the traditional term "heat" is put into quotation marks when it addresses the thermal energy. In this approach, entropy is a basic quantity. Thermoelectrics is an example par excellence to show the benefits of this philosophical perspective.

1.5. Entropy in Thermoelectrics

In the context of thermoelectrics, according to Boltzmann's principle, entropy is considered as a statistics-derived quantity when it is used to quantify the effect of spin and orbital degrees of freedom on the Seebeck coefficient in strongly correlated electron systems [43,44]. This, however, is a minor aspect. The approach by Clausius, to consider entropy as an energy-derived quantity does not play a significant role either.

In the so-called theory of thermodynamics of irreversible processes, as developed by Callen [26,27] and de Groot [28], it is rather the case that the thermal energy is derived from the entropy. Entropy is a fundamental quantity that is central to thermoelectrics. These texts can be read with great earning if entropy is considered as an indestructible substance-like quantity that is able to flow through the thermoelectric material and carries the thermal energy. The concept of energy carriers was developed by Falk et al. [45] and Herrmann [21].

However, the theory of thermodynamics of irreversible processes has the tendency to focus on the irreversibly produced excess entropy, but not on the entropy itself. Instead, energetic quantities are preferred. In §60 of his textbook, de Groot [28] presents an alternative presentation of thermoelectricity by the use of entropies of transfer, for which he has stated that the theory becomes somewhat more elegant compared to using energies of transfer. Unfortunately, he has not deepened this approach.

In a preceding paper [34], the author has shown that the rehabilitation of entropy into the theory by Callen [26,27] and de Groot [28] leads to a vivid description of thermoelectric devices. Like electrical charge carries the electrical energy, entropy carries the thermal energy. Thermal induction of an electrical current and electrical induction of a thermal current become understandable.

1.6. Aim of This Work

Like the preceding paper by the author [34], the present work aims to contribute to a better understanding of thermoelectrics by reconsidering it by treating entropy and electric charge as basic quantities of equal rank. This is semantically considered by naming the part of energy that flows together with entropy the thermal energy and part of energy flowing together with electrical charge the electrical energy. The energy flux through the thermoelectric material can thus be divided into thermal power and electrical power. Power conversion, which is in the focus of this article, implies that the system under consideration is not in equilibrium, but instead flown through by substance-like quantities. For the case of thermoelectric materials, these are entropy, electric charge, and energy.

By recalling the historical development of the perception of entropy, obstacles are identified, which have hindered the recognition of its important role in the field of thermoelectrics. The confused traditional approach and the use of model devices are avoided. Both power conversion and the efficiency of power conversion are accessed quantitatively for a thermoelectric material apart from a device. New physical insight into thermoelectrics is gained on the level of the thermoelectric material rather than on the device level. On the material's voltage–electrical current curve, distinct working points are identified (see Table 1), which not only allow for quantification of the material's properties and performance under specific operational conditions, but also relate generator mode (thermal-to-electrical power conversion) and entropy pump mode (electrical-to-thermal power conversion) of the same material to each other.

Table 1. Working points on the voltage–electrical current curve of a thermoelectric material in both operational modes, as addressed in this work.

Abbreviation	Working Point	Operational Mode
MCEP	Maximum (power) conversion efficiency point	entropy pump mode
EICP	Entropy conductivity inversion point	entropy pump mode
OC	(electrical) open circuit	generator mode
MCEP	(see above)	generator mode
MEPP	Maximum (electrical) power point	generator mode
SC	(electrical) short circuit	generator mode

The results are worked out in detail, and the outcome from the formalism is graphically illustrated and explained. The simplicity of thermoelectrics is clarified. The findings are linked to the outcome of the traditional approach to thermoelectrics and state-of-the-art thermoelectric materials.

2. Results

2.1. Categories

The results section is categorized, as follows.

- Section 2.2: Coupling currents of entropy and charge in thermoelectric materials
- Section 2.3: Material's voltage–electrical current and electrical power–electrical current characteristics
- Section 2.4: Material's thermal conductivity–electrical current characteristics
- Section 2.5: Thermoelectric material in generator mode
- Section 2.5.1: Working point for maximum electrical power
- Section 2.5.2: Thermal conductivity
- Section 2.5.3: Thermal power
- Section 2.5.4: Power conversion efficiency (thermal to electrical)
- Section 2.5.5: Working points for maximum conversion efficiency and maximum electrical power
- Section 2.6: Thermoelectric material in entropy pump mode
- Section 2.6.1: Power conversion efficiency (electrical to thermal)
- Section 2.6.2: Electrical and thermal power
- Section 2.7: Complete picture

2.2. Coupling Currents of Entropy and Charge in Thermoelectric Materials

When a thermoelectric material is simultaneously placed in a gradient of the electrochemical potential $\nabla\tilde{\mu}$ and a gradient of the temperature ∇T, electrical flux density $\mathbf{j}_q$, and entropy flux density $\mathbf{j}_S$ are observed [34,46].

$$\begin{pmatrix} \mathbf{j}_q \\ \mathbf{j}_S \end{pmatrix} = \begin{pmatrix} \sigma & \sigma \cdot \alpha \\ \sigma \cdot \alpha & \sigma \cdot \alpha^2 + \Lambda_{OC} \end{pmatrix} \cdot \begin{pmatrix} -\nabla\tilde{\mu}/q \\ -\nabla T \end{pmatrix} \tag{1}$$

With the classical thermodynamic potential gradients $\nabla\tilde{\mu}$ (per electric charge q) and ∇T being employed, the basic transport Equation (1) has the following structure.

$$\text{flux densities} \quad = \quad \text{material tensor} \quad \cdot \quad \text{potential gradients} \tag{2}$$

The thermoelectric material tensor in Equation (1) is composed of only three quantities, which are the isothermal electrical conductivity σ, the Seebeck coefficient α, and the entropy conductivity at electrical open circuit Λ_{OC} (i.e., at vanishing electrical current). In principle, all three quantities are tensors themselves, but, for homogenous materials, they are often treated as scalars.

The entropy conductivity Λ is related to the traditional "heat" conductivity λ by the absolute temperature T [32,34,37]. This, in principle, indicates that the traditional "heat" conduction is based on a more fundamental entropy conduction. The author proposes using the generic term thermal conductivity to address either the "heat" conductivity or the entropy conductivity [47,48].

$$\lambda = T \cdot \Lambda \tag{3}$$

It is emphasized that Equation (1) refers to a steady-state non-equilibrium situation. Instead of the quantities electric charge q and entropy S, their local flux densities appear. According to Falk [35], considering local flux densities allows addressing local energy conversion or better to say local power conversion. Because flowing quantities are involved, preference should be given to local power density. Remember, power is the flux of energy. Equation (1) allows for locally varying quantities to be considered, which can be expressed with the positional vector $\mathbf{r}$: $\mathbf{j}_q = \mathbf{j}_q(\mathbf{r})$, $\mathbf{j}_S = \mathbf{j}_S(\mathbf{r})$, $\sigma = \sigma(\mathbf{r})$, $\alpha = \alpha(\mathbf{r})$, $\Lambda_{OC} = \Lambda_{OC}(\mathbf{r})$, $\nabla \tilde{\mu} = \nabla \tilde{\mu}(\mathbf{r})$, $\nabla T = \nabla T(\mathbf{r})$. Of course, the thermodynamic potentials are locally varying when gradients are present: $\tilde{\mu} = \tilde{\mu}(\mathbf{r})$, $T = T(\mathbf{r})$.

However, if the local variation of all quantities in Equation (1) is neglected, a simplified formulation of the transport equation can be observed [34,49,50]. If a further weak temperature dependence is assumed for the electron chemical potential μ (i.e., $\frac{\partial \mu}{\partial T} \approx 0$), the temperature dependence of the electrochemical potential $\tilde{\mu} = \mu + q \cdot \varphi$ is only in the electrical potential φ. With $\nabla \mu / q \approx 0$ follows $\nabla \tilde{\mu} / q = \nabla \mu / q + \nabla \varphi \approx \nabla \varphi$. The assumption of constant gradients (i.e., linear potential curves) allows for them to be substituted by the difference of the respective potential along the thermoelectric material of length L: $\nabla \varphi \to -\Delta \varphi / L$, $\nabla T \to -\Delta T / L$. Furthermore, for a thermoelectric material of cross-sectional area A, the local flux densities can be replaced by the integrative currents of electrical charge and entropy, respectively: $\mathbf{j}_q \to I_q / A$, $\mathbf{j}_S \to I_S / A$. Subsequently, the transport equation follows as:

$$\begin{pmatrix} I_q \\ I_S \end{pmatrix} = \frac{A}{L} \cdot \begin{pmatrix} \sigma & \sigma \cdot \alpha \\ \sigma \cdot \alpha & \sigma \cdot \alpha^2 + \Lambda_{OC} \end{pmatrix} \cdot \begin{pmatrix} \Delta \varphi \\ \Delta T \end{pmatrix} \tag{4}$$

Equation (4) describes the coupling of currents of electrical charge I_q and entropy I_S in the thermoelectric material, which causes the occurence of either an electrically-induced entropy current [51] (Peltier effect) or a thermally-induced electrical current [52,53] (Seebeck effect). Note that Equation (4) describes these effects in a thermoelectric material, which is schematically shown in Figure 1, apart from a device.

2.3. Material's Voltage—Electrical Current and Electrical Power—Electrical Current Characteristics

Different working conditions of the thermoelectric material in this article are discussed with reference to the voltage–electrical current curve, which is derived from Equation (4) as Equation (5). Remember that the voltage $\Delta \varphi$ is the electrical potential difference along the thermoelectric material.

$$\Delta \varphi = -\alpha \cdot \Delta T + \frac{I_q}{\frac{A}{L} \cdot \sigma} \tag{5}$$

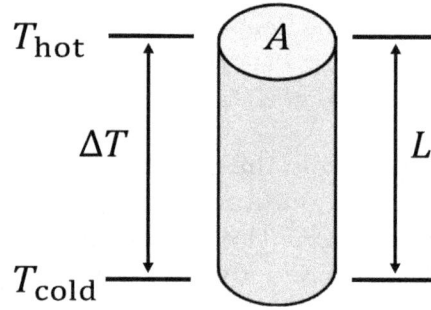

Figure 1. This paper discusses characteristics of a thermoelectric material of cross-sectional area A and length L when exposed to a temperature difference $\Delta T = T_{\text{hot}} - T_{\text{cold}}$ between a hot reservoir at T_{hot} and a cold reservoir at T_{cold}.

According to Equation (5), the voltage–electrical current characteristics is a line, which has the material's electrical resistance $R = \frac{1}{\frac{A}{L} \cdot \sigma}$ as its slope. This line is only determined by the voltage $\Delta\varphi_{\text{OC}}$ under electrically open-circuited conditions (i.e., at zero electrical current) and the electrical current I_{SC} at electrically short-circuited conditions (i.e., at zero voltage). The OC is of practical importance for the measurement of temperature using thermocouples.

$$\Delta\varphi_{\text{OC}} = -\alpha \cdot \Delta T \tag{6}$$

$$I_{q,\text{SC}} = \frac{A}{L} \cdot \alpha \cdot \sigma \cdot \Delta T \tag{7}$$

Obviously, the sign of the Seebeck coefficient α determines the sign of both the voltage $\Delta\varphi_{\text{OC}}$ under electrically short-circuited conditions and the electrical current $I_{q,\text{SC}}$ under electrically short-circuited conditions. Thus, the voltage–electrical current characteristics of p-type ($\alpha > 0$) or n-type ($\alpha < 0$) conductors differ from each other by principle (cf. Appendix A).

To discuss the materials independently of the sign of the Seebeck coefficient, the absolute of the voltage $|\Delta\varphi|$ is plotted in Figure 2 versus the absolute value of the electrical current $|I_q|$. In order to diminish Ohmic losses, the electrical resistance $R = \frac{1}{\frac{A}{L} \cdot \sigma}$ must be reduced, which, for the given geometry, requires the electrical conductivity σ to be increased.

To make the discussion independent from even the material parameters and temperature difference ΔT, the normalized electrical current i and normalized voltage u, as normalized to electrically short-circuited and open-circuited conditions, respectively, are considered in subsequent sections.

$$i = \frac{I_q}{I_{q,\text{SC}}} = \frac{I_q}{\frac{A}{L} \cdot \alpha \cdot \sigma \cdot \Delta T} \tag{8}$$

$$u = \frac{\Delta\varphi}{\Delta\varphi_{\text{OC}}} = \frac{\Delta\varphi}{-\alpha \cdot \Delta T} = 1 - i \tag{9}$$

The electrical power P_{el} is determined by the product of voltage and electrical current as given by Equation (10). It increases linearly with the electrical current, but it is parabolically damped at high electrical currents due to the limited electrical conductivity (Ohmic dissipation [54]).

$$
\begin{aligned}
P_{\text{el}} &= \Delta\varphi \cdot I_q = \left(-\alpha \cdot \Delta T + \frac{I_q}{\frac{A}{L} \cdot \sigma} \right) \cdot I_q \\
&= -\alpha \cdot \Delta T \cdot I_q + \frac{I_q^2}{\frac{A}{L} \cdot \sigma} \\
&= -\frac{A}{L} \cdot \sigma \cdot \alpha^2 \cdot (\Delta T)^2 \cdot (i - i^2)
\end{aligned}
\tag{10}
$$

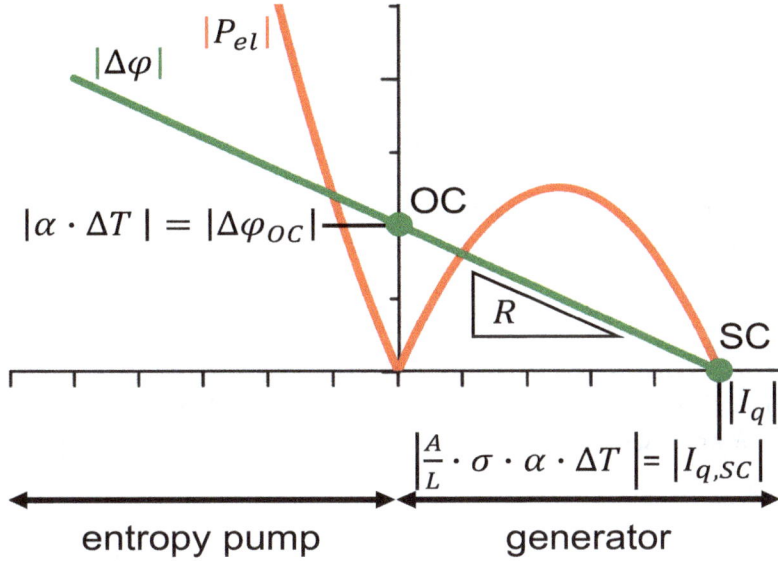

Figure 2. Absolute voltage $|\Delta\varphi|$ – electrical current $|I_q|$ curve (green), with slope given by the electrical resistance $R = \frac{1}{\frac{A}{L}\cdot\sigma}$, and the absolute electrical power $|P_{el}|$ – electrical current $|I_q|$ curve (red) for a thermoelectric material. Here, $\Delta T = \frac{T_{hot} - T_{cold}}{T_{hot}}$ is the temperature difference along the thermoelectric material of cross-sectional area A and length L. These quantities, together with the (isothermal) electrical conductivity σ and the Seebeck coefficient α, determine the electrical current I_{SC} under electrically short-circuited conditions. The voltage $\Delta\varphi_{OC}$ under electrically open-circuited conditions is determined by the Seebeck coefficient and the temperature difference. Generator mode refers to a positive sign and entropy pump mode to a negative sign of the electrical power (cf. Appendix A).

The absolute of the electrical power $|P_{el}|$ is plotted in Figure 2 versus the absolute value of the electrical current $|I_q|$ to discuss the thermoelectric materials independent of the sign of the Seebeck coefficient.

It is obvious from Figure 2 that the electrical power to be put into the material in entropy pump mode may distinctly exceed the electrical power that can be gained in generator mode if the material is applied to the same temperature difference.

2.4. Material's Thermal Conductivity—Electrical Current Characteristics

From Equation (4), the entropy current I_S flowing through the material is obtained. It depends on not only the temperature difference ΔT but also the Peltier effect that is associated with the thermally induced electrical current I_q, which can be expressed by the normalized electrical current i as given in Equation (8).

$$
\begin{aligned}
I_S &= \tfrac{A}{L} \cdot \Lambda_{OC} \cdot \Delta T + \alpha \cdot I_q \\
&= \tfrac{A}{L} \cdot \Lambda_{OC} \cdot \Delta T + \tfrac{A}{L} \cdot \sigma\alpha^2 \cdot i \cdot \Delta T \\
&= \tfrac{A}{L} \cdot \left(\Lambda_{OC} + \sigma\alpha^2 \cdot i\right) \Delta T \\
&= \tfrac{A}{L} \cdot \Lambda \cdot \Delta T
\end{aligned}
\tag{11}
$$

From Equation (11), it follows that the thermal conductivity, expressed here by the entropy conductivity Λ, is dependent on the electrical current i.

$$
\Lambda = \Lambda(i) = \Lambda_{OC} + \sigma\alpha^2 \cdot i
\tag{12}
$$

When compared to electrically open-circuited conditions, the power factor $\sigma\alpha^2$ gives an additional contribution to the entropy conductivity, which increases linearly with the electrical current. Under electrically short-circuited conditions (SC, i.e., $i = 1$), the entropy conductivity reaches its maximum value.

$$\Lambda_{SC} = \Lambda_{OC} + \sigma \alpha^2 \tag{13}$$

Under electrically short-circuited conditions, the electrical potential is spatially constant (i.e., its gradient vanishes: $\nabla \varphi = 0$). Note that the entropy conductivity at electrical short circuit Λ_{SC}, as given by Equation (13), is identical to tensor element M_{22} of the thermoelectric material tensor in the transport Equation (4).

To discuss the characteristics of the entropy conductivity in a general manner, it is normalized to its value under electrically open-circuited conditions:

$$\tilde{\Lambda} = \tilde{\Lambda}(i) = \frac{\Lambda}{\Lambda_{OC}} = 1 + \frac{\sigma \alpha^2}{\Lambda_{OC}} \cdot i = 1 + zT \cdot i \tag{14}$$

In Equation (14), a figure-of-merit zT has been identified, which only depends on the three material parameters σ, α and Λ_{OC}, which make up the material tensor of Equation (4).

$$zT = \frac{\sigma \cdot \alpha^2}{\Lambda_{OC}} \tag{15}$$

Equation (14) is visualized in Figure 3 for some hypothetical thermoelectric materials with $zT = 0.1, 0.5, 1, 2, 4$ and 8. Working points for electrically open-circuited (OC) conditions, maximum electrical power point (MEPP), and electrical short-circuited (SC) conditions are indicated on the voltage–electrical current curve. Note that the entropy conductivity inversion point (ECIP) is given by the negative reciprocal of the figure-of-merit $-1/zT$. Only for electrical currents being below the ECIP, effective entropy pump mode is reached with a negative entropy conductivity of the thermoelectric material. Only then, more entropy is pumped against the temperature difference than flows down it. Obviously, the measurements of the thermal conductivity of a thermoelectric material must refer to the working point on the voltage–electrical current curve.

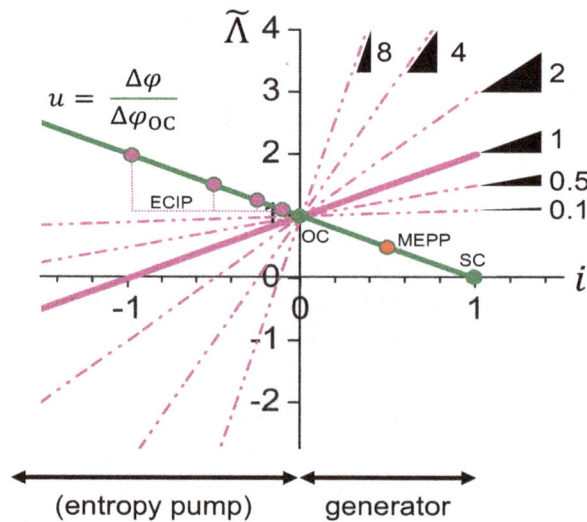

Figure 3. Normalized entropy conductivity $\tilde{\Lambda}$ as function of normalized electrical current i for some hypothetical thermoelectric materials. Depending on the figure-of-merit zT, the curves pivot through the working point for electrically open-circuited (OC) conditions. The figure-of-merit zT gives the slope of the curve and its negative reciprocal $-1/zT$ indicates the entropy conductivity inversion point (ECIP). For some thermoelectric materials, the respective ECIP is indicated as working point on the normalized voltage u–normalized electrical current i curve. Note that the ECIP for materials with $zT = 0.1.$ and $zT = 0.5$ is out of the applied scale. The term entropy pump mode is put into brackets because a net entropy current against the temperature difference will only occur if the magnitude of the electrical current is beyond the respective ECIP. For generator mode, the working points MEPP and SC are indicated.

2.5. Thermoelectric Material in Generator Mode

2.5.1. Working Point for Maximum Electrical Power

Remember, the characteristics of the thermoelectric material are all discussed for ΔT being different from zero, which implies non-isothermal conditions. It can be easily seen from Equation (10) that maximum electrical power output is obtained for half of the electrically short-circuited electrical current ($i_{\text{MEPP}} = \frac{1}{2}$, cf. Appendix B.1):

$$P_{\text{el, max}} = \mid P_{\text{el}} (i_{\text{MEPP}} = 0.5) \mid = \frac{1}{4} \cdot \frac{A}{L} \cdot \sigma \cdot \alpha^2 \cdot (\Delta T)^2 \tag{16}$$

To make the discussion independent from material parameters and temperature difference, the normalized electrical power p_{el}, as normalized to the maximum electrical power in generator mode, is plotted in Figure 4.

$$p_{\text{el}} = \frac{\mid P_{\text{el}} \mid}{P_{\text{el, max}}} = 4 \cdot \mid i - i^2 \mid \tag{17}$$

The maximum electrical power point (MEPP) is indicated on the normalized voltage–electrical current curve in Figure 4. It is clearly seen that the MEPP ($i_{\text{MEPP}} = 0.5$, $u_{\text{MEPP}} = 0.5$) is at half of the open-circuited voltage as well as at half of the electrically short-circuited electrical current, which also follows from Equation (9).

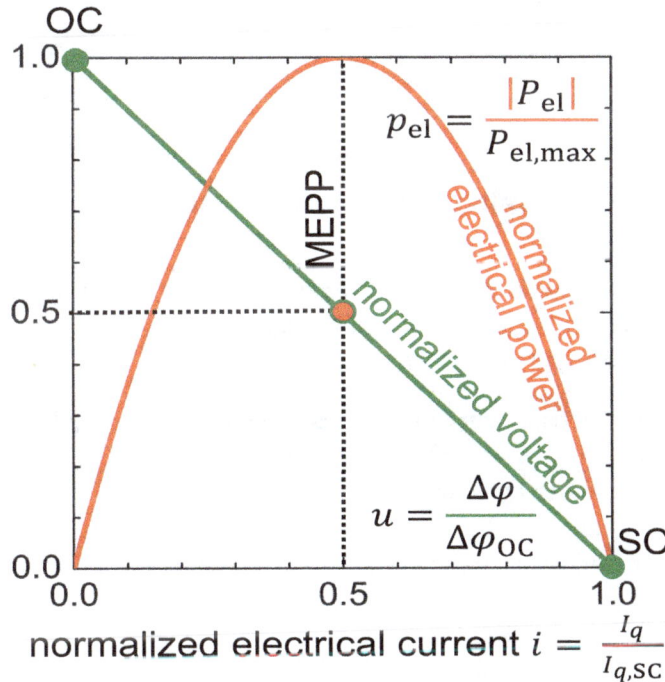

Figure 4. Normalized curves for both voltage u – electrical current i characteristics and electrical power p_{el}–electrical current i characteristics of a thermoelectric material when it is operated in generator mode. The working points open-circuited (OC), maximum electrical power point (MEPP), and short-circuited (SC) are indicated.

2.5.2. Thermal Conductivity

For the thermoelectric material being operated in generator mode, Equation (12) is graphically expressed in Figure 5. The electrically open-circuited entropy conductivity Λ_{OC} is purely dissipative, while the part of the entropy conductivity depending on the power factor $\sigma \cdot \alpha^2$ couples to the electrical current, and it fully contributes to the thermal-to-electric power conversion. Obviously, to maximize while the part of the entropy conductivity depending on the power factor $\sigma \cdot \alpha^2$ couples to the electrical current, and it fully contributes to the thermal-to-electric power conversion. Obviously, to maximize the electrical power at a given temperature difference, the power $\sigma \cdot \alpha^2$ must be maximized, which is in accordance with Equation (10).

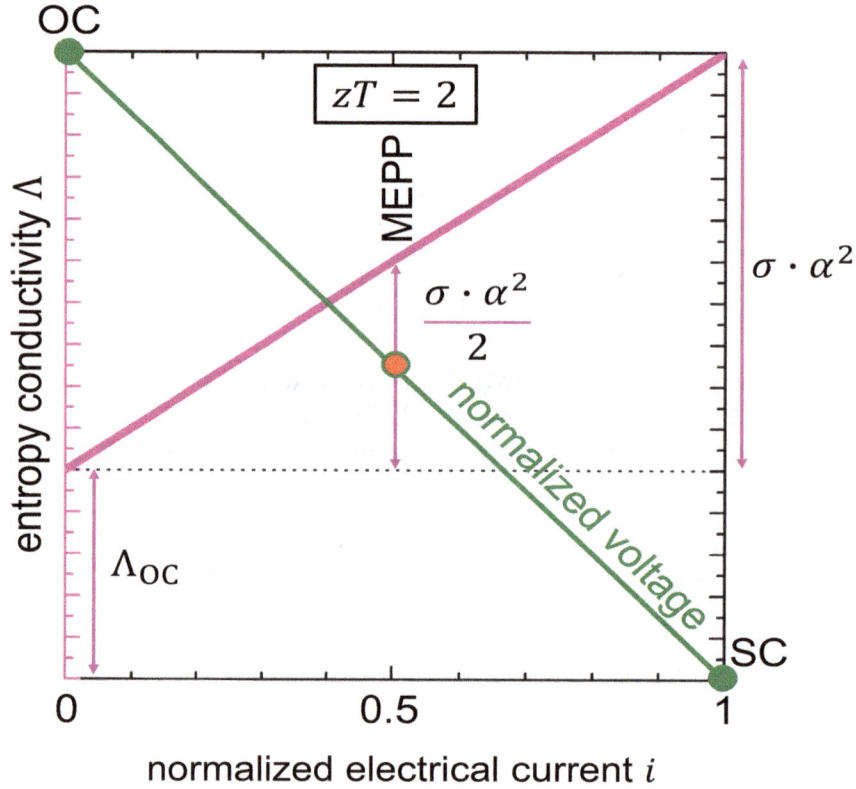

Figure 5. Entropy conductivity Λ as function of the normalized electrical current i for a thermoelectric material with $zT = 2$ in generator mode. The working points OC, MEPP, and SC are indicated on the normalized voltage–electrical current curve.

The thermally induced electrical current carries electrical energy, which, however, with increasing electrical current, is diminished by Ohmic losses due to the limited (isothermal) electrical conductivity σ as discussed above. At maximum electrical power, the entropy conductivity is increased by half of the power factor as compared to electrically open-circuited conditions. Under electrically short-circuited conditions, the entropy conductivity reaches its maximum (see Equation (13)).

2.5.3. Thermal Power

The thermal input power and the thermal output power depend on the electrical current i. According to Fuchs [33], the available thermal power P_{th} is determined by the fall of entropy down the temperature difference ΔT along the material.

$$P_{th} = I_S \cdot \Delta T = \Lambda \cdot (\Delta T)^2 = \frac{A}{L} \cdot \left(\Lambda_{OC} + \sigma \alpha^2 \cdot i \right) \cdot (\Delta T)^2 \tag{18}$$

Thus, the available thermal power, as given by Equation (18), depends on the electrical current in the same manner as the entropy conductivity in Figures 3 and 5.

2.5.4. Power Conversion Efficiency (Thermal to Electrical)

From Equations (10) and (18), the second-law power conversion efficiency for the thermoelectric material in generator mode is obtained:

$$
\eta_{\text{II,gen}} \quad = \left| \frac{P_{\text{el}}}{P_{\text{th,avail}}} \right| \quad = \frac{\frac{A}{L} \cdot \sigma \cdot \alpha^2 \cdot (\Delta T)^2 \cdot (i - i^2)}{\frac{A}{L} \cdot (\Lambda_{\text{OC}} + \sigma \alpha^2 \cdot i) \cdot (\Delta T)^2}
$$
$$
= \frac{i - i^2}{i + \frac{\Lambda_{\text{OC}}}{\sigma \cdot \alpha^2}}
$$
$$
= \frac{i - i^2}{i + \frac{1}{zT}}
\tag{19}
$$

Equation (19) is plotted in Figure 6 as solid blue curves for some hypothetical thermoelectric materials with different values of the figure-of-merit zT. Obviously, the figure-of-merit zT must be maximized in order to maximize the thermal-to-electrical power conversion efficiency at a given (thermally induced) electrical current.

Equation (19) can be read as the coupled thermal power being converted into electrical power with the constraint; however, with increasing electrical current, Ohmic dissipation gains overhead. As a result, the optimum power conversion efficiency is obtained at lower electrical current than the optimum electrical power output, and the working points for one or other task differ from each other, which can be seen in Figure 6.

According to Fuchs [33], the second-law efficiency $\eta_{\text{II,gen}}$ is related to the first-law efficiency $\eta_{\text{I,gen}}$ by Carnot's efficiency η_{C}.

$$
\eta_{\text{I,gen}} = \eta_{\text{C}} \cdot \eta_{\text{II,gen}} = \frac{T_{\text{hot}} - T_{\text{cold}}}{T_{\text{hot}}} \cdot \eta_{\text{II,gen}}
\tag{20}
$$

Carnot's efficiency η_{C} places a theoretical limit for the case in which the second-law efficiency $\eta_{\text{II,gen}} = 1$, which refers to the unrealistic case of vanishing dissipation. Nevertheless, the second-law efficiency $\eta_{\text{II,gen}}$ is the only material-dependent factor and has been used by Altenkirch [55] and Ioffe [56] in order to estimate the performance of thermoelectric materials by treating thermogenerators. It is worth noting that the entropy-based approach presented here allows for power conversion and its efficiency for a single thermoelectric material apart from a device to be discussed.

2.5.5. Working Points for Maximum Conversion Efficiency and Maximum Electrical Power

From the maximum of Equation (19), the maximum conversion efficiency point (MCEP) is obtained with the normalized electrical current $i_{\text{MCEP,gen}}$ being, as follows (cf. Appendix B.2):

$$
i_{\text{MCEP,gen}} \quad = \frac{1}{\sqrt{1 + zT} + 1}
\tag{21}
$$

At the MCEP, the maximum power conversion efficiency of the thermoelectric material in generator mode is then obtained, as follows (cf. Appendix B.2):

$$
\eta_{\text{II,gen,max}} \quad = \eta_{\text{II,gen}} \left(i_{\text{MCEP,gen}} \right) \quad = \frac{\sqrt{1 + zT} - 1}{\sqrt{1 + zT} + 1}
\tag{22}
$$

Equation (23), which shows the variation of the MCEP with varying $i_{\text{MCEP,gen}}$ due to varying zT, is plotted in Figure 6 as dotted blue line.

$$
\eta_{\text{II,gen,max}} \left(i_{\text{MCEP,gen}} \right) \quad = 1 - 2 \cdot i_{\text{MCEP,gen}}
\tag{23}
$$

Note that with increasing figure-of-merit zT, not only does the MCEP drift apart from the MEPP, but the electrical power output also decreases with respect to the MEPP (see Equation (16)), both of which can be seen in Figure 6 (cf. Appendix B.2).

$$P_{\text{el,MCEP}} = \frac{4 \cdot \sqrt{1+zT}}{\left(\sqrt{1+zT}+1\right)^2} \cdot P_{\text{el,max}} \tag{24}$$

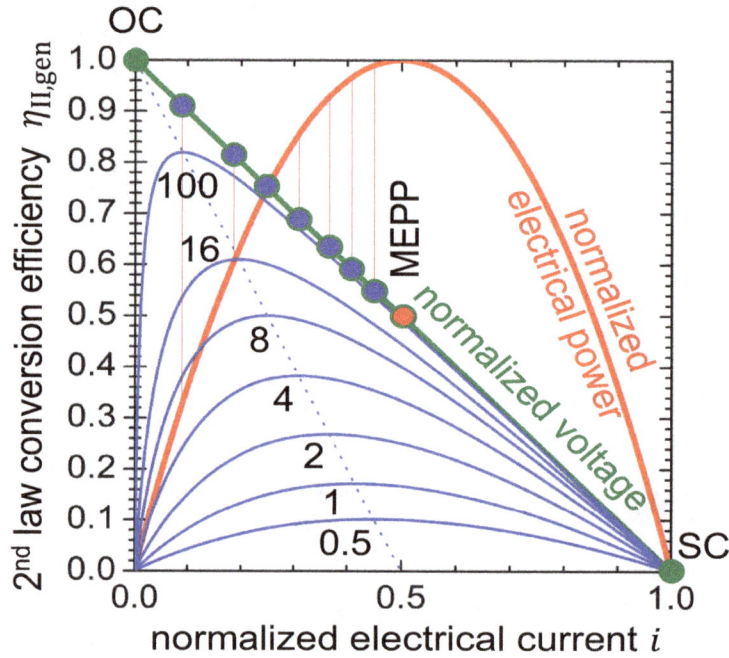

Figure 6. Thermal to electrical power conversion efficiency for some hypothetic materials with figure-of-merit zT varying from 0.5 to 100. Respective working points MCEP (blue) are indicated on the voltage–electrical current curve as well as the MEPP (red). Vertical lines indicate the electrical power output at the MCEP for the example materials. Note that the MCEP drifts apart from the MEPP with increasing figure-of-merit zT. The dashed line indicates the dependence of the MCEP with varying zT.

Obviously, with increasing figure-of-merit zT, the electrical power at the MCEP converges to zero. Figure 7 shows that a notable difference in electrical power output between MCEP and MEPP can be expected for thermoelectric materials with $zT > 0.3$ only (red curves). A notable difference in the power conversion efficiency of the thermoelectric material being operated in the MCEP or the MEPP can only be expected when $zT > 2$. This is also obvious from Table 2, which, for some hypothetical values of the material's figure-of-merit zT, gives values of the second-law power conversion efficiency at the working points under discussion. The 2nd law power conversion efficiency at the MEPP is obtained as follows (cf. Appendix B.1).

$$\eta_{\text{II,gen,MEPP}} = \eta_{\text{II,gen}}\left(i_{\text{MEPP}} = 0.5\right) = \tfrac{1}{2} \cdot \tfrac{zT}{zT+2} \tag{25}$$

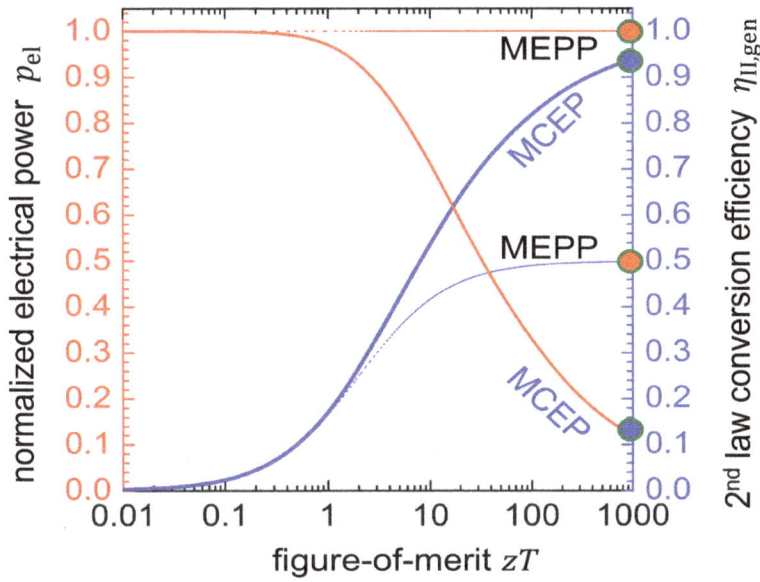

Figure 7. Electrical power output (red lines) and thermal-to-electrical power conversion efficiency (blue lines) for some hypothetic materials with figure-of-merit zT varying from 0.01 to 1000 when operated in two distinct working points, respectively. Solid lines refer to the MCEP and dashed lines refer to the MEPP.

It is worth noting that, for a thermoelectric material with $zT < 2$, there is no benefit from operating it apart from the MEPP.

Table 2. Second-law power conversion efficiency of a thermoelectric material at the MCEP in either entropy pump mode or generator mode and at the MEPP in generator mode for some hypothetical values of the figure-of-merit zT.

zT	Maximum 2nd Law Efficiency	2nd Law Efficiency at MEPP
0.1	0.02	0.02
0.5	0.1	0.1
1	0.17	0.17
1.5	0.23	0.21
2	0.27	0.25
2.5	0.30	0.28
3	0.33	0.3
3.5	0.36	0.32
4	0.38	0.33
8	0.5	0.4
16	0.61	0.44
32	0.70	0.47
100	0.82	0.49

2.6. Thermoelectric Material in Entropy Pump Mode

2.6.1. Power Conversion Efficiency (Electrical to Thermal)

Traditional approaches consider a coefficient of performance when addressing the performance of a thermoelectric cooling or heating device [56,57]. Analogously, a coefficient of performance COP of the thermoelectric material, when used in a cooler, can be considered. It is the thermal power removed from the cold side $T_{cold} \cdot I_S$ related to the electrical power (cf. Appendix C.1).

$$COP_{\text{cooler}} \quad =\left| \frac{T_{\text{cold}} \cdot I_S}{P_{\text{el}}} \right| \quad = \frac{T_{\text{cold}}}{\Delta T} \cdot \left| \frac{P_{\text{th}}}{P_{\text{el}}} \right|$$

$$= \frac{T_{\text{cold}}}{\Delta T} \cdot \eta_{\text{II,ep}} \tag{26}$$

If instead of a cooler, the thermoelectric material is used in a heater (see Fuchs [32], p. 135ff), the thermal power released to the hot side $T_{\text{hot}} \cdot I_S$ becomes the reference parameter, and the COP is then (cf. Appendix C.1):

$$COP_{\text{heater}} \quad =\left| \frac{T_{\text{hot}} \cdot I_S}{P_{\text{el}}} \right| \quad = \frac{T_{\text{hot}}}{\Delta T} \cdot \left| \frac{P_{\text{th}}}{P_{\text{el}}} \right|$$

$$= \frac{T_{\text{hot}}}{\Delta T} \cdot \eta_{\text{II,ep}} \tag{27}$$

$$= \frac{1}{\eta_C} \cdot \eta_{\text{II,ep}}$$

In both cases, Equations (26) and (27), the COP can be factorized into a temperature factor and the second-law efficiency for the thermoelectric material in entropy pump mode $\eta_{\text{II,ep}}$ (see Fuchs [32], p. 135ff). When the thermoelectric material is used in a heater (Equation (27)), the temperature factor is the inverse of Carnot's efficiency η_C [32]. The second-law efficiency for the thermoelectric material in entropy pump mode $\eta_{\text{II,ep}}$ relates the thermal power P_{th} that is needed to pump a certain entropy current from the cold side to the hot side to the electrical power P_{el} (cf. Appendix C.1).

$$\eta_{\text{II,ep}} \quad =\left| \frac{P_{\text{th}}}{P_{\text{el}}} \right| \quad = \frac{i + \frac{1}{zT}}{-i^2 + i} \tag{28}$$

The second-law efficiency for the thermoelectric material in entropy pump mode $\eta_{\text{II,ep}}$ only depends on the normalized electrical current i (i.e., working point on the voltage–electrical current curve) and the material's figure-of-merit zT. It can be used to assess the performance of the thermoelectric material when it is used to pump entropy, regardless of whether the purpose is cooling or heating.

Note that the second-law efficiency for the thermoelectric material in entropy pump mode $\eta_{\text{II,ep}}$ (Equation (28)) is the inverse of the second-law efficiency for the thermoelectric material in generator mode (Equation (19)). Because a net entropy current from the cold side to the hot side will only be obtained for negative entropy conductivity (see Equation (14) and Figure 3), here $\eta_{\text{II,ep}}$ will make sense only for the normalized electrical current being $i \le \frac{1}{zT}$. For this parameter range it is plotted in Figure 8 for some hypothetic thermoelectric materials with figure-of-merit zT between 0.5 and 100.

The maximum 2nd-law power conversion efficiency for a thermoelectric material operated in entropy pump mode is dependent on the material's figure-of-merit zT (cf. Appendix C.2):

$$\eta_{\text{II,ep,max}} \quad = \frac{\sqrt{1+zT}-1}{\sqrt{1+zT}+1} \tag{29}$$

It is obtained at a normalized electrical current $i_{\text{MCEP,ep}}$, which corresponds to the thermoelectric material's maximum conversion efficiency point (MCEP) in entropy pump mode (cf. Appendix C.2). Respective working points for some hypothetic thermoelectric materials are indicated on the voltage–electrical current curve presented in Figure 8.

$$i_{\text{MCEP,ep}} \quad = -\frac{1}{\sqrt{1+zT}-1} \tag{30}$$

The dependence of the maximum second-law efficiency on the electrical current is shown in Figure 8 as a hyperbolic line (cf. Appendix C.2).

$$\eta_{\text{II,ep,max}} \left(i_{\text{MCEP,ep}} \right) \quad = \frac{1}{1 - 2 \cdot i_{\text{MCEP,ep}}} \tag{31}$$

Obviously, an ideal thermoelectric material would have an infinite zT, but the MCEP converges then to the OC working point at vanishing electrical current and, thus, zero electrical power. On the contrary, for the limit of vanishing zT, the maximum second-law efficiency converges to zero at infinite magnitude of the electrical current.

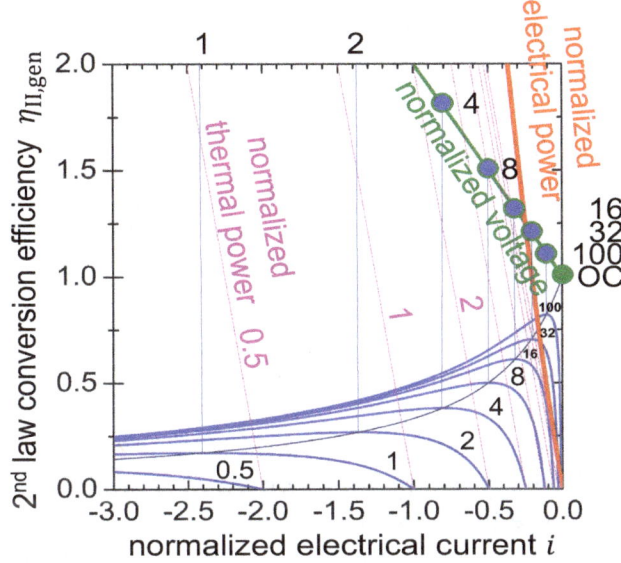

Figure 8. Electrical-to-thermal power conversion efficiency as a function of the reduced electrical current for some hypothetic materials with figure-of-merit zT varying from 0.5 to 100. Respective working points MCEP (blue) are indicated on the voltage–electrical current curve for $zT = 100, 32, 18, 8$ and 4. Further vertical lines (blue) indicate the MCEP for $zT = 2, 1$. The MCEP for $zT = 0.5$ is out of display. The hyperbolic curve indicates the dependence of the MCEP with varying zT. The red curve indicates electrical power–electrical current characteristics. The set of inclined parallel lines (magenta) indicate the thermal power–electrical current characteristics for the respective zT. All of the power curves are normalized to the MEPP in generator mode.

2.6.2. Electrical and Thermal Power

All of the power curves in Figure 8, for the thermoelectric material in entropy pump mode, are normalized to the MEPP in generator mode (see Figures 2 and 4) when the material is exposed to the same temperature difference ΔT. According to Equations (16) and (18), the normalized thermal power p_{th} in Figure 8 is given by a straight line that intersects the horizontal axis at $-\frac{1}{zT}$ and it has a slope of -4 (cf. Appendix C.3).

$$p_{th} \quad = \frac{|P_{th}|}{P_{el,max}} \quad = 4 \cdot \left| \frac{1}{zT} + i \right| \tag{32}$$

For different values of the figure-of-merit zT, a set of inclined parallel lines results. Only the lines for $zT = 0.5, 1$ and 2 are labelled in Figure 8. With increasing figure-of-merit zT, the normalized thermal power curve approaches the normalized electrical power curve, which is in accordance with the increasing power conversion efficiency. However, when the thermoelectric material is operated in its MCEP, the thermal power will decrease with increasing figure-of-merit zT, which becomes obvious when Equation (30) is combined with Equation (32) (cf. Appendix C.2).

$$p_{th,MCEP} \quad = p_{th}\left(i_{MCEP,ep}\right) \quad = 4 \cdot \frac{\sqrt{1+zT}}{zT} \tag{33}$$

The normalized thermal power at MCEP would be steeply curved in Figure 8, with the data point out of scale for $zT < 8$, but has been skipped for clarity. Instead, relevant values for the MCEP are listed in Table 3, together with the normalized electrical power and the normalized electrical current.

Table 3. Values of normalized electrical current $i_{\text{MCEP,ep}}$, normalized thermal power $p_{\text{th,MCEP}}$, and normalized electrical power $p_{\text{el,MCEP}}$ at the MCEP in entropy pump mode for some hypothetical values of the figure-of-merit zT. Values of the second law power conversion efficiency can be read from Table 2

zT	$i_{\text{MCEP,ep}}$	$p_{\text{th,MCEP}}$	$p_{\text{el,MCEP}}$
0.1	−20.49	41.95	1761.32
0.5	−4.45	9.80	97.01
1	−2.41	5.66	32.87
1.5	−1.72	4.22	19.67
2	−1.36	3.46	12.83
2.5	−1.48	2.99	10.77
3.0	−1	2.68	8.93
3.5	−0.89	2.42	7.56
4	−0.80	2.2	5.76
8	−0.50	1.5	3.00
16	−0.32	1.03	1.69
32	−0.21	0.71	1.02
100	−0.11	0.40	0.49

2.7. Complete Picture

With the approach chosen here, working points on the voltage–electrical current curve relate the power conversion properties of the thermoelectric material in generator mode and entropy pump mode to each other. Figure 9 illustrates the concise result for a hypothetical thermoelectric material with figure-of-merit $zT = 3.5$.

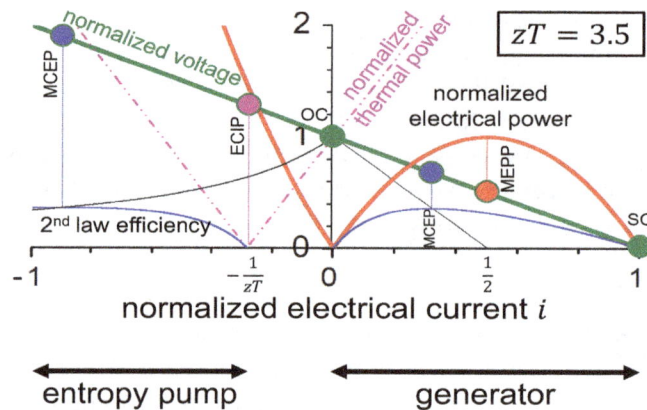

Figure 9. Related characteristics of a hypothetic thermoelectric material with figure-of-merit $zT = 3.5$ in entropy pump mode and generator mode: normalized voltage, normalized electrical power, normalized thermal power, and 2nd-law conversion efficiency as a function of the normalized electrical current. Different working points are indicated on the voltage–electrical current curve. Note that, for current state-of-the-art materials, the MCEP in entropy pump mode would be out of display (see Table 3).

For a given figure-of-merit zT, according to Equations (22) and (29), the values of the maximum 2nd-law conversion efficiency for both modes are identical. Some values are given in Table 2. In addition, values of the 2nd-law conversion efficiency at the MEPP in generator mode are given (see Equation (25)). Remember, the obtained power requires consideration of the absolute value of the electrical power, as determined by the power factor (see Equation (16)).

3. Materials and Methods

Detailed calculations, as given in Appendixs B and C, were made using pencil and paper. The manuscript was prepared using Latex in MikTex distribution. Figures were drawn with the aid of Microcal's Origin and Microsoft's PowerPoint.

4. Discussion

4.1. Remarks on the Use of Working Points

Traditionally, a thermoelectric device is considered and, in generator mode, the operational conditions are set by an external load resistance. The approach of this work, which uses working points on the material's voltage–electrical voltage curve, gives consistent results, which is explicitly shown in Appendix B.3. However, consideration of working points comes with the advantage that the contribution of individual thermoelectric materials in a device can be easily understood [58]. Moreover, the material's voltage–electrical voltage curve directly relates generator mode and entropy pump mode.

4.2. Remarks on the Altenkirch-Ioffe Model

Due to the prominence of the Altenkirch-Ioffe model [55,56], it is worth comparing it to the model, which has been introduced in this work. A comparison of important quantities described by the model of this work and the Altenkirch-Ioffe model is shown in Figure 10.

Remember, Equation (4) has been derived for a thermoelectric material apart from a device. Furthermore, a constant temperature gradient has been assumed, which means a constant slope of the temperature profile, which then connects the hot side at T_{hot} and the cold side at T_{cold} by a straight line (solid line in Figure 10a). The further assumption of a temperature-independent entropy conductivity Λ_{OC} at electrical open-circuit is plotted in Figure 10b as a solid line. As a consequence of these assumptions, at a given electrical current (including electrically open-circuited conditions), the entropy current will carry the highest energy current at the hot side of the thermoelectric material. When advancing through the thermoelectric material to lower temperatures, the entropy current cannot further carry all thermal energy ("heat"), which then needs to be dissipated. Following Walstrom's approach [59], thermal energy is assumed to be dissipated transversally together with instantaneously produced excess entropy as its carrier. It is important to emphasize that excess entropy leaves the thermoelectric material in directions transversal to the flow of the entropy inserted at the hot side. The ability to conduct thermal energy is decreased with decreasing temperature, which is reflected in a decreasing "heat" conductivity, as plotted in Figure 10c as a solid line.

Traced back to Altenkirch [55] and Ioffe [56], often a model is discussed that considers a two-leg thermogenerator and assumes constant "heat" conductivity. Concerning the thermoelectric material, the model is purely one-dimensional and does not allow for transversal dissipation of entropy and energy. All dissipation has to be considered parallel or antiparallel to the flow of entropy and thermal energy along the thermoelectric material. In fact, only the parallel option (i.e., down the temperature gradient) remains physically meaningful. Under electrically open-circuited conditions (i.e., vanishing electrical current), the temperature profile can still be linear. However, Heikes and Ure [60] have shown that, in the presence of a thermally-induced electrical current, the temperature profile is flattened at the hot side and steeply sloping at the cold side, which is shown in Figure 10a as a dashed line. As a consequence of the curved temperature profile and the constant "heat" conductivity (see dashed line in Figure 10c), the "heat" flux is diminished at the hot side (thermal energy input) and increased at the cold side (thermal energy output). The change in the temperature profile is such that, as compared to the zero electrical current situation, the thermal energy input is diminished by half of the Joule "heat" and the thermal energy release at the cold side is increased by half of the Joule "heat", as shown by Heikes and Ure [60,61]. This is to account for the dissipation of thermal energy being parallel to the flow of entropy and thermal energy. As a consequence, when compared to electrically open-circuited conditions, the thermoelectric material would be thermally less transparent when an electrical current flows.

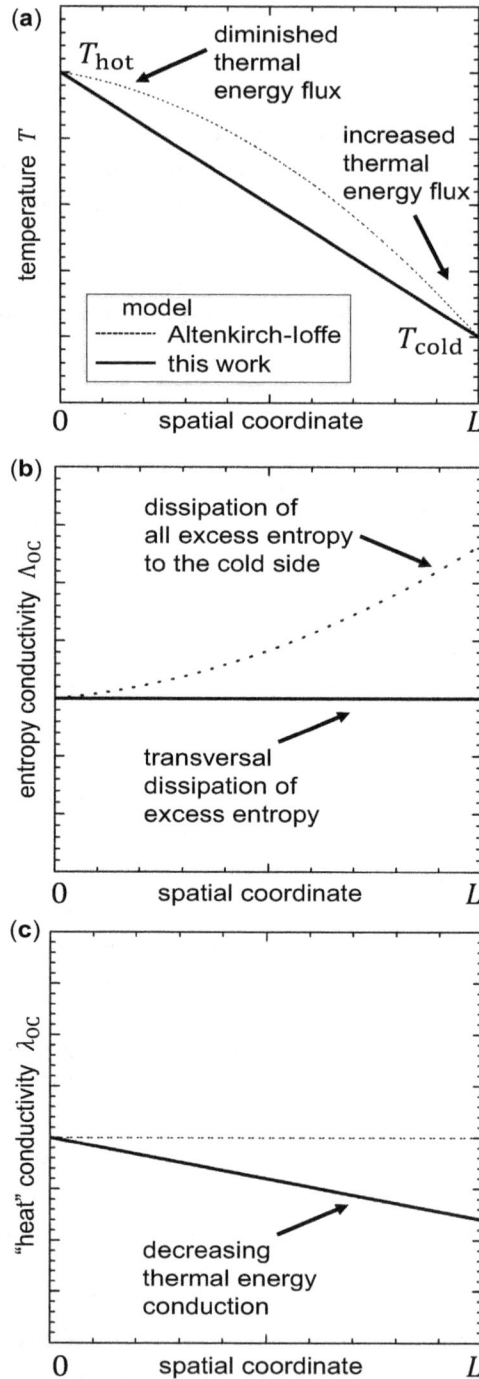

Figure 10. Comparison of the model of this work (constant entropy conductivity) to the Altenkirch-Ioffe model [33,55,56,60] (constant "heat" conductivity) with the schematic profiles of the following quantities over the thermoelectric material when the material is carrying a (thermally induced) electrical current: (a) temperature T; (b) electrically open-circuited entropy conductivity Λ_{OC}; and, (c) electrically open-circuited "heat" conductivity λ_{OC}. Note that profiles are not drawn to scale.

In contrast, the model of this work predicts the thermoelectric material to become thermally more transparent with increasing electrical current, which is reflected in the then reversible increased entropy conductivity $\Lambda(i)$ (see Equations (12) and (14)). In the author's opinion, this is an important characteristic of thermoelectric materials, which is fully embezzled in the traditional model.

In the Altenkirch-Ioffe model, all the excess entropy and excess thermal energy are dissipated to the cold side, which is reflected in an irreversible increase of the entropy conductivity along the thermoelectric material, as visualized in Figure 10b. The aforementioned assumption introduces a ratio of T_{hot}/T_{cold} into the formula for the 2nd-law efficiency at the MCEP (see Appendixs B.4 and B.5 for a device in generator mode; see Appendixs C.4 and C.5 for a device in entropy pump mode). Ioffe [56] has shown that the deviation from Equation (22) (generator mode) or Equation (29) (entropy pump mode), however, is only a few per cent when the efficiency itself is small. In other words, for a small temperature difference ΔT, both of the models give nearly the same results.

It must be emphasized that both of the models rely on very special assumptions and, thus, cannot claim general validity [62]. In this sense, all of the results have to be considered semi-quantitatively when it comes to real thermoelectric materials and devices. More general considerations, as provided by Equation (1), need to consider the local variation of thermoelectric parameters but are beyond the scope of this work. Heikes and Ure [60] and Gryasnov et al. [63] have considered the local variation of thermoelectric parameters to some extent. However, the advantage of the model of this work is not only to consider the thermoelectric material apart from a device, but also to clearly separate the dissipation of entropy and thermal energy from the reversible thermoelectric coupling. The simplicity of thermoelectrics is manifested.

4.3. Remarks on Narducci's Model

Narducci has put the question "Do we really need high thermoelectric figures of merit?" and found in his calculations that, when considering constant ΔT, the electrical power output of a two-leg thermogenerator device at the MEPP increases with increased thermal conductivity (see Narducci [64], Figure 2). The situation that is discussed by Narducci corresponds to a decreasing figure-of-merit (i.e., $zT \rightarrow 0$ limit) with the electrical power converging to what we have obtained here as $P_{el, max}$ (see Equation (16)). In light of this work, it becomes obvious that the MCEP and the MEPP of the thermoelectric material(s) then merge (see Figures 6 and 7).

4.4. Remarks on Λ_{OC}

In the model applied in this work, the electrically open-circuited entropy conductivity Λ_{OC} originates only from non-charge transporting excitations of the solid (mostly phonons). Here, the contribution from electrons to the entropy conductivity solely originates from the power factor (see Figures 3 and 5). Subsequently, distinguishing contributions from electrons and phonons to the thermal conductivity is straightforward (see Ioffe [56], p. 44) and has been coined the "phonon glass–electron crystal" (PGEC) concept by Slack [65]. In this case, Λ_{OC} is identical to the phonon contribution to the entropy conductivity.

However, as mentioned by Ioffe (see Ioffe [56], p. 46), in materials with charge carriers of both signs (electrons and holes from multiple bands), the situation is more intricate. Subsequently, important electronic contribution to the thermal conductivity can be expected for vanishing net flux of charge. In other words, the electrically open-circuited entropy conductivity Λ_{OC} has contributions from both phonons and electrons. The application of the empirical Wiedemann-Franz law to describe the relationship between thermal and electrical conductivity is questionable for these materials [48,56]. In practice, this is the case for many semiconductors and metals. To improve the thermoelectric properties of these materials, it is not sufficient to reduce the phonon contribution by the PGEC concept. In addition, electronic band engineering is required in order to diminish the electron contribution to Λ_{OC}. The theory in this work can be easily extended to treat this case by introducing a second type of charge carrier into Equations (1) and (4).

4.5. Remarks on Figure-of-Merit zT

In this work, the figure-of-merit has been introduced in context with the entropy conductivity (cf. Equation (14)) to underline that it is the dimensionless ratio of two entropy conductivities. Initially, the thermoelectric figure-of-merit was introduced by Ioffe [56] as a parameter $z = \frac{\sigma \cdot \alpha^2}{\lambda_{OC}}$ in the treatment of a thermogenerator referring to the "heat" conductivity. In subsequent treatment, Ioffe has taken into account the medium temperature $\overline{T}$ of the device and elucidated the thermoelectric material's figure-of-merit to be $z\overline{T} = \frac{\sigma \cdot \alpha^2}{\lambda_{OC}} \cdot \overline{T}$, which has subsequently been widely used as zT. With this formulation of the figure-of-merit, researchers often have been confused by the intensive variable temperature $\overline{T}$ showing up explicitly besides material parameters [66]. It is seen as a persistent residual outcome of the historical dispute between Ostwald and Boltzmann (see Section 1.2) that it has not been realized that the use of entropy conductivity Λ instead of the "heat" conductivity λ makes the figure-of merit depend on three material parameters only, which all implicitly depend on temperature (see Equations (3) and (15)).

The author has used zT to be consistent with the conventional nomenclature of the thermoelectric community. All of the formulas in this article, which contain the figure-of-merit, however, would look more straightforward if zT were to be substituted by a single letter, for instance, f as used by Zener [67].

$$f = \frac{\sigma \cdot \alpha^2}{\Lambda_{OC}} = \frac{\sigma \cdot \alpha^2}{\lambda_{OC}} \cdot T = zT \tag{34}$$

4.6. Remarks on State-of-the-Art and Emerging Thermoelectric Materials

It is worth noting that, for a thermoelectric material with $zT < 2$, there is no benefit from operating it apart from the MEPP (see Figure 6, Figure 7 and Table 2). In this context, it is important to perceive that current state-of-the-art materials hardly exceed a zT value of 2. The values listed in Table 4 are peak values. Among the materials of Table 4, PbTe$_{0.7}$S$_{0.3}$-2.5%K has a peak zT of 2.2 at 923 K and a record high average zT of 1.56 in the temperature interval of 300–900 K [68]. Conclusively, the tracking of the MEPP [69], but not of the MCEP, is reported for thermogenerators. However, for the application of emerging thermoelectric materials with further improved figure-of-merit, and thus more distant working points, tracking of the MCEP might become relevant.

Table 4. Maximum figure-of-merit zT_{max} and corresponding power factor $\sigma \cdot \alpha^2$ of some state-of-the-art and emerging thermoelectric materials at temperature T with indication of conduction type.

Material	Type	zT_{max}	$\sigma \cdot \alpha^2$ [μWcm^{-1}K^{-2}]	T [K]	Ref.
(Bi$_{0.25}$Sb$_{0.75}$)$_2$Te$_3$	p	1.05	43	323	[70]
FeNb$_{0.8}$Ti$_{0.2}$Sb	p	1.10	53	973	[48,71]
Hf$_{0.6}$Zr$_{0.4}$Hf$_{0.25}$NiSn$_{0.995}$Sb$_{0.005}$	n	1.20	47	900	[48,72]
Bi$_2$(Te$_{0.94}$Se$_{0.06}$)$_3$ (0.017 wt.% Te, 0.068 wt.% I)	n	1.25	57	298	[73]
(Bi$_{0.25}$Sb$_{0.75}$)$_2$Te$_3$ (8wt.% Te)	p	1.27	58	298	[73]
nano (Bi$_{0.25}$Sb$_{0.75}$)$_2$Te$_3$	p	1.4	38	373	[70]
ZrCoBi$_{0.65}$Sb$_{0.15}$Sn$_{0.20}$	p	1.42	38	973	[48,74]
FeNb$_{0.88}$Hf$_{0.12}$Sb	p	1.45	51	1200	[48,75]
Bi$_{0.88}$Ca$_{0.06}$Pb$_{0.06}$CuSeO	p	1.5	8	873	[48,76]
β-Cu$_{2-x}$Se	p	1.5	12	1000	[77]
Ti$_{0.5}$Zr$_{0.25}$Hf$_{0.25}$NiSn$_{0.998}$Sb$_{0.002}$Se	n	1.5	62	700	[48,78]
Mg$_3$Sb$_{1.48}$Bi$_{0.4}$Te$_{0.04}$	n	1.65	13	725	[79]
Ba$_{0.08}$La$_{0.05}$Yb$_{0.04}$Co$_4$Sb$_{12}$	n	1.7	51	850	[80]
Mg$_{3.175}$Mn$_{0.025}$Sb$_{1.5}$Bi$_{0.49}$Te$_{0.01}$	n	1.71	20	700	[48,81]
B-doped Si$_{80}$Ge$_{20}$ + YSi$_2$	p	1.81	39	1073	[48,82]
Cu$_{2-y}$S$_{1/3}$Se$_{1/3}$Te$_{1/3}$	p	1.9	8	1000	[83]
AgPb$_m$SbTe$_{2+m}$	n	2.2	11	800	[84]
PbTe$_{0.7}$S$_{0.3}$-2.5%K	p	2.2	14	923	[68]
PbTe-4%SrTe-2%Na	p	2.2	24	915	[85]
Ge$_{0.89}$Sb$_{0.1}$In$_{0.01}$Te	p	2.3	37	650	[86]
PbTe-8%SrTe	p	2.5	30	923	[87]
SnSe single crystal's b-axis	p	2.6	10	923	[88]
β-Cu$_2$Se/CuInSe$_2$ (1% In)	p	2.6	12.5	850	[89]
SnSe$_{0.97}$Br$_{0.03}$ single crystal's a-axis	n	2.8	9	773	[90]

The benefit of an increased figure-of-merit zT will be an increased power conversion efficiency at the MEPP anyway. Figure 6, Figure 7, and Table 2 indicate that the material's second-law power conversion efficiency at the MEPP will not exceed the value of 0.5 (see also Equation (25)). Interestingly, this value corresponds to the lower limit of the Curzon-Ahlborn efficiency of a Carnot engine operated at its MEPP [91,92]. At the MEPP, a real thermoelectric material will always be operated at less than half of the Carnot efficiency.

4.7. Remarks on the Importance of the Power Factor and Choice of Materials for Thermogenerators

Because normalized curves are discussed in this work, one might lose sight of the fact that the power factor $\sigma \cdot \alpha^2$ is at least as important as the figure-of-merit zT. According to Equation (16), it rules over the maximum achievable absolute electrical power when the thermoelectric material is operated in generator mode at MCEP. For a material with high zT (e.g., 100), the electrical power is much lower at the MCEP compared to the MEPP (Figures 6 and 7). This is because, at the low electrical current of the MCEP, the thermoelectric material is less permeable to entropy when compared to the MEPP (see Figure 5). Thus, less thermal power is available to be thermoelectrically converted into electrical power. The amount of useful thermal power depends on the power factor and the electrical current (see the second summand in Equation (18)).

The open-circuited entropy conductivity Λ_{OC} causes a thermoelectrically-inactive bypass, which eventually leads the temperature difference ΔT, which squared determines the maximum electrical power in Equation (16), to drop. To provide large ΔT, the open-circuited entropy conductivity Λ_{OC} should be kept small. Here, in addition to a high power factor $\sigma \dot{\alpha}^2$, the figure-of-merit zT comes into play, which relates the aforementioned contributions to the entropy conductivity (see Equation (12) and Equation (15)). The materials that are listed in Table 4 represent those with the highest values of the figure-of-merit reported thus far. In the author's opinion, the most interesting materials are those that also have a high power factor of at least 30 μWcm^{-1}K^{-2}.

A high electrical conductivity σ is also advantageous, as already mentioned in Section 2.3. The choice of materials can easily be made with the help of type-1 Ioffe plots [56] ($\sigma\alpha^2 - \sigma$) and type-2 Ioffe plots ($\Lambda_{OC} - \sigma$) [56,93], which have been recently revitalized on the example of current thermoelectric materials [47,48,94]. The reader is referred to Fuchs [32] (p. 135ff) for further details.

4.8. Remarks on the Second-Law Power Conversion Efficiency vs. Coefficient of Performance for Entropy Pumps

While the upper limit of the coefficient of performance will depend on temperature conditions, as involved in the Carnot efficiency η_C (Equation (27)) or the temperature factor $\frac{T_{cold}}{\Delta T}$ (Equation (26)), the upper limit for the second-law efficiency is fixed to unity (i.e., $\eta_{II,ep} \leq 1$). The unity value of the second-law efficiency refers to an ideal material. While the coefficient of performance is related to a floating scale, the second-law efficiency allows for the estimation of how far from ideal a thermoelectric material is. Another advantage is that the second-law efficiency in Equation (28) only depends on the figure-of-merit and the electrical current and, thus, allows for evaluation of the performance of the thermoelectric material apart from specific temperature conditions, as well as independent from use in a cooler or a heater.

Note that, according to Equations (29) and (22), the maximum second-law efficiency of a thermoelectric material is identical in entropy pump mode and generator mode:

$$\eta_{II,ep,max} \quad = \eta_{II,ep}\left(i_{MCEP,ep}\right) \quad = \eta_{II,gen}\left(i_{MCEP,gen}\right) \quad = \eta_{II,gen,max} \tag{35}$$

This is also apparent from Figure 9.

4.9. Remarks on the Choice of Materials for Entropy Pumps

Remember, electrical and thermal power in Figure 8 are normalized to the MEPP in generator mode (see Equations (17) and (32)). Thus, the absolute thermal power in entropy pump mode is determined by the material's power factor $\sigma \cdot \alpha^2$ (see Equation (16)). A low open-circuited entropy conductivity Λ_{OC} is desired to prevent the thermoelectrically inactive fall of entropy along the temperature difference ΔT, which would make it difficult to maintain the ΔT. Thus, in addition to a high power factor $\sigma \cdot \alpha^2$, a high figure-of-merit zT is favourable, which relates the aforementioned quantities (see Equation (15)).

Operating the thermoelectric material in entropy pump mode requires good performance at ambient temperature and below (e.g., for cooling 150–300 K) or above (e.g., for heating 300–400 K). Among the materials listed in Table 4, only bismuth telluride-based materials fulfil all requirements; and, they are the current materials of choice for the mentioned applications and are conclusively found in commercial devices.

According to Figure 8, emerging materials with improved figure-of-merit at a power factor comparable to bismuth telluride-based materials would have the benefit that comparable thermal power could be pumped from the cold to hot side at a lower electrical current and electrical power.

5. Conclusions

Treating entropy and electrical charge as basic quantities allows for a concise description of thermoelectric transport phenomena (entropy, charge, thermal energy, and electrical energy) and it is the key to comprehensibility. The basic transport equation involves classical thermodynamic potentials (temperature and electrical potential) and enables the identification of a thermoelectric material tensor. On the material's voltage–electrical current cure, distinct working points can be identified, which allow for consideration of the power conversion and its efficiency of the thermoelectric material apart from a device. The power depends on the power factor, and the conversion efficiency depends on the figure-of-merit zT. A clear physical meaning is given to the power factor as the part of the entropy conductivity that couples to the electrical current. The thermal conductivity, expressed here as entropy conductivity, depends on the electrical current and becomes negative when the thermoelectric material is operated in entropy pump mode. The dimensionless figure-of-merit zT is the ratio of two entropy conductivities, the one under electrically open-circuited conditions and the one that couples to the electrical current. The performance of the thermoelectric material in generator mode and entropy pump mode are related to each other and they can be considered on a single voltage–electrical current curve apart from a device.

Acknowledgments: The author is grateful to Jürgen Caro for his continuous encouragement and sustainable support. The author is grateful to Mario Wolf and Richard Hinterding for critical reading of the manuscript and their suggestions.

Abbreviations

The following abbreviations are used in this manuscript:

ECIP   Entropy Conductivity Inversion Point
MCEP   Maximum Conversion Efficiency Point (either in generator mode or entropy pump mode)
MEPP   Maximum Electrical Power Point (in generator mode)
OC   (Electrical) Open Circuit
SC   (Electrical) Short Circuit

Symbols

The following symbols are used in this manuscript:

Geometry

A	cross-sectional area of thermoelectric material
L	length of thermoelectric material

Material properties

α	Seebeck coefficient
f	figure-of-merit (as proposed by Zener [67])
λ	"heat" conductivity
λ_{OC}	"heat" conductivity under electrically open-circuited (OC) conditions
Λ	entropy conductivity
Λ_{OC}	entropy conductivity under electrically open-circuited (OC) conditions
Λ_{SC}	entropy conductivity under electrically open-circuited (SC) conditions
$\tilde{\Lambda}$	normalized entropy conductivity
M_{22}	tensor element (of the thermoelectric material tensor)
R	electrical resistance (of thermoelectric material)
σ	isothermal electrical conductivity
z	thermoelectric factor (as introduced by Ioffe [56])
zT	figure-of-merit (as introduced by Ioffe [56])
zT_{max}	maximum figure-of-merit

Thermodynamic potentials

μ	chemical potential
$\tilde{\mu}$	electrochemical potential ($\tilde{\mu} = \mu + q \cdot \varphi$)
$\nabla\tilde{\mu}$	gradient of the electrochemical potential
$\nabla\tilde{\mu}/q$	gradient of the electrochemical potential per electric charge ($\nabla\tilde{\mu}/q = \nabla\mu/q + \nabla\varphi$)
φ	electrical potential
$\nabla\varphi$	gradient of the electrical potential
$\Delta\varphi$	difference of electrical potential (along the thermoelectric material)
$\Delta\varphi_{OC}$	voltage under electrically open-circuited (OC) conditions
T	absolute temperature
T_{cold}	temperature of the thermoelectric material at its cold side
T_{hot}	temperature of the thermoelectric material at its hot side
∇T	gradient of the temperature
ΔT	difference of temperature (along the thermoelectric material)
u	normalized voltage
u_{MEPP}	normalized voltage at the maximum electrical power point (MEPP)

Fluxes

A	cross-sectional area of thermoelectric material
L	length of thermoelectric material
i	normalized electrical current
$i_{MCEP,ep}$	normalized electrical current at the maximum conversion efficiency point (MCEP) in entropy pump mode
$i_{MCEP,gen}$	normalized electrical current at the maximum conversion efficiency point (MCEP) in generator mode
i_{MEPP}	normalized electrical current at the maximum electrical power point (MEPP)
I_q	electrical current
$I_{q,SC}$	electrical current at electrically short-circuited (SC) conditions
I_S	entropy current
$\mathbf{j}_q$	electrical flux density
$\mathbf{j}_S$	entropy flux density
q	electric charge
S	entropy

Performance

COP_{cooler}	coefficient of performance of the thermoelectric material when used in a cooler
COP_{heater}	coefficient of performance of the thermoelectric material when used in a heater
$\eta_{L,gen}$	first-law power conversion efficiency of the thermoelectric material in generator mode
$\eta_{II,gen}$	second-law power conversion efficiency of the thermoelectric material in generator mode
$\eta_{II,gen,max}$	maximum second-law power conversion efficiency of the thermoelectric material in generator mode
$\eta_{II,ep}$	second-law power conversion efficiency of the thermoelectric material in entropy pump mode
$\eta_{II,ep,max}$	maximum second-law power conversion efficiency of the thermoelectric material in entropy pump mode
η_C	Carnot's efficiency
p_{el}	normalized electrical power
P_{el}	electrical power, needed for lifting electrical charge (generator mode) or made available by the fall of electric charge (entropy pump mode); simplified called output (generator mode) or input (entropy pump mode), when the electrical potential on one side of the thermoelectric material is set to zero
$P_{el, max}$	maximum electrical power output of the thermoelectric material in generator mode (at the MEPP)
$P_{el,MCEP}$	electrical power output, of the thermoelectric material in generator mode, at the MCEP
P_{th}	thermal power, made available by the fall of entropy (generator mode) or needed for lifting entropy (entropy pump mode)

Appendix A. Voltage–Electrical Current and Electrical Power–Electrical Current Characteristics: *p*- and *n*-Type Materials

The voltage–electrical current characteristics (green curve) and the electrical power–electrical current characteristics (red curve) of a thermoelectric material with either *p*-type or *n*-type conduction are given in Figure A1.

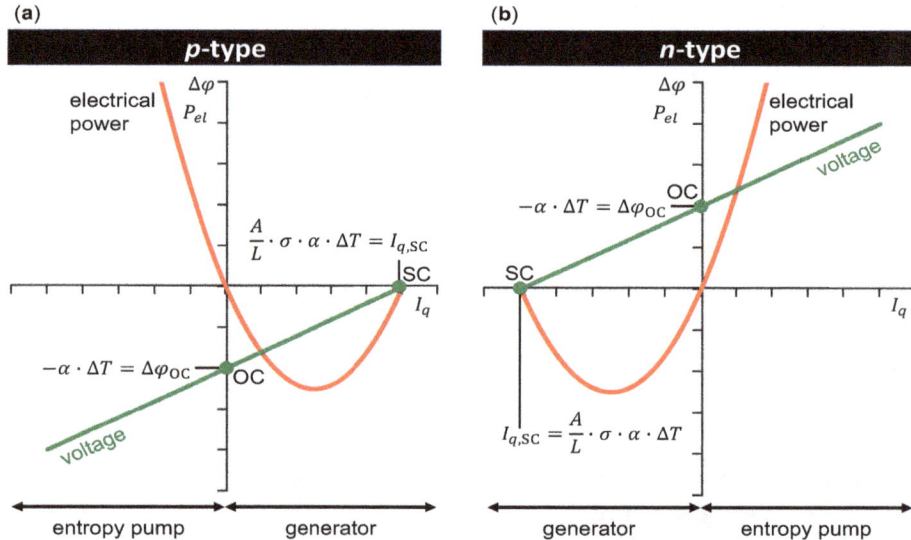

Figure A1. Voltage $\Delta\varphi$ – electrical current I_q characteristics (green curves) and electrical power P_{el} – electrical current characteristics I_q (red curves) for materials with: (**a**) Seebeck coefficient α being positive, which refers to *p*-type conduction and (**b**) Seebeck coefficient α being negative, which refers to *n*-type conduction. Here, $\Delta T = \frac{T_{hot}-T_{cold}}{T_{hot}}$ is the temperature difference along a thermoelectric material of length L and cross-sectional area A. These quantities, together with the (isothermal) electrical conductivity σ and the Seebeck coefficient, determine the electrical current I_{SC} under electrical short-circuited conditions. The voltage $\Delta\varphi_{OC}$ under electrical short-circuited conditions is determined by the Seebeck coefficient and the temperature difference. When the electrical power P_{el} is negative (electrical power output), the material is in generator mode (thermal-to-electrical power conversion). When the electrical power P_{el} is positive (electrical power input), the material is in entropy pump mode (electrical-to-thermal power conversion).

Appendix B. Thermal-to-Electrical Power Conversion: Calculations and Established Models

Appendix B.1. Maximum Electrical Power Point (MEPP): Material in Generator Mode

The MEPP is found by looking for the vanishing first derivative of the electrical power.

$$
\begin{aligned}
0 \quad = \frac{\partial P_{el}}{\partial i} \quad &= -\frac{A}{L} \cdot \sigma \cdot \alpha^2 \cdot (\Delta T)^2 \cdot \frac{\partial}{\partial i}\left(i - i^2\right) \\
&= -\frac{A}{L} \cdot \sigma \cdot \alpha^2 \cdot (\Delta T)^2 \cdot (1 - 2 \cdot i)
\end{aligned}
\tag{A1}
$$

The derivative vanishes if the term in the brackets vanishes, and the normalized current at the MEPP is as follows.

$$
i_{MEPP} = \frac{1}{2}
\tag{A2}
$$

At the MEPP, the maximum electrical power is obtained as follows.

$$
\begin{aligned}
P_{el,max} \quad = P_{el}\left(i_{MEPP}\right) \quad &= -\frac{A}{L} \cdot \sigma \cdot \alpha^2 \cdot (\Delta T)^2 \cdot \left(i_{MEPP} - i_{MEPP}^2\right) \\
&= -\frac{A}{L} \cdot \sigma \cdot \alpha^2 \cdot (\Delta T)^2 \cdot \left(\frac{1}{2} - \left(\frac{1}{2}\right)^2\right) \\
&= -\frac{A}{L} \cdot \sigma \cdot \alpha^2 \cdot (\Delta T)^2 \cdot \left(\frac{1}{2} - \frac{1}{4}\right) \\
&= -\frac{1}{4} \cdot \frac{A}{L} \cdot \sigma \cdot \alpha^2 \cdot (\Delta T)^2
\end{aligned}
\tag{A3}
$$

The 2nd-law power conversion efficiency at the MEPP is then obtained as follows.

$$
\begin{aligned}
\eta_{II,gen,MEPP} \quad = \eta_{II,gen}\left(i_{MEPP}\right) \quad &= \frac{i_{MEPP} - i_{MEPP}^2}{i_{MEPP} + \frac{1}{zT}} \\
&= \frac{\frac{1}{2} - \left(\frac{1}{2}\right)^2}{\frac{1}{2} + \frac{1}{zT}} \\
&= \frac{\frac{1}{2} - \frac{1}{4}}{\frac{1}{2} + \frac{1}{zT}} \\
&= \frac{1}{4} \cdot \frac{1}{\frac{1}{2} + \frac{1}{zT}} \\
&= \frac{zT}{4} \cdot \frac{1}{\frac{zT}{2} + 1} \\
&= \frac{1}{2} \cdot \frac{zT}{zT + 2}
\end{aligned}
\tag{A4}
$$

Appendix B.2. Maximum Conversion Efficiency Point (MCEP): Material in Generator Mode

The 2nd-law power conversion efficiency for a thermoelectric material operated in generator mode is obtained as follows.

$$
\eta_{II,gen} = \left|\frac{P_{el}}{P_{th}}\right| = \frac{\alpha \cdot I_q \cdot \Delta T - \frac{I_q^2}{\frac{A}{L} \cdot \sigma}}{\alpha \cdot I_q \cdot \Delta T + \frac{A}{L} \cdot \Lambda_{OC} \cdot (\Delta T)^2} = \frac{\frac{I_q}{I_{q,SC}} - \left(\frac{I_q}{I_{q,SC}}\right)^2}{\frac{I_q}{I_{q,SC}} + \frac{\Lambda_{OC}}{\sigma \cdot \alpha^2}}
\tag{A5}
$$

Substituting in Equation (A5) the dimensionless normalized electrical current $i = \frac{|I_q|}{|I_{q,SC}|}$ and the figure-of-merit $zT = \frac{\sigma \cdot \alpha^2}{\Lambda_{OC}}$, the 2nd law power conversion efficiency can be written as follows.

$$\eta_{II,gen} = \frac{i - i^2}{i + \frac{1}{zT}} \tag{A6}$$

The maximum power conversion efficiency point (MCEP) can then be found by the first derivative to vanish.

$$\begin{aligned} 0 &= \frac{\partial \eta_{II,gen}}{\partial i} \\[2mm] &= \frac{\partial}{\partial i}\left(\frac{i - i^2}{i + \frac{1}{zT}}\right) \\[2mm] &= \frac{(1 - 2 \cdot i)\cdot\left(i + \frac{1}{zT}\right) - \left(i - i^2\right)\cdot 1}{\left(i + \frac{1}{zT}\right)^2} \\[2mm] &= \frac{i + \frac{1}{zT} - 2\cdot i^2 - 2\cdot\frac{1}{zT}\cdot i - i + i^2}{\left(i + \frac{1}{zT}\right)^2} \\[2mm] &= \frac{-i^2 - 2\cdot\frac{1}{zT}\cdot i + \frac{1}{zT}}{\left(i + \frac{1}{zT}\right)^2} \end{aligned} \tag{A7}$$

The derivative vanishes when the numerator vanishes.

$$i^2 + 2\cdot\frac{1}{zT}\cdot i - \frac{1}{zT} = 0 \tag{A8}$$

This quadratic equation has two solutions, from which only one gives a positive-definite normalized current i at the maximum conversion efficiency point (MCEP).

$$\begin{aligned} i_{MCEP,gen} &= \frac{\sqrt{1 + zT} - 1}{zT} \\[2mm] &= \frac{\left(\sqrt{1 + zT} - 1\right)\cdot\left(\sqrt{1 + zT} + 1\right)}{zT\cdot\left(\sqrt{1 + zT} + 1\right)} \\[2mm] &= \frac{1 + zT + \sqrt{1 + zT} - \sqrt{1 + zT} - 1}{zT\cdot\left(\sqrt{1 + zT} + 1\right)} \\[2mm] &= \frac{zT}{zT\cdot\left(\sqrt{1 + zT} + 1\right)} \\[2mm] &= \frac{1}{\sqrt{1 + zT} + 1} \end{aligned} \tag{A9}$$

At the MCEP, the maximum power conversion efficiency of the thermoelectric material in generator mode is then obtained as follows.

$$
\begin{aligned}
\eta_{\text{II,gen,max}} \quad &= \eta_{\text{II,gen}}\left(i_{\text{MCEP,gen}}\right) &&= \frac{i_{\text{MCEP,gen}}-i_{\text{MCEP,gen}}^2}{i_{\text{MCEP,gen}}+\frac{1}{zT}} \\[2mm]
&= i_{\text{MCEP,gen}} \cdot \frac{1-i_{\text{MCEP,gen}}}{i_{\text{MCEP,gen}}+\frac{1}{zT}} \\[2mm]
&= i_{\text{MCEP,gen}} \cdot \frac{\frac{1}{i_{\text{MCEP,gen}}}-1}{1+\frac{1}{i_{\text{MCEP,gen}}\cdot zT}} \\[2mm]
&= \frac{1}{\sqrt{1+zT}+1} \cdot \frac{\frac{1}{\frac{1}{\sqrt{1+zT}+1}}-1}{1+\frac{1}{\frac{1}{\sqrt{1+zT}+1}\cdot zT}} \\[2mm]
&= \frac{1}{\sqrt{1+zT}+1} \cdot \frac{\sqrt{1+zT}+1-1}{1+\frac{\sqrt{1+zT}+1}{zT}} \\[2mm]
&= \frac{1}{\sqrt{1+zT}+1} \cdot \frac{zT\cdot\sqrt{1+zT}}{zT+\sqrt{1+zT}+1} &&\text{(A10)} \\[2mm]
&= \frac{1}{\sqrt{1+zT}+1} \cdot \frac{zT\cdot\sqrt{1+zT}}{1+zT+\sqrt{1+zT}} \\[2mm]
&= \frac{1}{\sqrt{1+zT}+1} \cdot \frac{zT}{\sqrt{1+zT}+1} \\[2mm]
&= \frac{zT}{\left(\sqrt{1+zT}+1\right)^2} \\[2mm]
&= \frac{1+zT-1}{\left(1+\sqrt{1+zT}\right)^2} \\[2mm]
&= \frac{\left(\sqrt{1+zT}+1\right)\cdot\left(\sqrt{1+zT}-1\right)}{\left(1+\sqrt{1+zT}\right)^2} \\[2mm]
&= \frac{\sqrt{1+zT}-1}{\sqrt{1+zT}+1}
\end{aligned}
$$

By combining Equation (A9) and Equation (A10), the dependence of the maximum second-law efficiency on the electrical current can be shown to be linear.

$$
\begin{aligned}
\eta_{\text{II,gen,max}}\left(i_{\text{MCEP,gen}}\right) \quad &= \frac{\sqrt{1+zT}-1}{\sqrt{1+zT}+1} \\[2mm]
&= \frac{\sqrt{1+zT}+1-2}{\sqrt{1+zT}+1} \\[2mm]
&= \frac{\frac{1}{i_{\text{MCEP,gen}}}-2}{\frac{1}{i_{\text{MCEP,gen}}}} &&\text{(A11)} \\[2mm]
&= 1-2\cdot i_{\text{MCEP,gen}}
\end{aligned}
$$

The electrical power at the MCEP is as follows.

$$
\begin{aligned}
P_{el,MCEP} \quad &= P_{el}\left(i_{MCEP,gen}\right) \quad = -\tfrac{A}{L} \cdot \sigma \cdot \alpha^2 \cdot (\Delta T)^2 \cdot \left(i_{MCEP,gen} - i^2_{MCEP,gen}\right) \\[2em]
&= -\tfrac{A}{L} \cdot \sigma \cdot \alpha^2 \cdot (\Delta T)^2 \cdot \left(\frac{1}{\sqrt{1+zT}+1} - \frac{1}{\left(\sqrt{1+zT}+1\right)^2}\right) \\[2em]
&= -\tfrac{A}{L} \cdot \sigma \cdot \alpha^2 \cdot (\Delta T)^2 \cdot \left(\frac{\sqrt{1+zT}+1}{\left(\sqrt{1+zT}+1\right)^2} - \frac{1}{\left(\sqrt{1+zT}+1\right)^2}\right) \\[2em]
&= -\tfrac{A}{L} \cdot \sigma \cdot \alpha^2 \cdot (\Delta T)^2 \cdot \frac{\sqrt{1+zT}}{\left(\sqrt{1+zT}+1\right)^2} \\[2em]
&= -\tfrac{1}{4} \cdot \tfrac{A}{L} \cdot \sigma \cdot \alpha^2 \cdot (\Delta T)^2 \cdot \frac{4 \cdot \sqrt{1+zT}}{\left(\sqrt{1+zT}+1\right)^2} \\[2em]
&= P_{el,max} \cdot \frac{4 \cdot \sqrt{1+zT}}{\left(\sqrt{1+zT}+1\right)^2}
\end{aligned}
\tag{A12}
$$

Appendix B.3. Comparison to Power Conversion Efficiency after Fuchs: Thermogenerator Device

By accepting temperature and entropy as primitive quantities, Fuchs [33] has created aggregate dynamical models of a Peltier device. Suggesting the Peltier device to function analogously to a battery, he has derived linear voltage-electrical current characteristics and identified the only two dissipative processes, which are the diffusion of electric charge and the diffusion of entropy. For the case of the device being operated as a thermogenerator, Fuchs [33] has derived its 2nd-law efficiency by the ratio of useful to available power and expressed the efficiency with respect to the internal resistance of the device R_{TEG} and an external load resistance R_{ext}.

$$
\eta_{II,TEG} = \frac{R_{ext}}{R_{TEG} + (R_{TEG} + R_{ext}) \cdot \frac{1}{zT}} \cdot \frac{R_{TEG}}{R_{TEG} + R_{ext}}
\tag{A13}
$$

For a given figure-of-merit zT, the 2nd-law efficiency of the device has its maximum at.

$$
R_{ext} = \sqrt{1 + zT} \cdot R_{TEG}
\tag{A14}
$$

Thus, the maximum 2nd-law power conversion efficiency is as follows.

$$
\begin{aligned}
\eta_{\text{II,TEG,max}} &= \frac{\sqrt{1+zT}\cdot R_{\text{TEG}}}{R_{\text{TEG}}+\left(R_{\text{TEG}}+\sqrt{1+zT}\cdot R_{\text{TEG}}\right)\cdot\frac{1}{zT}} \cdot \frac{R_{\text{TEG}}}{R_{\text{TEG}}+\sqrt{1+zT}\cdot R_{\text{TEG}}} \\[2mm]
&= \frac{\sqrt{1+zT}}{1+\left(1+\sqrt{1+zT}\right)\cdot\frac{1}{zT}} \cdot \frac{1}{1+\sqrt{1+zT}} \\[2mm]
&= \frac{\sqrt{1+zT}}{zT+\left(1+\sqrt{1+zT}\right)} \cdot \frac{zT}{1+\sqrt{1+zT}} \\[2mm]
&= \frac{\sqrt{1+zT}}{1+zT+\sqrt{1+zT}} \cdot \frac{zT}{1+\sqrt{1+zT}} \\[2mm]
&= \frac{1}{\sqrt{1+zT}+1} \cdot \frac{zT}{1+\sqrt{1+zT}} \\[2mm]
&= \frac{zT}{\left(1+\sqrt{1+zT}\right)^{2}} \\[2mm]
&= \frac{1+zT-1}{\left(1+\sqrt{1+zT}\right)^{2}} \\[2mm]
&= \frac{\left(\sqrt{1+zT}+1\right)\cdot\left(\sqrt{1+zT}-1\right)}{\left(1+\sqrt{1+zT}\right)^{2}} \\[2mm]
&= \frac{\sqrt{1+zT}-1}{\sqrt{1+zT}+1}
\end{aligned}
\tag{A15}
$$

Of note, Fuchs has neglected the Joule "heat", which would only have a small impact when the device is operated in generator mode. Note that Equation (A15) is equivalent to what has been obtained in this work for a thermoelectric material apart from a device (cf. Equation (A10)).

Appendix B.4. Comparison to Power Conversion Efficiency after Altenkirch: Thermogenerator Device

Altenkirch [55] has estimated the power conversion efficiency for a thermogenerator (called thermopile at that time), which has been assumed to be made of two legs of dissimilar materials. For a small temperature difference along the device, which will cause only a small thermally-induced electrical current and allows neglect the Joule heating as well as the Thomson effect, he has derived his Equation (4) for the 1st-law power conversion efficiency. Altenkirch [55] has factorized the 1st law power conversion efficiency into the Carnot efficiency and what we call here the 2nd-law power conversion efficiency η_{II}. The latter has been of the following form.

$$
\begin{aligned}
\eta_{\text{II,TEG}} &= \frac{zT\cdot\frac{R_{\text{ext}}}{R_{\text{TEG}}}}{zT\cdot\left(1+\frac{R_{\text{ext}}}{R_{\text{TEG}}}\right)+\left(1+\frac{R_{\text{ext}}}{R_{\text{TEG}}}\right)^{2}} \\[2mm]
&= \frac{zT\cdot R_{\text{ext}}\cdot R_{\text{TEG}}}{zT\cdot\left(R_{\text{TEG}}^{2}+R_{\text{TEG}}\cdot R_{\text{ext}}\right)+\left(R_{\text{TEG}}+R_{\text{ext}}\right)^{2}} \\[2mm]
&= \frac{R_{\text{ext}}\cdot R_{\text{TEG}}}{R_{\text{TEG}}\cdot\left(R_{\text{TEG}}+R_{\text{ext}}\right)+\left(R_{\text{TEG}}+R_{\text{ext}}\right)^{2}\cdot\frac{1}{zT}} \\[2mm]
&= \frac{R_{\text{ext}}}{R_{\text{TEG}}+\left(R_{\text{TEG}}+R_{\text{ext}}\right)\cdot\frac{1}{zT}} \cdot \frac{R_{\text{TEG}}}{R_{\text{TEG}}+R_{\text{ext}}}
\end{aligned}
\tag{A16}
$$

Here, Altenkirchs's nomenclature has been substituted by $\frac{R_{\text{ext}}}{R_{\text{TEG}}} = x$ and $zT = 10^{7}\cdot\eta'$. In his treatment, the factor 10^{7} appeared due to the use of the calorie as the energy units, and "η'" was called the effective thermopower of the device, which however contained the Seebeck coefficient multiplied with the square root of the ratio of specific thermal and specific electrical conductivities of the thermoelectric materials involved. Equation (A16) is equivalent to the result observed by Fuchs (cf. Equation (A13)).

Subsequently, Altenkirch derived the efficiency to be maximized for the following.

$$x = \frac{R_{\text{ext}}}{R_{\text{TEG}}} = \sqrt{1 + zT} \tag{A17}$$

Note that Equation (A17) is equivalent to the result obtained by Fuchs (cf. Equation (A14)).

For the thermoelectric generator (TEG), Altenkirch derived the maximum 2nd-law power conversion efficiency $\eta_{\text{II,TEG,max}}$ to be (see Altenkirch [55], Equation (5)) as follows.

$$\eta_{\text{II,TEG,max}} = \frac{\sqrt{1+zT}-1}{\sqrt{1+zT}+1} \tag{A18}$$

Note that Equation (A18) is equivalent to the result obtained by Fuchs (cf. Equation (A15)).

Even though Altenkirch did not use the term figure-of-merit (compare Altenkirch [55], Figure 3), he plotted the maximum 2nd law power conversion efficiency $\eta_{\text{II,TEG,max}}$ as a function of $x = \frac{R_{\text{ext}}}{R_{\text{TEG}}}$ for different values of his "η'", which despite a dimensionless factor has been identified with zT. In the plot, he indicated the shift of the MCEP with varied figure-of-merit.

Altenkirch extended his approach by considering the impact of the Thomson effect on the power conversion efficiency. Moreover, he added remarks on the rate of thermal power exchange of the device with a hot reservoir and cold reservoir and its impact on the effective temperature difference along the device.

Appendix B.5. Comparison to Power Conversion Efficiency after Ioffe: Thermogenerator Device

Ioffe [56] has considered a thermocouple in which legs of materials 1 and 2 of equal length are joined by a metallic bridge. The Seebeck coefficient of the device has been estimated from those of the two legs: $\alpha = |\alpha_1| + |\alpha_2|$. From equal length and the individual values of the electrical resistivities (ρ_1, ρ_2), "heat" conductivities ($\lambda_{\text{OC,1}}, \lambda_{\text{OC,2}}$) and cross-sectional area, he has calculated the total electrical resistance R_{TEG} and thermal conductance of the device K_{TEG} (see Ioffe [56], p. 36). To calculate the efficiency of thermal-to-electrical power conversion of the device, he has neglected the Thomson "heat". Furthermore, he made an assumption regarding the Joule "heat" (see Ioffe [56], p. 38): "Of the total Joule 'heat' $I_q^2 \cdot R_{\text{TEG}}$ generated in the thermoelement, half passes to the hot junction, returning the power $\frac{1}{2} \cdot I_q^2 \cdot R_{\text{TEG}}$ and the rest is transferred to the cold junction." As a result, the temperatures of the hot T_{hot} and cold junction T_{cold} appear in the maximum second-law efficiency.

$$\eta_{\text{II,TEG,max}} = \frac{\sqrt{1+z\overline{T}}-1}{\sqrt{1+z\overline{T}}+\frac{T_{\text{hot}}}{T_{\text{cold}}}} \tag{A19}$$

The aforementioned argument, which was probably inspired by Altenkirch's [55] article, is based on misunderstanding the dissipation, which in the author's opinion is thermal energy to leave the system together with produced entropy. The entropy, and thus the thermal energy, will not have driving force to flow to higher temperature. Anyway, following the argument, the thermal input power is diminished by half of the dissipated Joule "heat". In this work, it has been outlined that the effect of Joule "heat" would be a diminished thermal power supply due to a changed temperature profile (cf. Section 4.2 in the the main text).

Neglecting the Joule "heat", Ioffe has derived the following equation (see Ioffe [56], p. 40).

$$\eta_{\text{II,TEG,max}} = \frac{\sqrt{1+z\overline{T}}-1}{\sqrt{1+z\overline{T}}+1} \tag{A20}$$

Note that this is equivalent to what has been obtained in this work for a thermoelectric material apart from a device.

In the factor z, which Ioffe deduced (see Ioffe [56], p. 39), the cross-sectional areas A_1 and A_2 and length L cancel out, so it depends only on the thermoelectric properties of both materials but not their dimensions.

$$z = \frac{\alpha^2}{K_{\text{TEG}} \cdot R_{\text{TEG}}} = \frac{\alpha^2}{\left(\sqrt{\lambda_{\text{OC},1} \cdot \rho_1} + \sqrt{\lambda_{\text{OC},2} \cdot \rho_2}\right)^2} = \frac{\alpha^2}{\left(\sqrt{\frac{\lambda_{\text{OC},1}}{\sigma_1}} + \sqrt{\frac{\lambda_{\text{OC},2}}{\sigma_2}}\right)^2} \tag{A21}$$

In the case that the electrical conductivities ($\sigma = \sigma_n = \sigma_p$) and "heat" conductivities ($\lambda_{\text{OC}} = \lambda_{\text{OC},1} = \lambda_{\text{OC},2}$) are equal in both legs of the device, respectively, Equation (A21) becomes the following.

$$z = \frac{\sigma \cdot \alpha^2}{\lambda_{\text{OC}}} \tag{A22}$$

Ioffe used Equation (A22) when discussing a thermoelectric cooler (see Ioffe [56], p. 100) but derived an equivalent expression – using the thermal conductance instead of the thermal conductivity – when discussing the thermogenerator (see Ioffe [56], p. 38ff.). Anyway, in Equations (A19) and (A20) for the maximum power conversion efficiency, there appears not the factor z but this factor multiplied with the average temperature $\overline{T}$.

$$z\overline{T} = z \cdot \overline{T} = z \cdot \frac{T_{\text{hot}} + T_{\text{cold}}}{2} \tag{A23}$$

Because of Ioffe's Equations (A21)–(A23), the figure-of-merit of a thermoelectric material is currently termed $z\overline{T}$ or zT.

Appendix C. Electrical-to-Thermal Power Conversion: Calculations and Established Models

Appendix C.1. Power Conversion Efficiency

When the thermoelectric material is used in a cooler, the coefficient of performance COP is the ratio of the thermal power removed from the cold side $T_{\text{cold}} \cdot I_S$ related to the electrical power P_{el}.

$$
\begin{aligned}
COP_{\text{cooler}} \quad &= \left| \frac{T_{\text{cold}} \cdot I_S}{P_{\text{el}}} \right| \quad = \left| \frac{T_{\text{cold}} \cdot I_S}{P_{\text{th}}} \right| \cdot \left| \frac{P_{\text{th}}}{P_{\text{el}}} \right| \quad = \frac{T_{\text{cold}}}{\Delta T} \cdot \left| \frac{P_{\text{th}}}{P_{\text{el}}} \right| \\[2mm]
&= \frac{T_{\text{cold}}}{\Delta T} \cdot \frac{\frac{A}{L} \cdot (\Lambda_{\text{OC}} + \sigma\alpha^2 \cdot i) \cdot (\Delta T)^2}{\frac{A}{L} \cdot \sigma \cdot \alpha^2 \cdot (\Delta T)^2 \cdot (i - i^2)} \\[2mm]
&= \frac{T_{\text{cold}}}{\Delta T} \cdot \frac{\frac{\Lambda_{\text{OC}}}{\sigma\alpha^2} + i}{i - i^2} \\[2mm]
&= \frac{T_{\text{cold}}}{\Delta T} \cdot \frac{i + \frac{1}{zT}}{-i^2 + i} \\[2mm]
&= \frac{T_{\text{cold}}}{\Delta T} \cdot \eta_{\text{II,ep}}
\end{aligned}
\tag{A24}
$$

When the thermoelectric material is used in a heater, the coefficient of performance COP is the ratio of the thermal power released to the hot side $T_{\text{hot}} \cdot I_S$ related to the electrical power P_{el}.

$$
\begin{aligned}
COP_{\text{heater}} \quad &= \left| \frac{T_{\text{hot}} \cdot I_S}{P_{\text{el}}} \right| \quad = \left| \frac{T_{\text{hot}} \cdot I_S}{P_{\text{th}}} \right| \cdot \left| \frac{P_{\text{th}}}{P_{\text{el}}} \right| \quad = \frac{T_{\text{hot}}}{\Delta T} \cdot \left| \frac{P_{\text{th}}}{P_{\text{el}}} \right| \\[2mm]
&= \frac{T_{\text{hot}}}{\Delta T} \cdot \frac{i + \frac{1}{zT}}{-i^2 + i} \\[2mm]
&= \frac{T_{\text{hot}}}{\Delta T} \cdot \eta_{\text{II,ep}} \\[2mm]
&= \frac{\eta_{\text{II,ep}}}{\eta_{\text{C}}}
\end{aligned}
\tag{A25}
$$

The second-law efficiency for the thermoelectric material in entropy pump mode $\eta_{\text{II,ep}}$ is as follows.

$$
\begin{aligned}
\eta_{\text{II,ep}} \quad &= \left| \frac{P_{\text{th}}}{P_{\text{el}}} \right| \quad = \frac{\frac{A}{L} \cdot \left(\Lambda_{\text{OC}} + \sigma\alpha^2 \cdot i \right) \cdot (\Delta T)^2}{\frac{A}{L} \cdot \sigma \cdot \alpha^2 \cdot (\Delta T)^2 \cdot (i - i^2)} \\[2mm]
&= \frac{\frac{\Lambda_{\text{OC}}}{\sigma\alpha^2} + i}{i - i^2} \\[2mm]
&= \frac{i + \frac{1}{zT}}{-i^2 + i}
\end{aligned}
\tag{A26}
$$

The electrical power P_{el} used in Equations (A24)–(A26) is available by the fall of electric charge along the electrical potential difference $\Delta\varphi$. It drives the pumping of entropy from the material's cold side to its hot side. The thermal power $P_{\text{th}} = \Delta T \cdot I_S = T_{\text{hot}} \cdot I_S - T_{\text{cold}} \cdot I_S$ is needed for lifting entropy along the temperature difference ΔT. Some illustration is given in Figure A2.

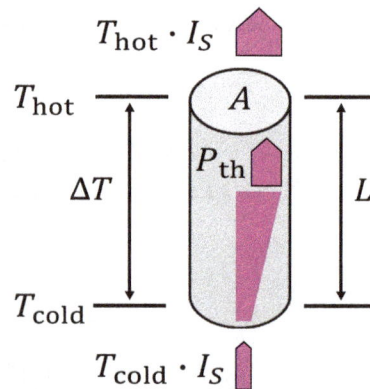

Figure A2. When the thermoelectric material is operated in entropy pump mode, electrical power P_{el}, which is available by the fall of electric charge along $\Delta\varphi$, drives the pumping of entropy from the cold side to hot side. The thermal power $P_{\text{th}} = \Delta T \cdot I_S = T_{\text{hot}} \cdot I_S - T_{\text{cold}} \cdot I_S$ for lifting entropy along the temperature difference ΔT adds to the thermal power removed from the cold side $T_{\text{cold}} \cdot I_S$ to give the thermal power released to the hot side $T_{\text{hot}} \cdot I_S$. Different width of arrows refers to different magnitudes of thermal power at the opposite sides of the material, which is due to thermoelectric power conversion.

Appendix C.2. Maximum Conversion Efficiency Point (MCEP): Material in Entropy Pump Mode

The maximum power conversion efficiency point (MCEP) follows when the first derivative of the 2nd-law power conversion efficiency, as given by Equation (A26), vanishes.

$$
\begin{aligned}
0 &= \frac{\partial \eta_{II,ep}}{\partial i} \\[2mm]
&= \frac{\partial}{\partial i}\left(\frac{i+\frac{1}{zT}}{-i^2+i}\right) \\[2mm]
&= \frac{1\cdot\left(-i^2+i\right)-\left(i+\frac{1}{zT}\right)\cdot\left(-2\cdot i+1\right)}{\left(-i^2+i\right)^2} \\[2mm]
&= \frac{-i^2+i+2\cdot i^2+\frac{2}{zT}\cdot i-i-\frac{1}{zT}}{\left(-i^2+i\right)^2} \\[2mm]
&= \frac{i^2+\frac{2}{zT}\cdot i-\frac{1}{zT}}{\left(-i^2+i\right)^2}
\end{aligned}
\tag{A27}
$$

The derivative vanishes when the numerator vanishes.

$$
i^2 + \tfrac{2}{zT}\cdot i - \tfrac{1}{zT} = 0
\tag{A28}
$$

The quadratic Equation (A28) has two solutions.

$$
\begin{aligned}
i_{1,2} &= -\tfrac{1}{zT} \pm \sqrt{\left(\tfrac{1}{zT}\right)^2 + \tfrac{1}{zT}} \\[2mm]
&= -\tfrac{1}{zT} \pm \tfrac{1}{zT}\cdot\sqrt{1+zT}
\end{aligned}
\tag{A29}
$$

From the two solutions shown in Equation (A29) only one fulfils the requirement $i \leq -\tfrac{1}{zT}$ for the material's maximum conversion efficiency point (MCEP) in entropy pump mode. Thus, the normalized electrical current at the maximum conversion efficiency point (MCEP) is obtained as follows.

$$
\begin{aligned}
i_{MCEP,ep} &= -\tfrac{1}{zT} - \tfrac{1}{zT}\cdot\sqrt{1+zT} \\[2mm]
&= -\frac{1+\sqrt{1+zT}}{zT} \\[2mm]
&= -\frac{\sqrt{1+zT}+1}{zT} \qquad ! \\[2mm]
&= -\frac{\sqrt{1+zT}+1}{1+zT-1} \\[2mm]
&= -\frac{\sqrt{1+zT}+1}{\left(\sqrt{1+zT}+1\right)\cdot\left(\sqrt{1+zT}-1\right)} \\[2mm]
&= -\frac{1}{\sqrt{1+zT}-1}
\end{aligned}
\tag{A30}
$$

The maximum 2nd-law power conversion efficiency for a thermoelectric material operated in entropy pump mode is then as follows.

$$
\eta_{\mathrm{II,ep,max}} = \eta_{\mathrm{II,ep,max}}\left(i_{\mathrm{MCEP,ep}}\right) = \frac{i_{\mathrm{MCEP,ep}} + \frac{1}{zT}}{-i_{\mathrm{MCEP,ep}}^2 + i_{\mathrm{MCEP,ep}}}
$$

$$
= \frac{-\frac{\sqrt{1+zT}+1}{zT} + \frac{1}{zT}}{-\left(\frac{\sqrt{1+zT}+1}{zT}\right)^2 - \frac{\sqrt{1+zT}+1}{zT}}
$$

$$
= \frac{zT}{zT} \cdot \frac{-\sqrt{1+zT}-1+1}{-\frac{1}{zT}\cdot\left(\sqrt{1+zT}+1\right)^2 - \left(\sqrt{1+zT}+1\right)}
$$

$$
= \frac{1}{-\left(\sqrt{1+zT}+1\right)} \cdot \frac{-\sqrt{1+zT}}{\frac{1}{zT}\cdot\left(\sqrt{1+zT}+1\right)+1}
$$

$$
= \frac{zT}{\sqrt{1+zT}+1} \cdot \frac{\sqrt{1+zT}}{\left(\sqrt{1+zT}+1\right)+zT}
$$

$$
= \frac{zT}{\sqrt{1+zT}+1} \cdot \frac{\sqrt{1+zT}}{1+zT+\sqrt{1+zT}} \tag{A31}
$$

$$
= \frac{zT}{\sqrt{1+zT}+1} \cdot \frac{1}{\sqrt{1+zT}+1}
$$

$$
= \frac{1+zT-1}{\sqrt{1+zT}+1} \cdot \frac{1}{\sqrt{1+zT}+1}
$$

$$
= \frac{\left(\sqrt{1+zT}+1\right)\cdot\left(\sqrt{1+zT}-1\right)}{\sqrt{1+zT}+1} \cdot \frac{1}{\sqrt{1+zT}+1}
$$

$$
= \frac{\sqrt{1+zT}-1}{\sqrt{1+zT}+1}
$$

By combining Equations (A30) and (A31), the dependence of the maximum second-law efficiency on the electrical current can be shown to be hyperbolic.

$$
\eta_{\mathrm{II,ep,max}}\left(i_{\mathrm{MCEP,ep}}\right) = \frac{\sqrt{1+zT}-1}{\sqrt{1+zT}+1}
$$

$$
= \frac{\sqrt{1+zT}-1}{\sqrt{1+zT}-1+2}
$$

$$
= \frac{\frac{-1}{i_{\mathrm{MCEP,ep}}}}{\frac{-1}{i_{\mathrm{MCEP,ep}}}+2} \tag{A32}
$$

$$
= \frac{1}{1-2\cdot i_{\mathrm{MCEP,ep}}}
$$

The normalized thermal power (cf. Appendix C.3) at the MCEP is obtained by combining Equations (A35) and (A30).

$$
\begin{aligned}
p_{\text{th,MCEP}} \quad = p_{\text{th}}\left(i_{\text{MCEP,ep}}\right) \quad &= 4 \cdot \left| \frac{1}{zT} - \frac{1}{\sqrt{1+zT}-1} \right| \\
&= \frac{4}{zT} \cdot \left| 1 - \frac{zT}{\sqrt{1+zT}-1} \right| \\
&= \frac{4}{zT} \cdot \left| 1 - \frac{1+zT-1}{\sqrt{1+zT}-1} \right| \\
&= \frac{4}{zT} \cdot \left| 1 - \frac{\left(\sqrt{1+zT}+1\right)\cdot\left(\sqrt{1+zT}-1\right)}{\sqrt{1+zT}-1} \right| \\
&= \frac{4}{zT} \cdot \left| 1 - \left(\sqrt{1+zT}+1\right) \right| \qquad\qquad\text{(A33)} \\
&= \frac{4}{zT} \cdot \left| 1 - \sqrt{1+zT} - 1 \right| \\
&= \frac{4}{zT} \cdot \left| -\sqrt{1+zT} \right| \\
&= 4 \cdot \left| -\frac{\sqrt{1+zT}}{zT} \right| \\
&= 4 \cdot \frac{\sqrt{1+zT}}{zT}
\end{aligned}
$$

The absolute thermal power at the MCEP in entropy pump mode, which is related to the MEPP in generator mode, is thus the following:

$$
\begin{aligned}
\left| P_{\text{th,MCEP}} \right| \quad &= p_{\text{th,MCEP}} \cdot P_{\text{el,max}} \\
&= 4 \cdot \frac{\sqrt{1+zT}}{zT} \cdot P_{\text{el,max}}
\end{aligned}
\qquad\qquad\text{(A34)}
$$

Appendix C.3. Normalized Thermal Power

The normalized thermal power p_{th} is obtained as follows.

$$
\begin{aligned}
p_{\text{th}} \quad &= \frac{|P_{\text{th}}|}{P_{\text{el,max}}} \\
&= \frac{\left| \frac{A}{L} \cdot \left(\Lambda_{\text{OC}} + \sigma\alpha^2 \cdot i\right) \cdot (\Delta T)^2 \right|}{\frac{1}{4} \cdot \frac{A}{L} \cdot \sigma \cdot \alpha^2 \cdot (\Delta T)^2} \\
&= 4 \cdot \left| \frac{\Lambda_{\text{OC}}}{\sigma\alpha^2} + i \right| \\
&= 4 \cdot \left| \frac{1}{zT} + i \right|
\end{aligned}
\qquad\qquad\text{(A35)}
$$

Appendix C.4. Comparison to Power Conversion Efficiency after Altenkirch: Thermoelectric Cooler Device

For a thermoelectric cooler made of two legs of dissimilar thermoelectric materials (called a thermopile) in steady-state condition, Altenkirch [57] has derived an expression for the minimum electrical power input related to a given cooling power (see Altenkirch [57], Equation (12)), which factorizes into a Carnot-type factor $\frac{T_{\text{hot}} - T_{\text{cold}}}{T_{\text{cold}}}$ and the reciprocal of what he called the dissipation factor for the electro-thermal device. It must be emphasized that the Carnot-type factor introduced by Altenkirch is different from Carnot's efficiency because it relates the temperature difference $T_{\text{hot}} - T_{\text{cold}}$

to the temperature of the cold side T_{cold} instead of the hot side T_{hot}. This is due to the thermal energy current removed from the cold side being related to the electrical power input.

When Altenkirch's nomenclature is substituted by $zT = 10^7 \cdot \eta'$, his dissipation factor (see Altenkirch [57], Equation (13)) for the thermoelectric cooler (TEC), which is the device-related analogue of what we here call the maximum 2nd-law power conversion efficiency for a thermoelectric material operated in entropy pump mode $\eta_{\text{II,ep,max}}$, becomes as follows.

$$\eta_{\text{II,TEC,max}} \quad = \quad \frac{\sqrt{1+zT}-\frac{T_{\text{hot}}}{T_{\text{cold}}}}{\sqrt{1+zT}+1} \tag{A36}$$

Altenkirch [57] states that, for small temperature differences (i.e., $\frac{T_{\text{hot}}}{T_{\text{cold}}} \approx 1$), the maximum 2nd-law power conversion efficiency for thermoelectric cooler $\eta_{\text{II,TEC,max}}$ becomes the following.

$$\eta_{\text{II,TEC,max}} \quad = \quad \frac{\sqrt{1+zT}-1}{\sqrt{1+zT}+1} \tag{A37}$$

Altenkirch's result of Equation (A37) for a device is identical to the maximum 2nd-law power conversion efficiency for a thermoelectric material operated in entropy pump mode $\eta_{\text{II,ep,max}}$ as obtained in this work (see Equation (A31)).

Appendix C.5. Comparison to Power Conversion Efficiency after Ioffe: Thermoelectric Cooler Device

For a thermoelectric cooler made of two legs of dissimilar thermoelectric materials, Ioffe [56] (see Ioffe [56], p. 99) has derived a maximum coefficient of performance COP, which he factorized into the inverse of a Carnot-type factor $\frac{T_{\text{cold}}}{T_{\text{hot}}-T_{\text{cold}}}$ and what we here call the maximum 2nd-law efficiency $\eta_{\text{II,ep,max}}$. After Ioffe [56], the device-related analogue of the latter has been as follows.

$$\eta_{\text{II,TEC,max}} \quad = \quad \frac{\sqrt{1+\frac{1}{2}\cdot z\cdot(T_{\text{hot}}+T_{\text{cold}})}-\frac{T_{\text{hot}}}{T_{\text{cold}}}}{\sqrt{1+\frac{1}{2}\cdot z\cdot(T_{\text{hot}}+T_{\text{cold}})}+1} \tag{A38}$$

In the case of small temperature difference (i.e., $\frac{T_{\text{cold}}}{T_{\text{hot}}} \approx 1$) and when identifying the average temperature $T = \frac{1}{2} \cdot (T_{\text{hot}} + T_{\text{cold}})$, it becomes identical to the result of this work for a thermoelectric material (see Equation (A31)).

References and Notes

1. Clausius, R. *Abhandlungen über die mechanische Wärmetheorie*; Friedrich Vieweg und Sohn: Braunschweig, Germany, 1864.
2. Clausius, R. Ueber verschiedene für die Anwendung bequeme Formen der Hauptgleichungen der mechanischen Wärmetheorie. *Poggendorffs Ann. Phys. Chem.* **1865**, *125*, 353–400. [CrossRef]
3. Clausius, R. *Mechanical Theory of Heat*; John van Voorst: London, UK, 1867.
4. Boltzmann, L. Über die Beziehung zwischen dem zweiten Hauptsatze der mechanischen Wärmetheorie und der Wahrscheinlichkeitsrechnung. *Wiener Berichte* **1877**, *76*, 373–435.
5. Boltzmann, L. *Wissenschaftliche Abhandlungen, Band 2*; J. A. Barth: Leipzig, Germany, 1909.
6. Flamm, D. Ludwig Boltzmann and his influence on science. *Stud. Hist. Phil. Sci.* **1983**, *14*, 225–278. [CrossRef]
7. Truesdell, C.A. *The Tragicomical History of Thermodynamics 1822-1854*; Springer: New York, NY, USA, 1980. [CrossRef]
8. Gibbs, J.W. On the equilibrium of heterogeneous substances. *Trans. Conn. Acad.* **1875**, *3*, 108–248. [CrossRef]
9. Gibbs, J.W. *The Collected Works of J. Willard Gibbs, Volume 1, Thermodynamics*; Longmans, Green and Co.: New York, NY, USA, 1928.

10. The monism of the "Energeticist" school should not be confused with "energetics" of the British pioneers in thermodynamics [11].

11. Smith, C. *The Science of Energy: A Cultural History of Energy Physics in Victorian Britain*; The University of Chicago Press: Chicago, IL, USA, 1998.

12. Ostwald, W. Studien zur Energetik: 2. Grundlinien der allgemeinen Energetik. *Berichte über die Verhandlungen der Königlich-Sächsischen Gesellschaft der Wissenschaften zu Leipzig, Mathematisch-Physische Klasse* **1892**, *44*, 211–237.

13. Helm, G. *Die Lehre von der Energie*; Arthur Felix: Leipzig, Germany, 1887.

14. Sommerfeld, A. Ludwig Boltzmann zum Gedächtnis. *Wien.-Chem.-Ztg.* **1944**, *3-4*, 25–28.

15. The aim here is not to discredit Wilhelm Ostwald or Ernst Mach who both have made remarkable contributions to science, but instead to give an understanding of how the actual perception of entropy in the scientific community has developed. Readers who are interested in more background information, including the impact of Mach's natural philosophy on the development of quantum mechanics, are referred to Flamm [6].

16. The late William Thomson [Lord Kelvin] wrote in 1906: "Young persons who have grown up in scientific work within the last fifteen years seem to have forgotten that energy is not an absolute existence. Even the Germans laugh on the 'Energetikers' [11]".

17. Boltzmann, L. Ein Wort der Mathematik an die Energetik. *Ann. Phys.* **1896**, *39*, 39–71. [CrossRef]

18. Boltzmann, L. Zur Energetik. *Ann. Phys.* **1896**, *39*, 595–598. [CrossRef]

19. Müller, I. Max Planck – a life for thermodynamics. *Ann. Phys.* **2008**, *17*, 73–87. [CrossRef]

20. Müller, I. Ein Leben für die Thermodynamik. *Physik Journal* **2008**, *7*, 39–45.

21. Herrmann, F. The Karlsruhe Physics Course. *Eur. J. Phys.* **2000**, *21*, 49–58. [CrossRef]

22. Jorda, S. Kontroverse um Karlsruher Physikkurs [Controverse about the Karlsruhe Physiscs Course]. *Phys. J.* **2013**, *12*, 6–7.

23. Strunk, C. *Moderne Thermodynamik—Band 2: Quantenstatistik aus experimenteller Sicht*, 2 ed.; De Gruyter: Berlin, Germany, 2018.

24. It is worth noting that there was another dispute. Max Planck, together with his scholar Ernst Zermelo, also agitated heavily against Ludwig Boltzmann's atomistic-statistical principle, which has been fought also in the pages of the Annalen der Physik [19,20] Only later did Planck became an aglow follower of Boltzmann, and Zermelo translated Gibb's book on statistical mechanics [19,20]. Planck used Boltzmann's principle to estimate the entropy of the electromagnetic field in his formulation of the spectrum of black body radiation [19,20].

25. Planck, M. *Über den zweiten Hauptsatz der mechanischen Wärmetheorie*; Theodor Ackermann: München, Germany, 1879. [CrossRef]

26. Callen, H. The application of Onsager's reciprocal relations to thermoelectric, thermomagnetic, and galvanomagnetic effects. *Phys. Rev.* **1948**, *489*, 414–418. [CrossRef]

27. Callen, H.B. *Thermodynamics—An Introduction to the Physical Theories of Equilibrium Thermostatics and Irreversible Thermodynamics*; John Wiley and Son: New York, NY, USA, 1960.

28. de Groot, G. *Thermodynamics of Irreversible Processes*, 1 ed.; North-Holland Publishing Company: Amsterdam, The Netherlands, 1951.

29. Callendar, H. The caloric theory of heat and Carnot's principle. *Proc. Phys. Soc. London* **1911**, *23*, 153–189. [CrossRef]

30. Carnot, S. *Réflexions sur la puissance motrice du feu*; Bachelier: Paris, France, 1824.

31. Strunk, C. *Moderne Thermodynamik – Band 1: Physikalische Systeme und ihre Beschreibung*, 2 ed.; De Gruyter: Berlin, Germany, 2018.

32. Fuchs, H.U. *The Dynamics of Heat—A Unified Approach to Thermodynamics and Heat Transfer*, 2 ed.; Springer: New York, NY, USA, 2010.

33. Fuchs, H.U. A direct entropic approach to uniform and spatially continuous dynamical models of thermoelectric devices. *Energy Harvest. Syst.* **2014**, *1*, 253–265. [CrossRef]

34. Feldhoff, A. Thermoelectric material tensor derived from the Onsager – de Groot – Callen model. *Energy Harvest. Syst.* **2015**, *2*, 5–13. [CrossRef]

35. Falk, G. *Physik—Zahl und Realität*, 1 ed.; Birkhäuser: Basel, Switzerland, 1990.

36. Job, G. *Neudarstellung der Wärmelehre—Die Entropie als Wärme*; Akademische Verlagsgesellschaft: Frankfurt, Germany, 1972.

37. Job, G.; Rüffler, R. *Physical Chemistry from a Different Angle*, 1 ed.; Springer: Berlin/Heidelberger, Germany, 2016.

38. Strunk [23,31] has shaped the conceptional approach by Falk [35] to put thermodynamics first and assign the statistical behaviour not to an ensemble, but to the individual quantum state itself. Strunk's approach solves the paradox of doubled statistics, which has been inherent to the traditional approach, and it overcomes attempts to interpret quantum statistics as a modified variant of Newtonian mechanics-based kinetic gas theory. Strunk [23,31] suggests to consider heat as to involve entropy and energy. In his approach, entropy is a basic quantity.

39. Neave, E. Joseph Black's lectures on the elements of chemistry. *Isis* **1936**, *25*, 372–390. [CrossRef]

40. Falk, G. Entropy, a resurrection of caloric—A look at the history of thermodynamics. *Eur. J. Phys.* **1985**, *6*, 108–115. [CrossRef]

41. Robison, J. *Lectures on the Elements of Chemistry—Delivered in the University of Edinburgh by the Late Joseph Black*; William Creech Edinburgh: Edinburgh, UK, 1803; Volume 1.

42. To not offend his readers, Strunk has chosen a slightly different point of view by stating that heat comprises entropy and energy and that its use should be avoided when it addresses thermal energy solely.

43. Koshibae, W.; Maekawa, S. Effects of spin and orbital degeneracy on the thermopower of strongly correlated systems. *Phys. Rev. Lett.* **2001**, *87*, 236603–1–236603–4. [CrossRef]

44. Wang, Y.; Rogado, N.S.; Cava, R.; Ong, O. Spin entropy as the likely source of enhanced thermopower in $Na_xCo_2O_4$. *Phys. Rev. Lett.* **2003**, *423*, 425–428. [CrossRef]

45. Falk, G.; Herrmann, F.; Schmid, G. Energy forms or energy carriers? *Am. J. Phys.* **1983**, *51*, 1074–1077. [CrossRef]

46. Treating a device, Fuchs has derived a corresponding equation [32,33].

47. Wolf, M.; Menekse, K.; Mundstock, A.; Hinterding, R.; Nietschke, F.; Oeckler, O.; Feldhoff, A. Low thermal conductivity in thermoelectric oxide-based multiphase composites. *J. Electron. Mater.* **2019**, *48*, 7551–7561. [CrossRef]

48. Wolf, M.; Hinterding, R.; Feldhoff, A. High power factor vs. high zT—A review of thermoelectric materials for high-temperature application. *Entropy* **2019**, *21*, 1058. [CrossRef]

49. Geppert, B.; Brittner, A.; Helmich, L.; Bittner, M.; Feldhoff, A. Enhanced flexible thermoelectric generators based on oxide-metal composite materials. *J. Electron. Mater.* **2017**, *46*, 2356–2365. [CrossRef]

50. A constant Seebeck coefficient α of the thermoelectric material being in a temperature gradient implies that the Thomson coefficient $\tau = T \cdot \frac{\partial \alpha}{\partial T} \approx 0$ is negligible.

51. Peltier, J.C.A. Nouvelles expériences sur la caloricité des courants électrique. *Ann. Chim. Phys.* **1834**, *56*, 371–386.

52. Seebeck, T.J. Magnetische Polarisation der Metalle und Erze durch Temperatur-Differenz. *Physicalische und medicinische Abhandlungen der königlichen Academie der Wissenschaften zu Berlin* **1822**, *1820–21*, 265–373.

53. Velmre, E. Thomas Johann Seebeck (1770–1831). *Proc. Estonian Acad. Sci. Eng.* **2007**, *13*, 276–282.

54. Ohmic losses are often referred to as Joule heating.

55. Altenkirch, E. Über den Nutzeffekt der Thermosäule. *Physikalische Zeitschrift* **1909**, *10*, 560–568.

56. Ioffe, A.F. *Semiconductor Thermoelements and Thermoelectric Cooling*, 1 ed.; Infosearch Ltd.: London, UK, 1957.

57. Altenkirch, E. Elektrothermische Kälteerzeugung und reversible elektrische Heizung. *Physikalische Zeitschrift* **1911**, *12*, 920–924.

58. Wolf, M.; Rybakov, A.; Feldhoff, A. Understanding and improving thermoelectric generators via optimized material working points. *Entropy* in preparation.

59. Walstrom, P. Satial dependence of thermoelectric voltages and reversible heats. *Am. J. Phys.* **1988**, *56*, 890–894. [CrossRef]

60. Heikes, R.R.; Ure, R.W. *Thermoelectricity: Science and Engineering*; Interscience Publishers: New York, NY, USA, 1961.

61. This has confused Ioffe [56], who misinterpreted the situation as an uphill "heat" flow: "Of the total Joule 'heat' $I_q^2 \cdot R_{TEG}$ generated in the thermoelement, half passes to the hot junction, returning the power $\frac{1}{2} \cdot I_q^2 \cdot R_{TEG}$ and the rest is transferred to the cold junction."

62. Fuchs, H.U. Personal communication, 8 December 2018.

63. Gryasnov, O.; Moizhes, B.; Nemchinskii, V. Generalized thermoelectric effectivness. *J. Tech. Phys.* **1978**, *48*, 1720–1728.

64. Narducci, D. Do we really need high thermoelectric figures of merit? A critical appraisal to the power conversion efficiency of thermoelectric materials. *Appl. Phys. Lett.* **2011**, *99*, 102104:1–102104:3. [CrossRef]

65. Slack, A. New Materials and Performance Limits for Thermoelectric Cooling. In *CRC Handbook of Thermoelectrics*; Rowe, D., Ed.; CRC Press: New York, NY, USA, 1994.

66. Goupil, C.; Seifert, W.; Zabrocki, K.; Müller, E.; Snyder, G.J. Thermodynamics of thermoelectric phenomena and applications. *Entropy* **2011**, *13*, 1481–1517. [CrossRef]

67. Zener, C. Putting electrons to work. *Trans. ASM* **1961**, *53*, 1052–1068.

68. Wu, H.J.; Zhao, L.D.; Zheng, F.S.; Wu, D.; Pei, Y.L.; Tong, X.; Kanatzidis, M.G. Broad temperature plateau for thermoelectric figure of merit $zT > 2$ in phase-separated $PbTe_{0.7}S_{0.3}$. *Nat. Commun.* **2014**, *5*, 4515. [CrossRef]

69. Risseh, A.E.; Nee, H.P.; Goupil, C. Electrical power conditioning system for thermoelectric waste heat recovery in commercial vehicles. *IEEE Trans. Transp. Electrif.* **2018**, *4*, 548–562. [CrossRef]

70. Poudel, B.; Hao, Q.; Ma, Y.; Lan, Y.; Minnich, A.; Yu, B.; Yan, X.; Wang, D.; Muto, A.; Vashaee, D.; et al. High-thermoelectric performance of nanostructured bismuth antimony telluride bulk alloys. *Science* **2008**, *320*, 634–638. [CrossRef]

71. He, R.; Kraemer, D.; Mao, J.; Zeng, L.; Jie, Q.; Lan, Y.; Li, C.; Shuai, J.; Kim, H.S.; Liu, Y.; et al. Power factor and output power density in *p*-type half-Heuslers $Nb_{1-x}Ti_xFeSb$. *Proc. Natl. Acad. Sci. USA* **2016**, *113*, 13576–13581. [CrossRef]

72. Chen, L.; Gao, S.; Zeng, X.; Mehdizadeh Dehkordi, A.; Tritt, T.; Poon, S. Uncovering high thermoelectric figure of merit in (Hf,Zr)NiSn half-Heusler alloys. *Appl. Phys. Lett.* **2015**, *107*, 041902. [CrossRef]

73. Yamashita, O.; Ochi, T.; Odahara, H. Effect of the cooling rate on the thermoelectric properties of p-type $(Bi_{0.25}Sb_{0.75})_2Te_3$ and n-type $Bi_2(Te_{0.94}Se_{0.06})_3$ after melting in the bismuth-telluride system. *Mater. Res. Bull.* **2009**, *44*, 1352–1359. [CrossRef]

74. Zhu, H.; He, R.; Mao, J.; Zhu, Q.; Li, C.; Sun, J.; Ren, W.; Wang, Y.; Liu, Z.; Tang, Z.; et al. Discovery of ZrCoBi based half-Heuslers with high thermoelectric conversion efficiency. *Nat. Commun.* **2018**, *9*, 1–9. [CrossRef] [PubMed]

75. Pei, Y.; Shi, X.; Lalonde, A.; Wang, H.; Chen, L.; Snyder, G. Convergence of electronic bands for high performance bulk thermoelectrics. *Nature* **2011**, *473*, 66–69. [CrossRef]

76. Liu, Y.; Zhao, L.D.; Zhu, Y.; Liu, Y.; Li, F.; Yu, M.; Liu, D.B.; Xu, W.; Lin, Y.H.; Nan, C.W. Synergistically optimizing electrical and thermal transport properties of BiCuSeO via a dual-doping approach. *Adv. Energy Mater.* **2016**, *6*, 1502423. [CrossRef]

77. Liu, H.; Shi, X.; Xu, F.; Zhang, L.; Zhang, W.; Chen, L.; Li, Q.; Uher, C.; Day, T.; Snyder, G.J. Copper ion liquid-like thermoelectrics. *Nat. Mater.* **2012**, *11*, 422–425. [CrossRef] [PubMed]

78. Shutoh, N.; Sakurada, S. Thermoelectric properties of the $Ti_x(Zr_{0.5}Hf_{0.5})_{1-x}NiSn$ half-Heusler compounds. *J. Alloy. Compd.* **2005**, *389*, 204–208. [CrossRef]

79. Zhang, J.; Song, L.; Pedersen, S.H.; Yin, H.; Hung, L.T.; Iversen, B.B. Discovery of high-performance low-cost n-type Mg_3Sb_2-based thermoelectric materials with multi-valley conduction bands. *Nat. Commun.* **2017**, *8*, 13901. [CrossRef]

80. Shi, X.; Yang, J.; Salvador, J.R.; Chi, M.; Cho, J.Y.; Wang, H.; Bai, S.; Yang, J.; Zhang, W.; Chen, L. Multiple-filled skutterudites: High thermoelectric figure of merit through separately optimizing electrical and thermal transports. *J. Am. Chem. Soc.* **2011**, *133*, 7837–7846. [CrossRef]

81. Chen, X.; Wu, H.; Cui, J.; Xiao, Y.; Zhang, Y.; He, J.; Chen, Y.; Cao, J.; Cai, W.; Pennycook, S.J.; et al. Extraordinary thermoelectric performance in n-type manganese doped Mg_3Sb_2 Zintl: High band degeneracy, tuned carrier scattering mechanism and hierarchical microstructure. *Nano Energy* **2018**, *52*, 246–255. [CrossRef]

82. Ahmad, S.; Singh, A.; Bohra, A.; Basu, R.; Bhattacharya, S.; Bhatt, R.; Meshram, K.N.; Roy, M.; Sarkar, S.K.; Hayakawa, Y.; et al. Boosting thermoelectric performance of p-type SiGe alloys through in-situ metallic YSi_2 nanoinclusions. *Nano Energy* **2016**, *527*, 282–297. [CrossRef]

83. Zhao, K.; Zhu, C.; Qiu, P.; Qiu, P.; Blichfeld, A.B.; Eikeland, E.; Ren, D.; Iversen, B.B.; Xu, F.; Shi, X.; et al. High thermoelectric performance and low thermal conductivity in $Cu_{2-y}S_{1/3}Se_{1/3}Te_{1/3}$ liquid-like materials with nanoscale mosaic structures. *Nano Energy* **2017**, *42*, 43–50. [CrossRef]

84. Hsu, K.F.; Loo, S.; Guo, F.; Chen, W.; Dyck, J.S.; Uher, C.; Hogan, T.; Polychroniadis, E.K.; Kanatzidis, M.G. Cubic $AgPb_mSbTe_{2+m}$: Bulk thermoelectric materials with high figure of merit. *Science* **2014**, *303*, 818–821. [CrossRef] [PubMed]

85. Biswas, K.; He, J.; Blum, I.D.; Wu, C.I.; Hogan, T.P.; Seidman, D.N.; Dravid, V.P.; Kanatzidis, M.G. High-performance bulk thermoelectrics with all-scale hierarchical architectures. *Nature* **2012**, *489*, 414–418. [CrossRef] [PubMed]

86. Hong, M.; Chen, Z.G.; Yang, L.; Zou, Y.C.; Dargusch, M.S.; Wang, H.; Zou, J. Realizing zT of 2.3 in $Ge_{1-x-y}Sb_xIn_yTe$ via Reducing the phase-transition temperature and introducing resonant energy doping. *Adv. Mater.* **2018**, *30*, 1705942. [CrossRef]

87. Tan, G.; Shi, F.; Hao, S.; Zhao, L.D.; Chi, H.; Zhang, X.; Uher, C.; Wolverton, C.; Dravid, V.P.; Kanatzidis, M.G. Non-equilibrium processing leads to record high thermoelectric figure of merit in PbTe-SrTe. *Nat. Commun.* **2016**, *7*, 12167. [CrossRef] [PubMed]

88. Zhao, L.D.; Lo, S.H.; Zhang, Y.; Sun, H.; Tan, G.; Uher, C.; Wolverton, C.; Dravid, V.P.; Kanatzidis, M.G. Ultralow thermal conductivity and high thermoelectric figure of merit in SnSe crystals. *Nature* **2014**, *508*, 373. [CrossRef]

89. Olvera, A.A.; Moroz, N.A.; Sahoo, P.; Ren, P.; Bailey, T.P.; Page, A.A.; Uher, C.; Poudeu, P.F.P. Partial indium solubility induces chemical stability and colossal thermoelectric figure of merit in Cu_2Se. *Energy Environ. Sci.* **2017**, *10*, 1668–1676. [CrossRef]

90. Chang, C.; Wu, M.; He, D.; Pei, Y.; Wu, C.F.; Wu, X.; Yu, H.; Zhu, F.; Wang, K.; Chen, Y. 3D charge and 2D phonon transports leading to high out-of-plane zT in n-type SnSe crystals. *Science* **2018**, *360*, 778–782. [CrossRef]

91. Curzon, F.; Ahlborn, B. Efficiency of a Carnot engine at maximum power output. *Am. J. Phys.* **1975**, *43*, 22–24. [CrossRef]

92. Leff, H.S. Thermal efficiency at maximum work ouptut: New results for old heat engines. *Am. J. Phys.* **1987**, *55*, 602–610. [CrossRef]

93. Ioffe (see [56], p. 45) has introduced it as "heat" conductivity–electrical conductivity plot ($\lambda_{OC} - \sigma$).

94. Bittner, M.; Kanas, N.; Hinterding, R.; Steinbach, F.; Räthel, J.; Schrade, M.; Wiik, K.; Einarsrud, M.A.; Feldhoff, A. A comprehensive study on improved power materials for high-temperature thermoelectric generators. *J. Power Sources* **2019**, *410–411*, 143–151. [CrossRef]

8

Adapted or Adaptable: How to Manage Entropy Production?

Christophe Goupil *,† **and Eric Herbert** †

Université de Paris, Laboratoire Interdisciplinaire des Energies de Demain (LIED), UMR 8236 CNRS, F-75013 Paris, France; eric.herbert@univ-paris-diderot.fr

* Correspondence: christophe.goupil@univ-paris-diderot.fr

† These authors contributed equally to this work.

Abstract: Adaptable or adapted? Whether it is a question of physical, biological, or even economic systems, this problem arises when all these systems are the location of matter and energy conversion. To this interdisciplinary question, we propose a theoretical framework based on the two principles of thermodynamics. Considering a finite time linear thermodynamic approach, we show that non-equilibrium systems operating in a quasi-static regime are quite deterministic as long as boundary conditions are correctly defined. The Novikov–Curzon–Ahlborn derivation applied to non-endoreversible systems then makes it possible to precisely determine the conditions for obtaining characteristic operating points. As a result, power maximization principle (MPP), entropy minimization principle (mEP), efficiency maximization, or waste minimization states are only specific modalities of system operation. We show that boundary conditions play a major role in defining operating points because they define the intensity of the feedback that ultimately characterizes the operation. Armed with these thermodynamic foundations, we show that the intrinsically most efficient systems are also the most constrained in terms of controlling the entropy and dissipation production. In particular, we show that the best figure of merit necessarily leads to a vanishing production of power. On the other hand, a class of systems emerges, which, although they do not offer extreme efficiency or power, have a wide range of use and therefore marked robustness. It therefore appears that the number of degrees of freedom of the system leads to an optimization of the allocation of entropy production.

Keywords: out of equilibrium thermodynamics; finite time thermodynamics; living systems

1. Introduction

The issue of energy conversion is the subject of historical debate. Without going back to its roots, let us mention the work initiated by Glansdorf and Prigogine, which placed at the center the question of entropy production in out-of-equilibrium systems, an issue that is still largely relevant [1,2]. This debate is itself part of an even broader debate that questions the operating points of the systems, considering mainly the maximization of entropy production (MEP), its minimization (mEP), or power maximization (MPP) [3,4]. One of the reasons why these questions do not find a general consensus today is that they are most often considered on very different systems, in particular in the definition of the boundary conditions of the device with its environment, considered immutable. The case of idealized mechanical systems is, from this point of view, much simpler, since, broadly speaking, the absence of any friction process means that the system interacts with its environment via a very limited number of degrees of freedom, which makes variational approaches relevant. On the contrary, it has long been accepted that there is no variational principle that governs the out-of-equilibrium steady state of a thermodynamic system [5]. This can be understood as an impossibility to establish a variational principle when the number of degrees of freedom diverges, which is obviously the case when the system is connected to a thermostat, and when dissipative processes occur. However, it is equally obvious that many out-of-equilibrium systems are perfectly deterministic in their evolution, and have a perfectly defined stationary state, as is the case, for example, for Kirchoff's networks in electronics. As a result, these systems, although not governed by a Lagrangian form and an associated variational principle, have a completely established stationary operating point, without any possible affirmation of an underlying minimization or maximization of the production of the entropy or the power.

These questions of power and finite time performance have been the subject of much work [6] particularly in thermoelectricity [7–11]. Without entering into these debates again, we propose an approach that provides a fairly generic framework for describing a complete thermodynamic system with perfectly established boundary conditions. In this article, we will limit ourselves to the case of locally linear machines, subscribing to Onsager's formalism. This formalism, based on the concept of local equilibrium, makes it possible to consider the thermodynamic potentials of the system, which are the intensive parameters. As a result, it becomes possible to derive a thermodynamics close to equilibrium, with, in particular, a rigorous choice of potentials that allow for obtaining the symmetry of the out-of-diagonal coefficients of the Onsager matrix. The stationary nature also requires that kinetic coefficients and boundary conditions of the system be constant or slowly variable compared to the characteristic relaxation time of entropy production and dissipation diffusion, thus guaranteeing both stationary processes and local equilibrium.

In this article, we consider the transport of energy and matter within a system, where the thermodynamic conversion is produced by coupling the energy and matter currents. By applying the first law of thermodynamics, both of these currents are conservative. By applying the second law, the energy, and sometimes the matter, used during the conversion process is subject to dispersion in the degrees of freedom accessible to the system. As a result, thermodynamics is based on both quantity and quality principles. Since the loss of quality is directly related to dispersion in the degrees of freedom, the search for processes to reduce their number has always been a guideline. It should be noted that, in the case of non-spontaneous processes, it is possible to consider a reduction in the degrees of freedom, but this operation requires the implementation of external processes. These processes offer other opportunities for energy dispersion, in greater proportions than those gained within the system. As a result, any physical process taking place over a finite period of time is the location of a compromise between the total energy used to carry out a process, and the energy actually converted for the needs to be covered. The process efficiency is therefore written as the ratio between the actually converted energy and the total energy supplied. We propose to consider energy conversion processes in a very generic form, in order to establish their main characteristics and constraints. In particular, we address the question of power and entropy production, insisting on the compromises they impose.

The question of adapting a device to the uses assigned to it then arises. In the case of single working point, the system may be designed to be as much adapted as is it possible. However, this single operating working point is a rare configuration, and realistic systems are asked to work in a given range of working points. Then, the concept of adaptability, or flexibility, arises, which enters into competition with the previous adapted concept. This problem of adaptation or adaptability concerns all thermodynamic systems, including, of course, living systems. Indeed, as soon as we define an envelope, we delimit the boundaries of a space occupied by a given device and the interactions of this device with the outside world. Considering the energy and matter budget at the borders of the device, we then characterize the relationship between the device and its environment. Since the processes take place over a finite period of time, it is important to consider an out-of-equilibrium description. In this paper, we consider an out-of-equilibrium thermodynamic description, driven by locally linear equations. We show that the intrinsic characteristics of the device, on the one hand, and the boundary conditions, on the other hand, totally determine the behavior of the system. It appears that the allocation of dissipation largely determines the possible ranges of use of an out-of-equilibrium thermodynamic system.

In terms of boundary conditions, we show that the real coupling conditions of a system with its environment are always located between the Dirichlet and Neumann boundaries, also called "stock" and "flow" boundary conditions. It should be noted that both pure stock and flow are extreme boundary conditions which can never being strictly reached. Between adaptable and adapted, the performances of thermodynamic systems are therefore the result of a compromise between intrinsic performance of a device and the coupling to the environment. This question of coupling to the environment is the subject of the first section of this article. In the following section, we describe the envisaged system in its most general form. The third section concerns the descriptions of the device at the heart of the system, while the fourth section describes its insertion into the complete system. The fifth section considers the different configurations that such a global system may encounter, and the consequences on the production of power, dissipation, and more generally, entropy. The article ends with concluding remarks.

2. System Description

2.1. Boundary Conditions

As indicated above, the system is composed of two sub-parts: a central zone, which we will call the device, and which is the place of thermodynamic conversion, on the one hand, and the boundary conditions, consisting of the source, and, on the other hand, the sink and the elements connecting it to the device. These elements allow for modifying at will the boundary conditions that condition the coupling of the device with the source and the sink, which is a central question for the optimization. Among the latter, we can distinguish systems whose intrinsic parameters are constant, as is the case for most machines, and systems, whose intrinsic parameters are subject to modification, as is the case for living or societal systems. These latter are subject to potential developments and evolution, which are not possible for the above-mentioned machines. By potential development, we consider the case of living systems, societies or organisms, which can, under conditions of energy and matter supply, develop, maintain, or regress.

In the case of systems under Neumann boundary conditions, the system is somehow fed by a constant current of energy and/or matter, which guarantees the maintenance of the system as much as it constrains its development. Under such conditions, the possible development of the system is limited by the value of the current of matter and/or energy. In the case of Dirichlet systems, there are no restrictions on access to the resource, except for the intrinsic limitations of the conversion device. As a result, the currents of energy and matter may diverge completely, if the characteristics of the device lend themselves to it. The same reasoning applies to the production and rejection of waste to the sink. Access to the resource and waste production are therefore both dependent on these boundary conditions. Let us consider, as an historical illustration, the situation of the industrial revolution, which saw the rise of the use of fossil energy [12]. The latter are by definition stock resources that lead the human societies to find themselves in Dirichlet conditions, as far as access to the resource is concerned. Concerning the waste rejected to the sink, the Dirichlet's condition has been the norm, as long as the planet has been considered a bottomless sink. On the other hand, if we consider the situation before the industrial revolution, it can be noted that the main resource for development, which is the food resource, was dependent on Neumann-type boundary conditions, due to the subjection to solar flux. Without going further into this illustration, which is beyond the scope of this article, we can nevertheless observe the importance of boundary conditions, both on the functioning of systems, but also for their possible evolutions. Indeed, in the case of boundary conditions of the Neumann type, there is no possibility of development, in the sense of increasing the current of energy and matter that feed the conversion device. Consequently, there is no possibility of any increase of the quantities. On the other hand, there are possibilities of increase of the quality because the conditions of coupling between energy and matter may change, as the history of life proved it.

On the other hand, in the case of Dirichlet boundary conditions, there is no limit to the increase in energy and matter currents, which could lead to their possible divergence. It should be noted that the actual Dirichlet conditions for the access to the energy for the human species are quite singular in the history of the living systems. In order to remain explicit and relatively simple to address, these questions need to be modeled in the most compact form possible. This why we propose to describe a generic thermodynamic machine in order to guarantee a general character to the developments of this article. Many extensions and refinements can be added, as for previous systems in the literature [6,10].

2.2. Thermodynamic Device

The proposed thermodynamic system is described in Figure 1. It consists of a reservoir providing the resource and a sink receiving the waste, with the respective potentials Π_1^R and Π_1^S fixed at constant values. Between these two reservoirs is the energy conversion device which is the place of coupling between a current of matter I_2, and a current of energy I_E. The energy current entering the system is associated with an incoming entropy current, I_1, with Π_1 its conjugated potential. In the case of a thermal system of heat current I_Q, temperature T and entropy current I_S, we would simply have $\Pi_1 I_1 = I_Q = T I_S$ so I_1 would be the classical entropy current. The current of matter is defined by I_2 and its conjugated potential Π_2. The energy currents budget finally writes $I_E = \Pi_1 I_1 + \Pi_2 I_2$. We recognize the fractions of dispersed energy, $\Pi_1 I_1$, and concentrated energy, $\Pi_2 I_2$, which are a generalization of the notions of heat and work extended to the case of non-thermal systems [13,14]. The coupling term between energy and matter is defined, under $I_2 = 0$ condition, as $\alpha = -(\delta \Pi_2 / \delta \Pi_1)_{I_2}$. The geometry of the system is given by its length L and its cross-section A. The two currents of energy and matter are then associated with two conductivities σ_1 and σ_2, which, at the integrated scale, behave like two resistive dipoles $R_{1/2} = \frac{1}{\sigma_{1/2}} \frac{L}{A}$. The connection of the conversion zone with the two reservoirs is defined by the coupling resistors R_+ and R_-, which allow the boundary conditions to be set, at will, between Dirichlet conditions ($R_+ = R_- = 0$), or Neumann conditions, where R_+ and R_- diverge. This type of configuration is not in itself new, and has already been used in specific systems [14,15]. In particular, it has been shown that, under these conditions, the way the system operates is partially governed by the feedback effects induced by boundary conditions. Some of this feedback can lead to the presence of oscillations. It should be noted that these processes do not violate the first principle in that they are not self-sustained oscillations, at least from an energy point of view. They do not violate the second principle either, since these structures are highly dissipative and are only maintained by a continuous supply of energy. It can also be noted that the incoming current of energy is used to produce a potential difference, which, if maintained, allows the circulation of the matter under the action of the thermodynamic force, which is defined from the gradient of the potential. This type of analysis of thermodynamic conversion has been used with success by Alicki in various systems [16,17]. This description of two coupled currents can, of course, be extended to a larger number of coupled currents without changing the spirit of the study.

As it is represented, the system is therefore quite generic. The main determinants of functioning are thus summarized by three terms, the capture of the resource, its conversion into a usable form, and the rejection of waste. It is clear that ideally the target is the one where the output power would be maximum and the amount of energy released would be minimal. The study of the limits to achieving this target is one of the objectives of this article. As the coupling parameter for the conversion, the α parameter is therefore central since it determines the system's ability to convert energy into a usable form. A naive picture may suggest that the largest possible α value necessarily leads to the most efficient system, but this is not correct, as we will see now.

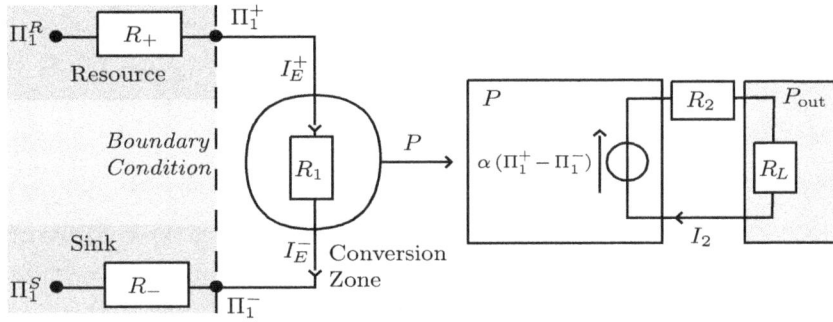

Figure 1. Schematic view of the generic system, with a resource and a sink, whose potential Π_1^R and Π_1^S are constant. The coupling of the conversion zone (circle) with the two reservoirs is ensured by the elements R_+ and R_-. As a result, the difference potential $\Pi_1^+ - \Pi_1^-$ is less than that between reservoir and sink. Power produced in the conversion zone (circle) is $P = -\alpha \Delta \Pi_1 = \Delta \Pi_2$. The internal resistance $R_2 = \frac{L}{A\sigma_2}$ gives rise to a dissipative contribution $R_2 I_2^2$. The R_L resistance is the output load, and the output power is $P_{\text{out}} = R_L I_2^2$.

3. Local Energy Conversion

3.1. Presentation

At the local level, energy conversion is produced by coupling the energy and matter currents flowing through the device. These currents are generated by the presence of differences between the two thermodynamic potentials Π_1 and Π_2. This local modeling is therefore based on the three parameters of conductivity associated with energy transport, σ_1, conductivity associated with matter transport, σ_2, and the coupling coefficient between the gradients of the two potentials, α. We deduce from this the formulation of local Onsager matrix, where $\nabla = \frac{d}{dx}$ is the spatial gradient, here reduced to $1D$ in order to simplify the description.

$$\begin{pmatrix} J_2 \\ J_E \end{pmatrix} = \begin{pmatrix} L_{11} & L_{12} \\ L_{21} & L_{22} \end{pmatrix} \begin{pmatrix} -\nabla\left(\frac{\Pi_2}{\Pi_1}\right) \\ \nabla\left(\frac{1}{\Pi_1}\right) \end{pmatrix}. \tag{1}$$

J_E and J_2 are the densities of the two currents, and are extensive and conservative quantities. Given the differential form $J_E = \Pi_1 J_1 + \Pi_2 J_2$, the equality of non-diagonal terms $L_{12} = L_{21}$ is insured according to the choice of the correct potentials $-\frac{\Pi_2}{\Pi_1}$ and $\frac{1}{\Pi_1}$ [18,19]. The four terms of the matrix are therefore reduced to three, σ_1, σ_2 and α, whose correspondences with the coefficients L_{ij} are

$$\sigma_1 = \frac{1}{\Pi_1^2}\left[\frac{L_{11}L_{22} - L_{21}L_{12}}{L_{11}}\right], \tag{2}$$

$$\sigma_2 = \frac{L_{11}}{\Pi_2}, \tag{3}$$

$$\alpha = -\frac{\Delta\Pi_2}{\Delta\Pi_1} = \frac{1}{\Pi_1}\frac{L_{12}}{L_{11}}. \tag{4}$$

In the absence of a matter gradient, the energy conductivity can be defined as $\sigma_{\Pi_2} = \sigma_1\left[1 + \alpha^2\sigma_2/\sigma_1\Pi_2\right]$. The figure of merit is then defined as

$$F_m = \frac{\alpha^2 R_1}{R_2}\Pi_2 = \frac{L_{12}^2}{L_{11}L_{22} - L_{21}L_{12}}. \tag{5}$$

It is known that the ratio σ_2/σ_1, therefore F_m, is a direct measure of the intrinsic capacity of energy conversion. F_m can be related to the ratio of the equivalent specific heats by the expression

$\gamma = \frac{C_{\Pi_2}}{C_{I_2}} = F_m + 1$. In their seminal paper, Kedem and Caplan derived the following expression of the coupling parameter between the two fluxes involved in the conversion process [13]:

$$q = \frac{L_{12}}{\sqrt{L_{11}L_{22}}} = \sqrt{\frac{F_m}{1 + F_m}} \tag{6}$$

an expression that explicitly includes the kinetic coefficients L_{ij}. The figure of merit and the coupling factor q are equivalent in terms of measure of the system performance: the larger their (absolute) values, the better the energy conversion system. This can be evidenced by the derivation of the local maximal efficiency of the conversion process in generator mode, η_{max}:

$$\eta_{max} = \left(\frac{1 + \sqrt{1 - q^2}}{q} \right)^2 = \frac{\sqrt{\gamma} - 1}{\sqrt{\gamma} + 1}. \tag{7}$$

3.2. Entropy Production and Efficiency

The volumetric entropy production rate is given by the summation of the force-flow products,

$$\dot{S} = J_2 \nabla \left(-\frac{\Pi_2}{\Pi_1} \right) + J_E \nabla \left(\frac{1}{\Pi_1} \right) = -\frac{1}{\Pi_1} \left[J_2 \nabla \Pi_2 + J_1 \nabla \Pi_1 \right]. \tag{8}$$

In the case of a reversible process $\dot{S} = 0$ so does $J_2 \nabla \Pi_2 + J_1 \nabla \Pi_1$. We get $-\frac{J_2 \nabla \Pi_2}{\Pi_1 J_1} = \frac{\nabla \Pi_1}{\Pi_1} = \eta_C$, where η_C is the Carnot efficiency. This leads to the general expression of the local efficiency,

$$\eta = -\frac{J_2 \nabla \Pi_2}{J_1 \Pi_1} < \eta_C. \tag{9}$$

Let us define the reduced current as

$$j = \frac{\alpha J_2}{J_1}, \tag{10}$$

which is the ratio between the entropy carried by the transport of the matter, divided by the total entropy transported. In the case of a reversible process, both terms are equal so $j = 1$ [20]. This expression shows three regions for the $\eta(j)$ meaning. For $0 < j < 1$, the device works as a generator. For $j < 0$ and $j > 1$, the device works as a receptor. For reasons of brevity, we will mainly deal with the generator configuration in this article.

Rewriting the Onsager matrix in more suitable form [21], we get

$$\begin{pmatrix} J_2 \\ \Pi_1 J_1 \end{pmatrix} = \begin{pmatrix} \sigma_2 & \alpha \sigma_2 \\ \alpha \Pi_1 \sigma_2 & \gamma \sigma_1 \end{pmatrix} \begin{pmatrix} -\nabla \Pi_2 \\ -\nabla \Pi_1 \end{pmatrix}. \tag{11}$$

Then,

$$j = \frac{\eta \alpha \Pi_1 \sigma_2 - j \alpha \sigma_2 \nabla \Pi_1}{\eta \alpha \Pi_1 \sigma_2 - \frac{j \gamma \sigma_1}{\alpha} \frac{\nabla \Pi_1}{\Pi_1}}. \tag{12}$$

Thus,

$$\eta = \eta_C j \frac{j \gamma - \frac{\alpha^2 \sigma_2}{\sigma_1} \Pi_1}{j \frac{\alpha^2 \sigma_2}{\sigma_1} \Pi_1 - \frac{\alpha^2 \sigma_2}{\sigma_1} \Pi_1}, \tag{13}$$

where $\gamma = \frac{\alpha^2 \sigma_2}{\sigma_1} \Pi_1 + 1$. After a few algebra, we get

$$\eta = \frac{\eta_C}{(\gamma - 1)} \frac{\gamma j^2 - (\gamma - 1) j}{j - 1} \tag{14}$$

η presents a maximum for $j_{opt} = 1 + \sqrt{\frac{1}{\gamma}}$ for a receptor mode, and $j_{opt} = 1 - \sqrt{\frac{1}{\gamma}}$ for a generator mode. Both optima reduce to $j = 1$ in the ideal case, when γ diverges, where we recover the Carnot efficiency. In this diverging case, the system do not present anymore dissipation production, and the equivalence between the receptor and generator modes is a proof of the absence of causality of the Carnot configuration. This absence of causality is another name for reversibility. We then recover the Kedem–Caplan expression of the maximal efficiency, $\eta_{max} = \eta_C \frac{\sqrt{\gamma}-1}{\sqrt{\gamma}+1}$ for the generator mode, and $\eta_{max} = \eta_C \frac{\sqrt{\gamma}+1}{\sqrt{\gamma}-1}$ for the receptor mode. Let us now plot the efficiency versus the reduced current, as reported in Figure 2.

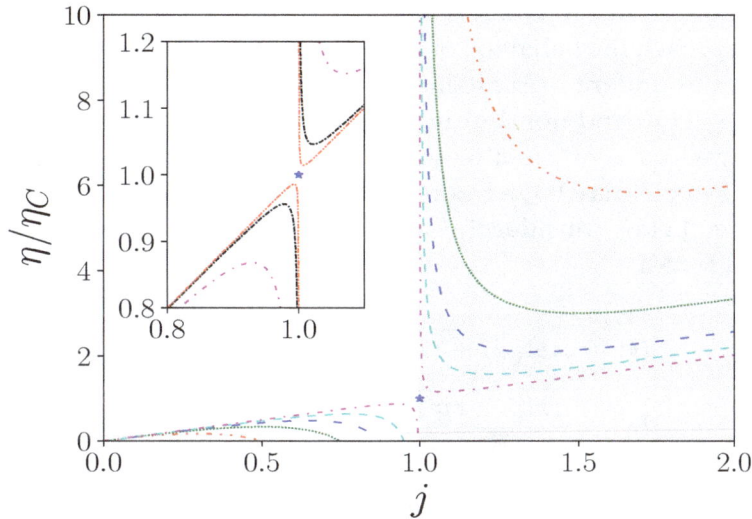

Figure 2. Normalized efficiency $\frac{\eta}{\eta_C}$ according to reduced current $j = \alpha J_2 / J_1$ with $\gamma = 2$ (red, dot dashed), 4 (green, dots), 8 (blue, loosely dashed), 20 (cyan, dashed), 2×10^2 (magenta, loosely dot dashed) in main figure, and $\gamma = 2 \times 10^2$ (magenta, loosely dot dashed), 2×10^3 (black, dot dashed), 2×10^4 (red, dot dot dashed) in inset. The grey area corresponds to the receptor mode (resp. generator mode). Note that the figure is symmetrical with respect to the Carnot point (blue star), which is never reached. This singular point defines the reversible configuration, where causality is broken.

As expected for the maximum performance achieved, η_{max} is an increasing function of the figure of merit. On the other hand, it also appears that the sensitivity to fluctuations in j becomes all the more important as η_{max} is important. This is confirmed by estimating the value of the slope in the vicinity of the maximum yield, which is $\partial \eta / \partial(j) \approx -2\eta_{max} F_m$. The larger the figure of merit, the steeper the slope. This local description allows us to conclude that the performance of the device is obtained at the cost of a constraint of stability of the operating points, directly driven by the value of the figure of merit. As an intrinsic quantity, the figure of merit defines the performance ceiling beyond which it cannot be exceeded. It is clear from the figure that the system defined by a high figure of merit exceeds in performance all the systems of lower figure of merit value. However, this result is strongly weighted by the fact that, for excursions of j around the optimal value, the efficiency falls rapidly. Then, it is not necessarily relevant to look for a device with a large figure of merit, without first inventorying the operating range that will be brought to run this device. For simplicity's sake, we have only dealt here with the case where the system works as a generator, which is obtained by $0 < j < 1$. It is clear that the same study can be carried out for the case where the system operates as a receptor, instead of working as a generator. This situation, well known for thermal machines, corresponds to heat pump operation. More broadly, and in the case of non-thermal machines, this case actually corresponds to the operation in recycling mode where the treated quantity undergoes regeneration. It should be noted that the expression of performance refers only to γ, and therefore to the figure of merit, without specifying any contribution from σ_1, σ_2 and α, respectively. The local level is totally blind to these issues so we now consider the situation of the entire system to see the relative contributions.

4. Global Conversion System

4.1. Presentation

In accordance with the diagram in Figure 1, the device of the conversion zone is connected to its reservoirs via the two resistors R_+ and R_-, which makes it possible to explore all boundary conditions. The presence of R_+ and R_- may lead to the pinching of the potential difference $\Pi_1^+ - \Pi_1^-$ according to the system operating point. More precisely, R_+ governs the limitation of access to the resource while R_- reflects possible saturation effects of waste disposal. This global model, although limited, makes it possible to approach the behavior of many systems, including living systems, depending on whether the resource is abundant or scarce, and whether waste disposal, including thermal waste, is easy or not. Living system and non-living systems differ from the fact that the energy current is never zero in living systems, so R_1 is always finite, and there is a non-zero resting point. On the contrary, a non-living system may have a zero resting point, with zero energy current, so R_1 may be infinite in these systems. Let us consider the set of the four equations that governs the functioning of the system (see Appendix A):

$$I_{E-} = \alpha \Pi_1^- I_2 + (1 - \varphi) R_2 I_2^2 + \frac{(\Pi_1^+ - \Pi_1^-)}{R_1}, \tag{15}$$

$$I_{E-} = \frac{(\Pi_1^- - \Pi_1^S)}{R_-}, \tag{16}$$

$$I_{E+} = \alpha \Pi_1^+ I_2 - \varphi R_2 I_2^2 + \frac{(\Pi_1^+ - \Pi_1^-)}{R_1}, \tag{17}$$

$$I_{E+} = \frac{(\Pi_1^R - \Pi_1^+)}{R_+}. \tag{18}$$

These equations have their origin in the integration of the local form described in the previous paragraph. These developments have been the subject of previous articles [14,22] , and will not be re-described here. φ controls the dissipation fraction that returned to the source or to the sink. In the following, we will choose $\varphi = 0$. This choice is not critical here since the effect of $\varphi = 0$ is driven by R_2, which is equal to zero.

4.2. Devices with Zero Resting Point

First of all, we consider that $R_2 = 0$ and R_1 diverge, in order to separate the contributions of entropy production and internal dissipation. R_2 governs the current of matter, so we therefore consider that this current may not be limited, so there is no intrinsic dissipation within the device. The figure of merit of the device is then infinite and we may expect to reach the ideal conditions and the Carnot efficiency. However, the classical discussion around the Carnot efficiency is based on pure Dirichlet boundary conditions, which is clearly not the case here, so we have to consider the new conditions introduced by the modification of the boundary conditions. In the present configuration of zero resting point systems, the general equations (see Appendix A) can be summarized as

$$I_{E-} = \frac{\Pi_1^S I_2}{\frac{1}{\alpha} - R_- I_2}, \tag{19}$$

$$I_{E+} = \frac{\Pi_1^R I_2}{R_+ I_2 + \frac{1}{\alpha}}, \tag{20}$$

with the output power given by $P = I_{E+} - I_{E-}$.

The plots in Figure 3 summarize the behavior of the global system. The output power presents a maximum and two zero values. The first value corresponds to the case where the efficiency reaches its maximum. This situation is obtained for $I_2 = 0$, so $I_{E-} = I_{E+} = P = 0$. This means that no matter or energy can flow through the system, which is a totally useless situation for a physical system. The second zero power value is reached for a current of matter $I_{2_{sc}}$, named the short-circuit current, by analogy with electronics. In this situation, the produced power is completely re-dissipated inside the system. $I_{2_{sc}}$ is therefore an ultimate operating point for the system, working as an energy generator. For a truly efficient operation, it is therefore necessary to try to push $I_{2_{sc}}$ to large values, which are obtained by getting as close as possible to Dirichlet conditions. In the general case, the approximate expression of this current is

$$I_{2sc} \approx \frac{1}{\alpha} \frac{\Delta\Pi_1}{R_-\Pi_1^S + R_+\Pi_1^R},\tag{21}$$

which confirms that Dirichlet's conditions where $R_- = R_+ \approx 0$ are to be sought, if accessible. Since the resting point here is zero, the power curve necessarily intercepts that of I_{E-}. Beyond this interception point, the system is in a situation where it releases more waste than it produces output power. We call *critical point* the point where $P = I_{E-}$, reached for I_{2cp}. The fact that power is not a monotonous function of I_2 is actually quite unexpected because, to the extent that $R_2 = 0$, the total absence of intrinsic viscosity should not lead to any limit to I_2. However, if we carry out a development at the first order of the expression of power we find

$$P \approx \left[\alpha\Delta\Pi_1 - \left(\Pi_1^R R_+ + \Pi_1^S R_-\right)\alpha^2 I_2\right] I_2\tag{22}$$

which clearly indicates the presence of a viscous friction term R_{fb},

$$R_{fb} \approx \alpha^2(\Pi_1^R R_+ + \Pi_1^S R_-)\tag{23}$$

which reduces the transport of the matter, even though the intrinsic viscosity, i.e., $\frac{1}{\sigma_2}$, associated with the transport of the matter, is zero. This additional dissipation is a pure feedback effect that is due to the presence of boundary conditions at the general limits where R_+ et R_- are non-zero. This additional dissipation can only be rendered null if $R_+ = R_- = 0$, i.e., a strict Dirichlet condition, which is, in reality, only very rarely observed. Note that the condition $\alpha = 0$ leads to the same result but it is useless because in this case the transport of energy and matter are fully decoupled, and the device does not convert the energy anymore. The conditions $R_2 = 0$ and $R_1 \to \infty$ determine the performance envelope for a system with an ideal conversion zone. In particular, it is noted that, although I_{E+} and I_{E-} are increasing functions of the current of matter I_2, the growth rate of the energy waste current I_{E-} always ends up reaching that of the energy current I_{E+} supplied to the system. In addition, even in the case of a system whose core is composed of an ideal device, ($R_2 = 0, R_1 \to \infty$), the increase in the current of matter inexorably leads to an increase in the current of waste in larger proportions to the rate of supply of resources. The only way out is to limit the current of matter to values below a threshold, which may be that of maximum power, maximum efficiency, minimum waste generation, or below the critical point. In the Figure 3, the response is given for two different values of the coupling parameter α. The influence of α is quite surprising. At first we observe that the lower is α and the lower are the output power and efficiencies, as expected for a lower conversion level of the energy. However, in the same time, the short-circuit current is strongly enhanced, opening the way to a large range of I_2 working points for the transport of the matter. This is due to the α^{-2} dependency of I_{2sc}. This leads to the conclusion that *the search for a very efficient system is in contradiction with the search for a very adaptable system.*

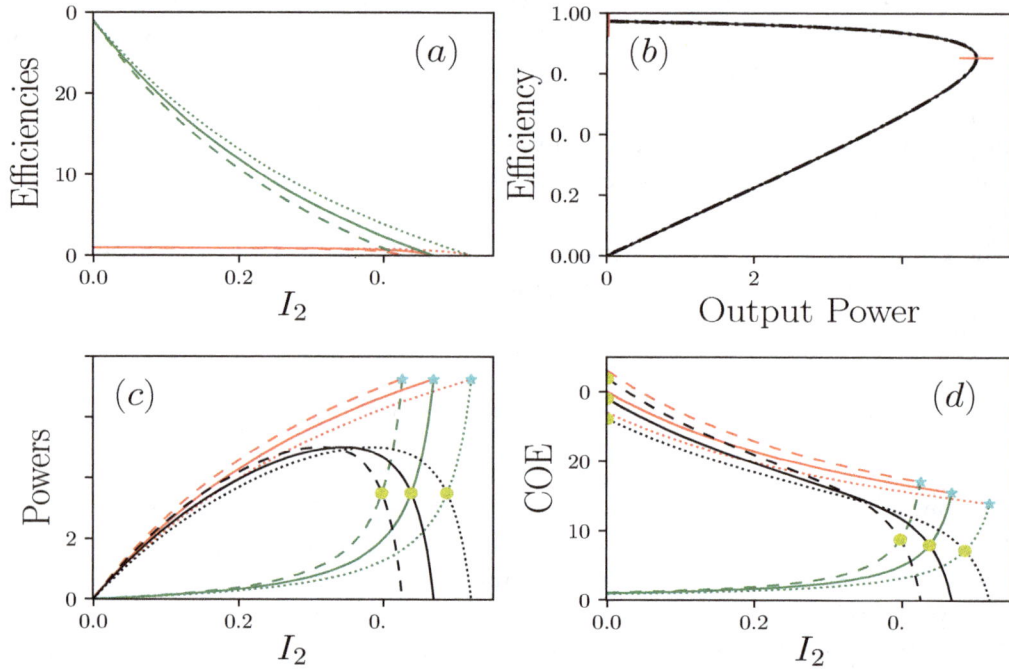

Figure 3. Representations of the powers I_{E+}, I_{E-}, with $R_1^{-1} = 0$, $R_2 = 0$ (and $P = P_{\text{out}}$), $R_+ = R_- = 2$, $\Pi_1^S = 1$, $\Pi_1^R = 30$. (a) shows efficiencies (resp. in red and green) $\eta_{+/-} = P_{\text{out}}/I_{2_{+/-}}$ in function of I_2, the current of matter. (b) is the efficiency in function of the power produced P_{out}. (c) show the power (resp. in red, green and black) I_{E+}, I_{E-} and P_{out} in function of I_2. (d) shows (resp. in red, green and black) $COE_{+/-} = I_{E_{+/-}}/I_2$ and $COE_{P_{\text{out}}} = P_{\text{out}}/I_2$ in function of I_2. Dotted lines are $\alpha = 0.9$, solid lines are $\alpha = 1$, dashed lines are $\alpha = 1.1$. In (c) and (d) cyan stars show short circuit situations I_{2sc}, yellow circles are critical points I_{2cp}. In (b) vertical and horizontal red lines are respectively maximal efficiency and maximal power.

Let us now focus on the issue of the trade-off between power efficiency and waste generation. The Figure 3a represents the curves of the production efficiency $\eta_{prod} = P/I_{E+}$ and the waste efficiency $\eta_{waste} = P/I_{E-}$. Note that η_{prod}, which is the traditional efficiency, is limited by the Carnot efficiency but η_{waste} is not, since it does not refer to the traditional expression of efficiency but is just an extension of the notations. η_{prod} is bounded by a zero value, which corresponds to zero power, and a maximum efficiency point, reported in Figure 3b. Between these two values, the system presents a maximum of the power, which absolutely does not coincide with the maximum efficiency. In this configuration the MPP or mEP operations are clearly disjointed as already mentioned [23,24]. Let us now consider the cost of carrying out a unitary process. By unitary process we consider a process standardized by the value of the associated transport of matter, i.e., the ratio between the energy currents and the matter current. We call this quantity Cost Of Energy, i.e., COE. This makes it possible to consider energy expenditures with regard to the associated matter transformation along a unitary displacements. In other words, COE can measure the amount of energy needed to be rejected as a waste, for displacing the matter from a unit length. This quantity is already known in biology as Cost Of Oxygen Transport (COT), where it has made it possible to qualify a unit displacement with regard to the energy released in the form of waste [25,26]. Here, we extend the notion in a more general form where COE is defined by COE_+ which is the cost of energy needed to feed the system, and COE_- which is the cost of waste energy that is rejected, so,

$$COE_{+/-} = \frac{I_{E+/-}}{I_2} \tag{24}$$

Note that the COE_+ is a strictly decreasing function of I_2 and COE_- is a strictly increasing function of I_2. This means that the cost of energy needed for a unitary process decrease when I_2 increases but, in the same time the amount of waste always increases. There is therefore no optimum to consider any minimization of the waste. In addition, it is important to note that the $R_1^{-1} = 0$ configuration is the only one that provides the strong coupling conditions, for which the energy and matter currents are roughly proportional [10]. In this case, the Onsager matrix has a zero determinant. This situation is an idealization of the transport of energy entirely achieved by the transport of matter. In other words, it is a question of considering that the behavior of out-of-equilibrium thermodynamics may be equivalently described by pure mechanics. This is obviously never fully encountered unless it is considered that a $\Delta\Pi_1$ difference can persist without an associated current of matter existing. This is the purpose of the following paragraph.

4.3. Devices with Non Zero Resting Point

The study of devices with non zero resting points concern the case of all systems for which a shutdown means death. Indeed, unlike a machine, all living systems are never totally shut down, and always keep a minimum operating point value , which we call basal, also known as a resting point. This situation corresponds to the case where R_1 has a finite value. While remaining, for the moment in the case where $R_2 = 0$, we can develop the main results from this configuration. The general equations of the system are given in Appendix B. In this situation, the efficiency, nor the power, can reach the previous values, as reported in the Figure 4. At the resting point $I_2 = 0$, the system is in its basal configuration where $P = 0$, so $I_{E+} = I_{E-} = B$ with,

$$B = \frac{\Delta\Pi_1}{R_+ + R_1 + R_-} \tag{25}$$

The typical response of systems with non zero resting points is given in the Figure 4. One can notice that the general shape is not strongly modified from the case of zero resting point configurations, except the presence of a non zero current of energy even at zero I_2 and a slight modification of the short-circuit point. Regardless of the reduction in efficiency introduced by the presence of R_1, the search for a system with a very low basal point requires to be located in a configuration close to Neumann conditions where R_+ and R_- have very large values. This is not problematic except that it requires the system to operate at low values of I_2, in order to limit the dissipation due to the term R_{fb}. There is therefore a fundamental contradiction between having a system with low resting power consumption and a system that can provide significant power. It is clear that a sober system, in the sense of its consumption at rest, is unsuited to the production of significant power, without leading to significant dissipation at high speed, or equivalently, high I_2. If such a power is sought, then it implies that the boundary conditions should be of Dirichlet like with $R_+ \approx R_- \approx 0$. However, in this case the system will have a necessarily high rest consumption. Compared to systems with a zero resting point, it can be seen that the maximum power operating point and maximum efficiency operating point tend to approach each other as R_1 increases. In this configuration, as can be derivated in [27], the feedback resistance is approximately given by

$$R_{fb} \approx \frac{\alpha^2 \langle \Pi_1 \rangle}{\frac{1}{R_+ + R_-} + \frac{1}{R_1}} = R^* \alpha^2 \langle \Pi_1 \rangle \tag{26}$$

where $R^* = \frac{(R_+ + R_-)R_1}{R_1 + R_+ + R_-}$ and $\langle \Pi_1 \rangle = \Pi_1^R/2 + \Pi_1^S/2$.

Compared to the previous configuration the dissipation introduced by the presence of R_{fb} can now be modified whatever are the boundary conditions because $R^* < Min(R_+ + R_-, R_1)$. More precisely, in the case of Neumann-like boundary conditions, there is a restriction to the value of R_1 where $R_1 \ll R_+ + R_-$ is expected. Under Dirichlet-like boundary conditions R_+ and R_- are small so there is no condition on R_1. Consequently, a system with a very low basal point, with large values of both (R_+, R_-) (Neumann like) and R_1 will suffer from a large R_{fb} and is then limited to very low I_2 currents. If the boundary conditions are more like Dirichlet conditions, then R_{fb} keeps low but the low basal level now imposes that R_1 strongly increases, which reduced both the available power P and the efficiency. Thus, we can see that there is no room for a powerful and efficient system working in all conditions. The main trade-off is between power and efficiency, but it ultimately extends beyond that.

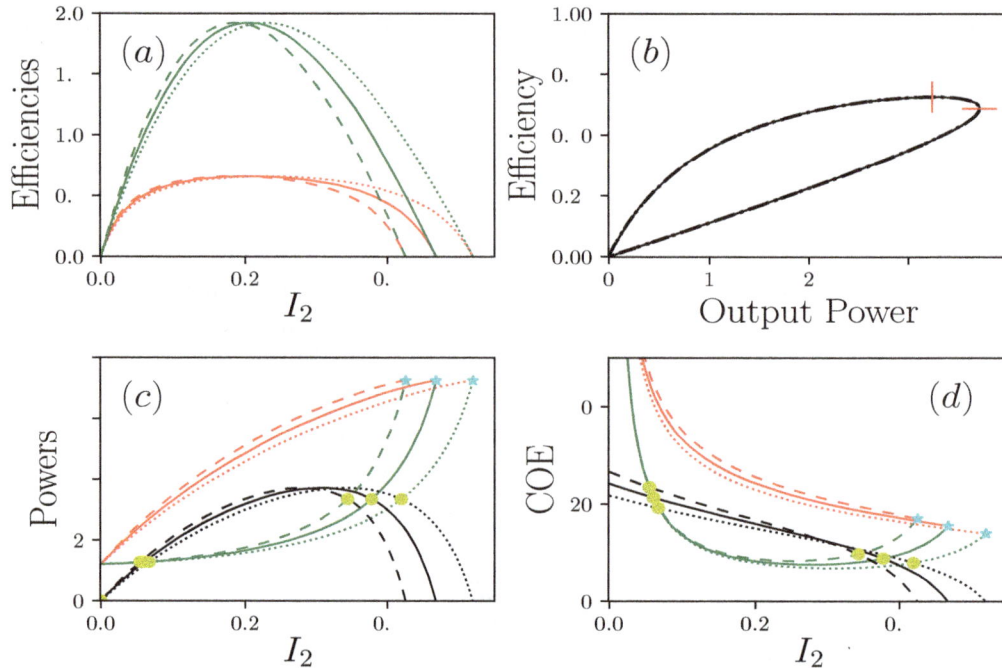

Figure 4. Representations of the powers I_{E+}, I_{E-} and P, with $R_1^{-1} = 0.05$, $R_2 = 0$ (and $P = P_{out}$), $R_+ = R_- = 2$, $\Pi_1^S = 1$, $\Pi_1^R = 30$. **(a)** shows efficiencies (resp. in red and green) $\eta_{+/-} = P_{out}/I_{2+/-}$ in function of I_2 the current of matter; **(b)** is the efficiency in function of the power produced P_{out}; **(c)** shows the power (resp. in red, green and black) I_{E+}, I_{E-} and P_{out} in function of I_2; **(d)** shows (resp. in red, green and black) $COE_{+/-} = I_{E+/-}/I_2$ and $COE_{P_{out}} = P/I_2$ in a function of I_2. Dotted lines are $\alpha = 0.9$, solid lines are $\alpha = 1$, dashed lines are $\alpha = 1.1$. In **(c)** and **(d)**, cyan stars show short circuit situations I_{2sc}, and yellow circles are critical points I_{2cp}. In **(b)**, vertical and horizontal red lines are respectively maximal efficiency and maximal power.

From a rather general point of view, the incoming energy current I_{E+} makes it possible to establish and maintain, thanks to the presence of R_1, a potential difference that permits the production of output work. On this point, we join the work of Alicki [16], who considers that the incoming energy current makes it possible to maintain a difference in potential, exactly as a pump would do. This situation is particularly described in the case of photovoltaic structures, with a difference in electrochemical potential [16], or in the case of muscles where the attachment and release cycles of actin and myosin structures lead to the maintenance of a force [28]. It should be noted that, depending on the position of the resting point, the power curve can intercept between zero and twice the I_{E+} curve. It can therefore be seen that, in the case of systems with a relatively low resting point, there may be an area for which the power produced is greater than the power released as a waste. More intriguing, this area can start with a non-zero value of I_2. In other words, there may be systems for which the situation $I_2 \neq 0$ leads to a proportionally smaller waste production than at rest. Systems with a non-zero resting point therefore present very different optima than non-living systems, whose zero resting point leads to minimizing power by stopping the machine. By using the definition $COE_- = I_{E-}/I_2$, we can plot its response according to I_2. It should be noted that the $COE-$ has a minimum value, which does not coincide with the maximum power point. This defines a new operating point for the system, which characterizes the situation where the system minimizes its production of waste.

An illustration of this can be given if we consider the motion of living systems. Let us consider that the task to be accomplished consists in moving the body over a unit distance, the question arises as to how fast this operation will lead to a minimum of waste, essentially in the form of heat and metabolic degradation products. It is clear that displacement here corresponds to the transport of matter, and is therefore assimilated to I_2 proportional to the speed of travel as previously said. There is an abundant amount of literature showing that there exists a minimum of the so-called $COT \equiv COE_-$ point for all animals for which movement appears to be favored when the COT is minimal [25,26]. As expected, see Figure 4, COE_- and $COE+$ curves have a common point at the short circuit point. We previously saw that Dirichlet's conditions, $R_+ = R_- = 0$, were those that minimized the feedback resistance R_{fb} and allowed for considering potentially a divergence of the current of matter and the output power. This simple observation shows that strict Dirichlet's conditions are simply nonphysical. Nevertheless, one can consider that this condition can be approached. However, the presence of R_+, R_- and R_1 in series shows that Dirichlet's condition is asymptotically obtained only if the ratios R_+/R_1 and R_-/R_1 are negligible, which imposes an important value for R_1, and therefore a high value of the basal power. *We therefore see the emergence of a paradox, which, seeking to minimize the dissipation due to R_{fb} leads to the constraint of high consumption at rest. The same system cannot therefore be both very powerful and very energy-efficient at its resting point.* We find here the generalization of a well-known situation, for example for the thermal engines of vehicles, in which the engine's displacement determines its ability to produce power, as well as its efficiency.

4.4. Internal Dissipation Devices

Let us now consider the introduction of the dissipative term R_2. The output power of the system is now represented by Figure 5. As a thermodynamic engine, the system provides a power $P = \alpha \left(\Pi_1^+ - \Pi_1^- \right) I_2$ as already defined. The efficiency of this part of the system is given by $\eta_2 = \frac{P - R_2 I_2^2}{P}$. Thus, the total efficiency of the system is

$$\eta_{sys} = \eta_1 \eta_2, \tag{27}$$

with $\eta_1 = \frac{P}{I_{E_+}}$. Compared to the previous configurations, both the power, the short-circuit current I_{sc}, and the efficiency are now reduced. The influence of R_2 appears to be always detrimental, which was not the case for R_1. It is clear that one should look for minimal R_2 if possible. In other words, in the expression of the figure of merit, there is a constraint on R_2. At first, both α and R_1 seem to be non-constrained, and the same figure of merit can be obtained for various values of the couple (α, R_1). Nevertheless, as we have mentioned, the present description shows that R_2 is linked in series with R_{fb}. Consequently, the constraint on R_2 can be relaxed to the condition $R_2 \ll R_{fb}$. According to the expression $R_{fb} \approx \frac{\alpha^2 \langle \Pi_1 \rangle}{\frac{1}{R_+ + R_-} + \frac{1}{R_1}}$, this leads to the condition $1 + \frac{R_1}{R_\Sigma} < \frac{\alpha^2 R_1}{R_2} \langle \Pi_1 \rangle$ where we recognize the figure of merit, so the condition becomes

$$1 + \frac{R_1}{R_\Sigma} < F_m, \tag{28}$$

where $R_\Sigma = R_+ + R_-$. According to the previous observation, the minimization of the dissipation occurring from the R_{fb} term imposes that $\frac{R_1}{R_\Sigma}$ should be large enough. Thus, we now get a supplementary condition for F_m. In this expression, the boundary conditions and the intrinsic performances of the device are considered together. Under Dirichlet conditions, $1 + \frac{R_1}{R_\Sigma}$ diverges so the system keeps its level of dissipation low only in the case of a very large figure of merit, and is forced to work at very low I_2 values. Under Neumann conditions, R_Σ diverges and then the condition on the figure of merit is then relaxed. Ideally, even when achieved asymptotically, one might want to achieve maximum power, as well as minimal waste production, combined with maximum efficiency. *We conclude that looking for maximum efficiency always leads to approaching the Carnot point, which is, even in an out-of-equilibrium description, the point where power production is canceled out.*

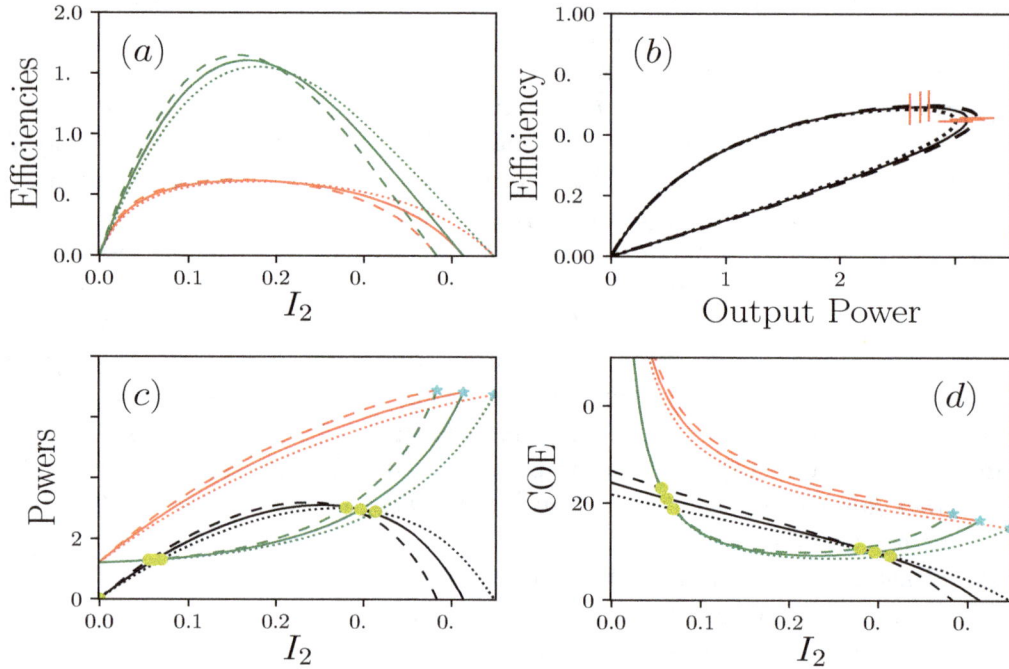

Figure 5. Different representations of the powers I_{E+}, I_{E-} and P, with $R_1^{-1} = 0.05$, $R_2 = 4$, $R_+ = R_- = 2$, $\Pi_1^S = 1$, $\Pi_1^R = 30$. (**a**) shows efficiencies (resp. in red and green) $\eta_{+/-} = P/I_{2_{+/-}}$ as a function of I_2 the current of matter. (**b**) is the efficiency in function of the power produced P. (**c**) shows the power (resp. in red, green and black) I_{E+}, I_{E-} and P_{out} in function of I_2. (**d**) shows (resp. in red, green and black) $COE_{+/-} = I_{E+/-}/I_2$ and $COE_{P_{\text{out}}} = P_{\text{out}}/I_2$ as a function of I_2. Dotted lines are $\alpha = 0.9$, solid lines are $\alpha = 1$, dashed lines are $\alpha = 1.1$. In (**c**) and (**d**), cyan stars show short circuit situations I_{2sc} and yellow circles are critical points I_{2cp}. In (**b**), vertical and horizontal red lines are respectively maximal efficiency and maximal power.

5. Entropic Point of View

The previous power budget analysis highlighted three classes of systems: systems with a zero resting point, systems with a non-zero resting point, and, finally, systems with an additional internal dissipation term R_2. Let us consider these three classes again from the entropic point of view.

5.1. Devices with Zero Resting Point

The production of entropy from the presence of R_- and R_+ is given respectively on both sides of the device by

$$\dot{S}_{E+} = I_{E+}\left(\frac{1}{\Pi_1^+} - \frac{1}{\Pi_1^R}\right) = \frac{\alpha^2 I_2^2 R_+}{1 + \alpha I_2 R_+}, \tag{29}$$

$$\dot{S}_{E-} = I_{E-}\left(\frac{1}{\Pi_1^S} - \frac{1}{\Pi_1^-}\right) = \frac{\alpha^2 I_2^2 R_-}{1 - \alpha I_2 R_-}. \tag{30}$$

The results are given in Figure 6.

There is clearly an asymmetry in the two entropy productions. Indeed, if the two contributions initially increase in a quadratic form with the current of matter, the contribution of the resource side, $\dot{S}_{E+}$, tends to a linear progression independent of the coupling condition R_+, while the contribution on the waste rejection side $\dot{S}_{E-}$ tends to diverge as soon as $I_2 \approx 1/\alpha R_-$. It is surprising to see that, in addition, this divergence is more marked as the coupling factor α between energy and matter is important. *There is therefore no other solution than to make R_- as small as possible, and therefore reject all the waste easily.* This is an additional constraint for the design of efficient systems.

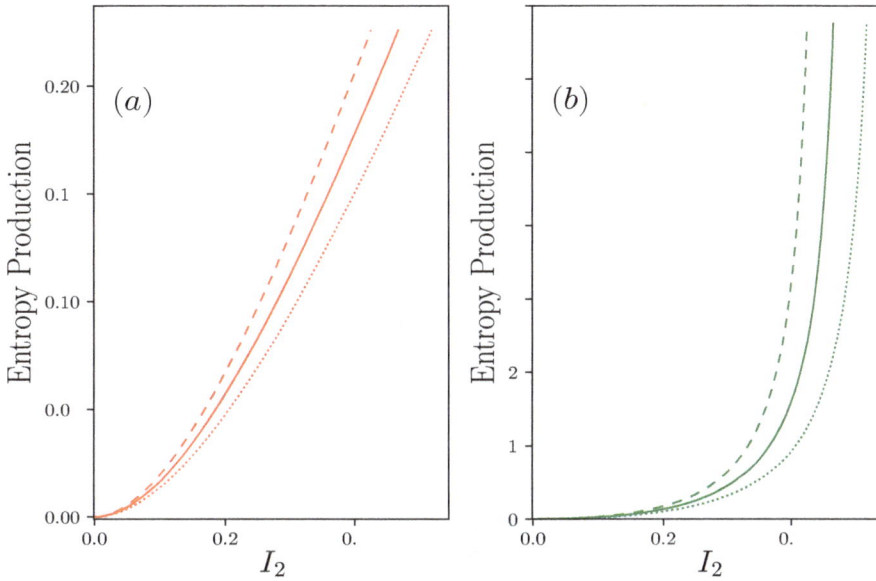

Figure 6. Evaluation of the entropy production with the same configuration as in Figure 3, $R_1^{-1} = 0$, $R_2 = 0$, $R_+ = R_- = 2$, $\Pi_1^S = 1$, $\Pi_1^R = 30$. (**a**) shows $\dot{S}_{E+}$ and (**b**) shows $\dot{S}_{E-}$, both in function of I_2 the current of matter—the same color and line-style code as in Figure 3.

5.2. Devices with Non-Zero Resting Points

Let us now look at the configuration of non-zero resting point systems, while keeping $R_2 \approx 0$. In this case, the general expressions become

$$\dot{S}_{E+} = I_{E+}\left(\frac{1}{\Pi_1^+} - \frac{1}{\Pi_1^R}\right) = \frac{R_+ I_{E+}^2}{\left(\Pi_1^R - R_+ I_{E+}\right)\Pi_1^R},\tag{31}$$

$$\dot{S}_{E-} = I_{E-}\left(\frac{1}{\Pi_1^S} - \frac{1}{\Pi_1^-}\right) = \frac{R_- I_{E-}^2}{\left(R_- I_{E-} + \Pi_1^S\right)\Pi_1^S}.\tag{32}$$

The results are given in Figure 7 where I_{E+} and I_{E-} are defined according to the Appendix B. We can see that the presence of R_1 reintroduces a significant symmetry between the two contributions to the entropy production. Moreover, the question of the importance of the quality of the coupling on the resource side, by minimizing R_+, or to the rejection side, by minimizing R_-, is now of equal importance.

5.3. Internal Dissipation Devices

For internally dissipated devices, the term R_2 produces a quadratic dissipation $R_2 I_2^2$. We have seen before that the presence of R_2 never brings any advantage in terms of energy conversion since it only contributes to lowering the power available at the output of the system. As this dissipation diffuses into the system, it is itself a source of entropy, as shown in Figure 8. At this stage, it is important to know how this dissipation occurs. In the case of some thermal systems, an analytical calculation can be carried out that leads to an equal distribution of this dissipation between the resource and the sink, i.e., $\varphi = 0.5$, see Appendix B in accordance with [22]. In other systems, such as muscles subjected to moderate stress, this dissipation is considered to be completely rejected into the sink ($\varphi = 0$) [14]. For some living systems, including homeothermic species, it is likely that a fraction of this dissipation is partially released, and partially used to maintain the central temperature of the body, leading to a value $\varphi \approx 1$, depending on outdoor conditions. One example is the case of vaso-dilatation and vasoconstriction of peripheral vessels, which is a solution for modulating the value of R_- and consequently reject less, or more, heat outside of the body.

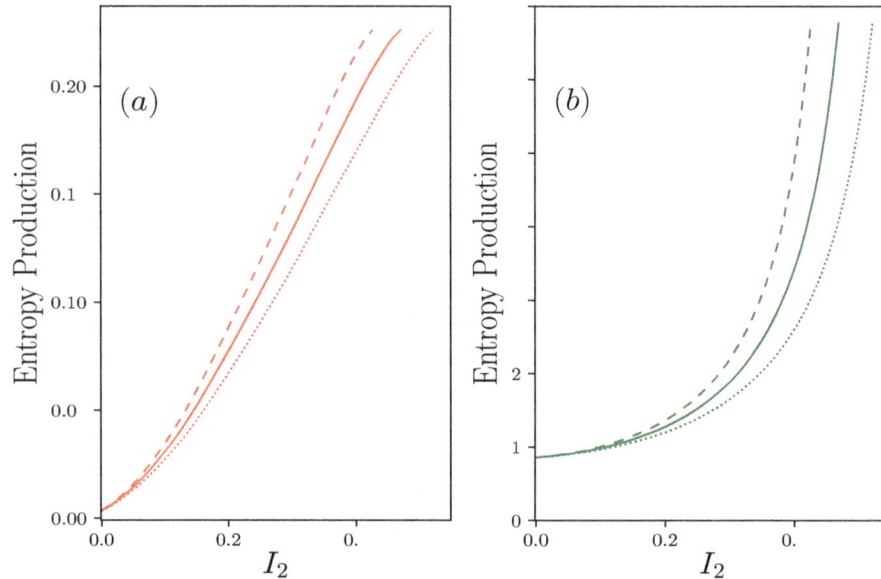

Figure 7. Evaluation of the entropy production with the same configuration as in Figure 4 with $R_1^{-1} = 0.05$, $R_2 = 0$, $R_+ = R_- = 2$, $\Pi_1^S = 1$, $\Pi_1^R = 30$. (**a**) shows $\dot{S}_{E+}$ and (**b**) shows $\dot{S}_{E-}$, both in function of I_2 the current of matter—the same color and line-style code as in Figure 3.

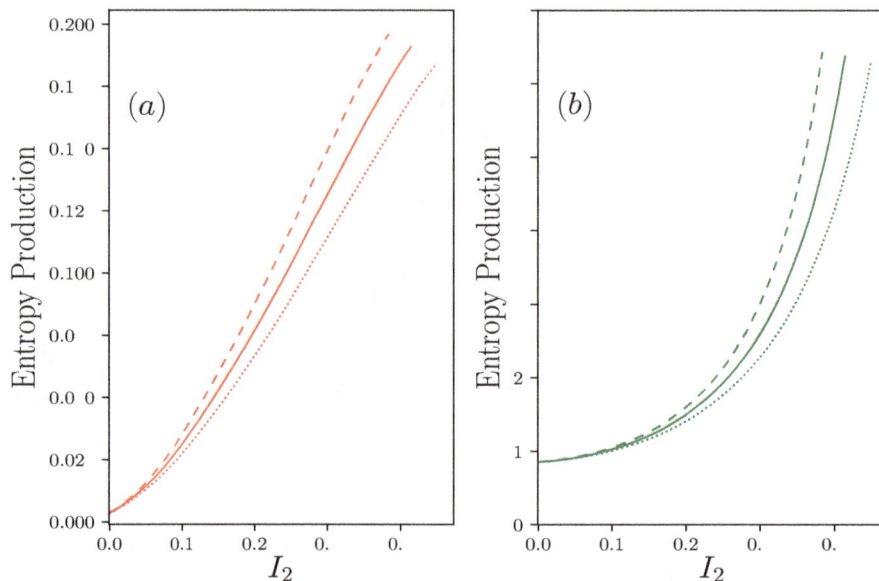

Figure 8. Plot of the entropy production with the same configuration as in Figure 5 with $R_1^{-1} = 0.05$, $R_2 = 4$, $R_+ = R_- = 2$, $\Pi_1^S = 1$, $\Pi_1^R = 30$. (**a**) shows $\dot{S}_{E+}$ and (**b**) shows $\dot{S}_{E-}$, both in function of I_2 the current of matter—the same color and linestyle code as in Figure 3.

6. Adaptable or Adapted?

The study of the behavior of a generic system composed of a conversion device, and the boundary conditions to the reservoirs, now allows us to establish several observations. First, the search for the best device, in terms of power and efficiency, can be summarized by the search for the largest figure of merit F_m. However, this result must be modulated by the fact that the value of F_m is determined by the set of the three parameters R_1, R_2, and α which, at this stage, do not present any constraints. In addition, few thermodynamic devices have a single operating point, but are generally expected to operate over a wide range of uses that principally means large range of I_2. In the precedent paragraph, we concluded that the greater the figure of merit, the smaller the effective operating range becomes. Indeed, for such a narrow range, the users must then conform quite strictly to that imposed by the value of the figure of merit of the device. This observation explains quite simply why the consumption observed by vehicle drivers is always larger than that reported by vehicle manufacturers, since the actual conditions of use never coincide with the test conditions. Similarly, the measured performance of equipment in dwellings, as well as the performance of the dwellings themselves, is below the expected performance during construction. This observation leads to the recommendation that devices intended to operate over a wide range of uses should not be designed solely on the basis of their maximum performance in terms of efficiency and power. Beyond this observation, the question arises of determining, within a system, which of the three parameters R_1, R_2, and α should be optimized as a priority. We can first conclude that, unless there are situations where dissipation is explicitly sought, R_2 must be systematically minimized. With regard to R_1, we have seen that its choice determines two categories of systems, depending on whether R_1 is zero or not. It must be noticed that $R_1 = 0$ is not possible for living systems because a resting point does exist until the death. In the category where $R_1 = 0$, the operating range of the system is limited by the feedback effects that introduce an excess dissipation term R_{fb}. Note that this term can be minimized if the boundary conditions are as close as possible to Dirichlet conditions. In this situation, the currents of matter I_2 and energy I_E may diverge. This situation has been that of our societies since the beginning of the industrial revolution [12], with coal, followed by an acceleration after the Second World War, due to the rise in oil consumption. The divergence of matter and energy currents is directly linked to an increase in the figure of merit, through an increased facilitation of the circulation of matters and energies, which is produced by a minimization of R_2, as well as an increase of α, i.e., technological progress that allows thermodynamic potentials to be more strongly coupled. A basic illustration of this increase is the performance of steam machines, which have gradually increased the ratio between outlet pressures and inlet temperatures [29]. The second category of system concerns the case where $R_1 \neq 0$. These systems are particular in that they consume energy, even in a resting situation. We can include living organisms and societies, but also machines, when the latter operate in the idle position, with no other power production than the maintenance of this idle. We have seen that, in this case, there are two categories of systems depending on whether we favor power production or low consumption at rest. These two categories are resolutely distinct and it is illusory to think of a system capable of producing a very high power, while maintaining a very low basic consumption. The choice of R_1, i.e., the dissipation at rest, is also decisive in the dissipation produced by feedback. The issues of minimization or maximization of efficiency and power are therefore part of a much broader framework than initially thought.

7. Discussion

We proposed a generic thermodynamic system model that allows for considering several situations of coupling of the energy and matter currents, as well as their conversions. At the local level, the intrinsic performance of the device that constitutes the core of the system was studied. It appears that the best intrinsic performance in terms of power and efficiency is obtained for the devices with the largest figure of merit, without specifying the respective contributions of the conductivities associated with the transport of energy or matter. However, the sensitivity of these devices to changes in the reduced current j shows that the intrinsically most efficient devices are also the most constraining because they require precise control of this reduced current, and therefore of energy and matter currents. At the scale of a complete system, the coupling to the external environment very strongly modifies the conclusions compared to the observations made at the local level. It is observed that behavior is mainly governed by the boundary conditions that connect the local system to the resource and the waste. The presence of boundary conditions such as Dirichlet or Neumann leads to a wide variety of behaviors. The ideal Dirichlet conditions are the only ones that do not lead to any feedback, and consequently conduct in the absence of limitations for the energy and matter currents. When the boundary conditions are between Dirichlet and Neumann, many possibilities then arise. The presence or absence of a resting point for the system strongly influences these possibilities in terms of power, but also in terms of waste production associated with the completion of a task. The concept of coefficient of energy cost, COE, is introduced, generalizing the classical COT already established for biological systems. Finally, it is observed that the internal dissipation produced by the presence of R_2 is always detrimental for both the efficiency and the power. Its only positive contribution is limited to cases where dissipation and entropy production are explicitly sought, as in the case of homeothermic animals.

Author Contributions: Conceptualization, C.G.; methodology, C.G. and E.H.; software, E.H.; validation, C.G. and E.H.; formal analysis, C.G.; writing—original draft preparation, C.G.; writing—review and editing, C.G. and E.H.; visualization, E.H.; supervision, C.G.; project administration, C.G.; funding acquisition, C.G. All authors have read and agreed to the published version of the manuscript.

Acknowledgments: The authors would like to thank Henni Ouerdane for enlightening discussions.

Appendix A

We consider the set of the four equations of the generic model:

$$I_{E-} = \alpha \Pi_1^- I_2 + (1-\varphi) R_2 I_2^2 + \frac{(\Pi_1^+ - \Pi_1^-)}{R_1}, \tag{A1}$$

$$I_{E-} = \frac{(\Pi_1^- - \Pi_1^S)}{R_-}, \tag{A2}$$

$$I_{E+} = \alpha \Pi_1^+ I_2 - \varphi R_2 I_2^2 + \frac{(\Pi_1^+ - \Pi_1^-)}{R_1}, \tag{A3}$$

$$I_{E+} = \frac{(\Pi_1^R - \Pi_1^+)}{R_+}. \tag{A4}$$

The φ term defines the fraction of the waste which is respectively rejected to the source and to the sink. This is a well known parameter in some thermal engines [22]. In the case of a living system, φ may define the ratio of heat rejected outside of the body and kept inside.

The resolution of the four equations gives

$$\begin{pmatrix} \Pi_1^- \\ \Pi_1^+ \end{pmatrix} = \frac{1}{AD - BC} \begin{pmatrix} D & -B \\ -C & A \end{pmatrix} \begin{pmatrix} \Pi_1^S + (1-\varphi) R_- R_2 I^2 \\ \Pi_1^R + \varphi R_+ R_2 I^2 \end{pmatrix}, \tag{A5}$$

with

$$A = 1 - \alpha R_- I + \frac{R_-}{R_1},$$

$$B = -\frac{R_-}{R_1},$$

$$C = -\frac{R_+}{R_1},$$

$$D = \alpha R_+ I + \frac{R_+}{R_1} + 1.$$

Appendix B

In the case of a system without dissipation, $R_2=0$, the general equations become

$$I_{E-} = \alpha \Pi_1^- I_2 + \frac{\Pi_1^+ - \Pi_1^-}{R_1}, \tag{A6}$$

$$\Pi_1^- = R_- I_{E-} + \Pi_1^S, \tag{A7}$$

$$I_{E+} = \alpha \Pi_1^+ I_2 + \frac{\Pi_1^+ - \Pi_1^-}{R_1}, \tag{A8}$$

$$\Pi_1^+ = \Pi_1^R - R_+ I_{E+}, \tag{A9}$$

which leads to

$$I_{E+} = \frac{\alpha I_2 \frac{R_1}{R_+} \Pi_1^S + \frac{\Delta \Pi_1}{R_+} + \alpha A I_2 \frac{R_1}{R_-} \Pi_1^R + A \frac{\Delta \Pi_1}{R_-}}{1 + AB}, \tag{A10}$$

$$I_{E-} = \frac{\alpha I_2 \frac{R_1}{R_-} \Pi_1^R + \frac{\Delta \Pi_1}{R_-} - \alpha B I_2 \frac{R_1}{R_+} \Pi_1^S - B \frac{\Delta \Pi_1}{R_+}}{1 + AB}, \tag{A11}$$

with

$$A = \left(\alpha I_2 R_1 \frac{R_-}{R_+} - \frac{R_1}{R_+} - \frac{R_-}{R_+} \right),$$

$$B = \left(\alpha I_2 R_1 \frac{R_+}{R_-} + \frac{R_+}{R_-} + \frac{R_1}{R_-} \right).$$

References

1. Kondepudi, D.; Kapcha, L. Entropy production in chiral symmetry breaking transitions. *Chirality* **2008**, *20*, 524–528. [CrossRef] [PubMed]
2. Grandy, W.T., Jr. *Entropy and the Time Evolution of Macroscopic Systems*; Oxford University Press: New York, NY, USA, 2008.
3. Martyushev, L.M.; Seleznev, V.D. Maximum entropy production principle in physics, chemistry and biology. *Phys. Rep.* **2006**, *426*, 1–45. [CrossRef]
4. Klein, M.J.; Meijer, P.H.E. Principle of Minimum Entropy Production. *Phys. Rev.* **1954**, *96*, 250–255. [CrossRef]
5. Lebon, G.; Jou, D. *Understanding Non-Equilibrium Thermodynamics: Foundations, Applications, Frontiers*; Springer: Berlin/Heidelberg, Germany, 2008.
6. Esposito, M.; Lindenberg, K.; Van den Broeck, C. Universality of Efficiency at Maximum Power. *Phys. Rev. Lett.* **2009**, *102*, 130602. [CrossRef]
7. Esposito, M.; Lindenberg, K.; Broeck, C.V.d. Thermoelectric efficiency at maximum power in a quantum dot. *EPL* **2009**, *85*, 60010. [CrossRef]
8. Balachandran, V.; Benenti, G.; Casati, G. Efficiency of three-terminal thermoelectric transport under broken time-reversal symmetry. *Phys. Rev. B* **2013**, *87*, 165419. [CrossRef]
9. Benenti, G.; Ouerdane, H.; Goupil, C. The thermoelectric working fluid: Thermodynamics and transport. *CR Phys.* **2016**, *17*, 1072–1083. [CrossRef]
10. Apertet, Y.; Ouerdane, H.; Goupil, C.; Lecoeur, P. Irreversibilities and efficiency at maximum power of heat engines: The illustrative case of a thermoelectric generator. *Phys. Rev. E* **2012**, *85*, 031116. [CrossRef]
11. Schmiedl, T.; Seifert, U. Efficiency at maximum power: An analytically solvable model for stochastic heat engines. *EPL* **2007**, *81*, 20003. [CrossRef]
12. Wrigley, E.A. Energy and the English Industrial Revolution. *Philos. Trans. R. Soc. A* **2013**, *371*, 20110568. [CrossRef]
13. Kedem, O.; Caplan, S.R. Degree of coupling and its relation to efficiency of energy conversion. *Trans. Faraday Soc.* **1965**, *61*, 1897–1911. [CrossRef]

14. Goupil, C.; Ouerdane, H.; Herbert, E.; Goupil, C.; D'Angelo, Y. Thermodynamics of metabolic energy conversion under muscle load. *New J. Phys.* **2019**, *21*, 023021. [CrossRef]

15. Goupil, C.; Ouerdane, H.; Herbert, E.; Benenti, G.; D'Angelo, Y.; Lecoeur, P. Closed-loop approach to thermodynamics. *Phys. Rev. E* **2016**, *94*, 032136. [CrossRef]

16. Alicki, R.; Gelbwaser-Klimovsky, D.; Jenkins, A. A thermodynamic cycle for the solar cell. *Ann. Phys.* **2017**, *378*, 71–87. [CrossRef]

17. Alicki, R.; Horodecki, M.; Horodecki, P.; Horodecki, R. Thermodynamics of Quantum Information Systems—Hamiltonian Description. *Open Syst. Inf. Dyn.* **2004**, *11*, 205–217. [CrossRef]

18. Onsager, L. Reciprocal Relations in Irreversible Processes. II. *Phys. Rev.* **1931**, *38*, 2265–2279. [CrossRef]

19. Onsager, L. Reciprocal Relations in Irreversible Processes. I. *Phys. Rev.* **1931**, *37*, 405–426. [CrossRef]

20. Apertet, Y.; Ouerdane, H.; Goupil, C.; Lecoeur, P. Revisiting Feynman's ratchet with thermoelectric transport theory. *Phys. Rev. E* **2014**, *90*, 012113. [CrossRef]

21. Goupil, C.; Seifert, W.; Zabrocki, K.; Müller, E.; Snyder, G.J. Thermodynamics of Thermoelectric Phenomena and Applications. *Entropy* **2011**, *13*, 1481–1517. [CrossRef]

22. Apertet, Y.; Ouerdane, H.; Goupil, C.; Lecoeur, P. From local force-flux relationships to internal dissipations and their impact on heat engine performance: The illustrative case of a thermoelectric generator. *Phys. Rev. E* **2013**, *88*, 022137. [CrossRef]

23. Novikov, I.I. The efficiency of atomic power stations (a review). *J. Nucl. Energy* **1958**, *7*, 125–128. [CrossRef]

24. Curzon, F.L.; Ahlborn, B. Efficiency of a Carnot engine at maximum power output. *Am. J. Phys.* **1975**, *43*, 22–24. [CrossRef]

25. Tucker, V.A. The Energetic Cost of Moving About: Walking and running are extremely inefficient forms of locomotion. Much greater efficiency is achieved by birds, fish—and bicyclists. *Am. Sci.* **1975**, *63*, 413–419. [PubMed]

26. Hoyt, D.F.; Taylor, C.R. Gait and the energetics of locomotion in horses. *Nature* **1981**, *292*, 239–240. [CrossRef]

27. Apertet, Y.; Ouerdane, H.; Glavatskaya, O.; Goupil, C.; Lecoeur, P. Optimal working conditions for thermoelectric generators with realistic thermal coupling. *EPL* **2012**, *97*, 28001. [CrossRef]

28. Huxley, A.F.; Simmons, R.M. Proposed Mechanism of Force Generation in Striated Muscle. *Nature* **1971**, *233*, 533–538. [CrossRef]

29. Jevons, W.S. *The Coal Question: An Inquiry Concerning the Progress of the Nation, and the Probable Exhaustion of Our Coal-Mines*; Macmillan: London, UK, 1866.

9

Non-Equilibrium Quantum Brain Dynamics: Super-Radiance and Equilibration in 2+1 Dimensions

Akihiro Nishiyama [1,*], Shigenori Tanaka [1,*] and Jack A. Tuszynski [2,3,4,*]

[1]   Graduate School of System Informatics, Kobe University, 1-1 Rokkodai, Nada-ku, Kobe 657-8501, Japan
[2]   Department of Oncology, University of Alberta, Cross Cancer Institute, Edmonton, AB T6G 1Z2, Canada
[3]   Department of Physics, University of Alberta, Edmonton, AB T6G 2J1, Canada
[4]   DIMEAS, Corso Duca degli Abruzzi, 24, Politecnico di Torino, 10129 Turin, TO, Italy
*   Correspondence: anishiyama@people.kobe-u.ac.jp (A.N.); tanaka2@kobe-u.ac.jp (S.T.);
     jackt@ualberta.ca (J.A.T.)

Abstract: We derive time evolution equations, namely the Schrödinger-like equations and the Klein–Gordon equations for coherent fields and the Kadanoff–Baym (KB) equations for quantum fluctuations, in quantum electrodynamics (QED) with electric dipoles in $2 + 1$ dimensions. Next we introduce a kinetic entropy current based on the KB equations in the first order of the gradient expansion. We show the H-theorem for the leading-order self-energy in the coupling expansion (the Hartree–Fock approximation). We show conserved energy in the spatially homogeneous systems in the time evolution. We derive aspects of the super-radiance and the equilibration in our single Lagrangian. Our analysis can be applied to quantum brain dynamics, that is QED, with water electric dipoles. The total energy consumption to maintain super-radiant states in microtubules seems to be within the energy consumption to maintain the ordered systems in a brain.

Keywords: non-equilibrium quantum field theory; quantum brain dynamics; Kadanoff–Baym equation; entropy; super-radiance

1. Introduction

Numerous attempts to understand memory in a brain have been made over one hundred years starting at the end of 19th century. Nevertheless, the concrete mechanism of memory still remains an open question in conventional neuroscience [1–3]. Conventional neuroscience is based on classical mechanics with neurons connected by synapses. However, we still cannot answer how limited connections between neurons describe mass excitations in a brain in classical neuron doctrine.

Quantum field theory (QFT) of the brain or quantum brain dynamics (QBD), is one of the hypotheses expected to describe the mechanism of memory in the brain [4–6]. Experimentally, several properties of memory, namely the diversity, the long-term but imperfect stability and nonlocality (Memory is diffused and non-localized in several domains in a brain. It does not disappear due to the destruction in a particular local domain. The term 'nonlocality' does not indicate nonlocality in entanglement in quantum mechanics.), are suggested in [7–9]. The QBD can describe these properties by adopting infinitely physically or unitarily inequivalent vacua in QFT, distinguished from quantum mechanics which cannot describe unitarily inequivalence. Unitarily inequivalence represents the emergence of the diversity of phases and allows the possibility of spontaneous symmetry breaking (SSB) [10–13]. The vacua or the ground states appearing in SSB describe the stability of the states. Furthermore, the QFT can describe both microscopic degrees of freedom and macroscopic matter [10].

To describe stored information, we can adopt the macroscopic ordered states in QFT with SSB involving long-range correlation via Nambu–Goldstone (NG) quanta. In 1967, Ricciardi and Umezawa proposed a quantum field theoretical approach to describe memory in a brain [14]. They adopted the SSB with long-range correlations mediated by NG quanta in QFT. Stuart et al. developed QBD by assuming a brain as a mixed system of classical neurons and quantum degrees of freedom, namely corticons and exchange bosons [15,16]. The vacua appearing in SSB, the macroscopic order, are interpreted as the memory storage in QBD. The finite number of excitations of NG modes represents the memory retrieval. Around the same time, Fröhlich proposed the application of a theory of electric dipoles to the study of biological systems [17–22]. He suggested a theory of the emergence of a giant dipole in open systems with breakdown of rotational symmetry of dipoles where dipoles are aligned in the same direction (the ordered states with coherent wave propagation of dipole oscillation in the Fröhrich condensate). In 1976, Davydov and Kislukha studied a theory of solitary wave propagation in protein chains, called the Davydov soliton [23]. It is found that the theory by Fröhlich and that by Davydov represent static and dynamical properties in the nonlinear Schödinger equation with an equivalent quantum Hamiltonian, respectively [24]. In the 1980s, Del Giudice et al. applied a theory of water electric dipoles to biological systems [25–28]. In particular, the derivation of laser-like behavior is a suggestive study. In the 1990s, Jibu and Yasue gave a concrete picture of corticons and exchange bosons, namely water electric dipole fields and photon fields [4,29–32]. The QBD is nothing but quantum electrodynamics (QED) with water electric dipole fields. When electric dipoles are aligned in the same directions coherently, the polaritons, NG bosons in SSB of rotational symmetry, emerge. The dynamical order in the vacua in SSB is maintained by long-range correlation of the massless NG bosons. In QED, the NG bosons are absorbed by photons and then photons acquire mass due to the Higgs mechanism and can stay in coherent domains. The massive photons are called evanescent photons. The size of a coherent domain is in the order of $50\,\mu m$. Furthermore, two quantum mechanisms of information transfer and integration among coherent domains are suggested. The first one is to use the super-radiance and the self-induced transparency via microtubules connecting two coherent domains [31]. Super-radiance is the phenomenon indicating coherent photon emission with correlation among not only photons but also atoms (or dipoles) [33–37]. The atoms (or dipoles) cooperatively decay in a short time interval due to correlation; coherent photons with intensity proportional to the square of the number of atoms (or dipoles) are emitted. The pulse wave photons in super-radiance propagate through microtubules without decay. Then the self-induced transparency appears, since microtubules are perfectly transparent in the propagation. The second one is to use the quantum tunneling effect among coherent domains surrounded by incoherent domains [32]. The effect is essentially equivalent to the Josephson effect between two superconducting domains separated by a normal domain. Del Giudice et al. studied this effect in biological systems [28]. In 1995, Vitiello has shown that a huge memory capacity can be realized by regarding a brain as an open dissipative system and doubling the degrees of freedom with mathematical techniques in thermo-field dynamics [38]. In dissipative model of a brain, each memory state evolves in classical deterministic trajectory like a chaos [39]. The overlap among distinct memory states is zero at any time in the infinite volume limit. However, finite volume effects allow states to overlap one another, which might represent association of memories [6]. In 2003, exclusion zone (EZ) water was discovered experimentally [40]. The properties of EZ water correspond to those of coherent water [41].

However, we have never seen the dynamical memory formations based on QBD at the physiological temperature in the presence of thermal effects written by quantum fluctuations. Hence, there are still criticisms related with the decoherence phenomena (We should use the mass of polaritons in estimating the critical temperature of ordered states, not that of water molecules themselves.)in memory formations in QBD [42]. So, we need to derive time evolution equations of coherent fields and quantum fluctuations and show numerical simulations of memory formation processes in non-equilibrium situations to check whether or not memory in QBD is robust against thermal effects. Futhermore, in 2012 Craddock et al. suggested the mechanism of memory coding in microtubules with phosphorylation by Ca^{2+} calmodulin kinase II [43]. It will be an interesting topic to investigate how water electric dipoles and evanescent photons are affected by phosphorylated microtubules.

The aim of this paper is to derive time evolution equations, namely the Schrödinger-like equations for coherent dipole fields, the Klein–Gordon equations for coherent photon fields, the Kadanoff–Baym equations for quantum fluctuations [44–46], with the two-particle-irreducible effective action technique with Keldysh formalism [47–51]. We derive both the equilibration for quantum fluctuations and the super-radiance for background coherent fields from the single Lagrangian in quantum electrodynamics (QED) with electric dipole fields. We arrive at the Maxwell–Bloch equations for the super-radiance by starting with QED with electric dipole fields in $2 + 1$ dimensions. When we consider electric fields in super-radiance, we only need two spatial dimensions, one axis for the amplitude and another axis for the propagation. Hence we have discussed the case in $2 + 1$ dimensions in this paper. By using our equations for super-radiance in this paper, we can describe information transfer via microtubules. Then, microtubule-associated proteins can make an important contribution to information transfer with interconnections among microtubules. We also derive the Higgs mechanism and the tachyonic instability for coherent fields in the Klein–Gordon equation for coherent electric fields. In two energy level approximation for electric dipole fields, namely with the ground state and the first excited states, the Higgs mechanism appears in normal population in which the probability amplitude in the ground state is larger than that in the first excited states. The penetrating length in the Meissner effect due to the Higgs mechanism is $6.3\,\mu m$ derived by using coefficients in $2 + 1$ dimensions and the number density of liquid water molecules in $3 + 1$ dimensions. On the other hand, the tachyonic instability appears in inverted population in which the probability amplitudes in the first excited states are larger than that in the ground state. Then the electric field increases exponentially while the system is in inverted population. The increase stops at times when normal population is realized. Our analysis also contains the dynamics of quantum fluctuations in non-equilibrium cases. We also derive the Kadanoff–Baym equations for quantum fluctuations with the leading-order self-energy in the coupling expansion. The Kadanoff–Baym equations describe the entropy producing dynamics during equilibration as shown in the proof of the H-theorem. Entropy production stops when the Bose–Einstein distribution is realized. By combining time evolution equations (the Klein–Gordon equations for coherent electric fields and the Schrödinger-like equations for coherent electric dipole fields) and the Kadanoff–Baym equations for quantum fluctuations, we can describe the dynamical behavior of dipoles with thermal effects written by quantum fluctuations. Our analysis will be applied to memory formation processes in QBD. In particular, by extending our method to the case in open systems (networks), we can also trace dynamical memory recalling processes with excitations of particles in coherent domains via quantum tunneling processes, which are described by the Kadanoff–Baym equations. We can perform the simulations of the dynamical recalling processes in QBD with our equations to understand our thinking processes.

This paper is organized as follows. In Section 2, we introduce the two-particle-irreducible effective action in the closed-time path contour to describe non-equilibrium phenomena and derive time evolution equations. In Section 3, we introduce a kinetic entropy current in the first order of the gradient expansion, and show the H-theorem in the leading-order approximation of the coupling expansion. In Section 4, we show the time evolution equations, the conserved total energy and the potential energy in spatially homogeneous systems in an isolated system. In Section 5, we derive the super-radiance by analyzing the time evolution equations for coherent fields. In Section 6, we discuss our results. In Section 7, we provide the concluding remarks. In the Appendix A, we show how quantum fluctuations appear as additional terms in the Klein–Gordon equations. In this paper, the labels $i, j = 1$ and 2 represent x and y directions in space, the labels $a, b, c, d = 1, 2$ represent two contours in the closed-time path, the labels $\alpha = -1, 1$ represent the angular momentum of electric dipoles. The speed of light, the Planck constant divided by 2π and the Boltzmann constant are set to be 1 in this paper. We adopt the metric tensor $\eta^{\mu\nu} = \text{diag}(1, -1, -1)$ with $\mu, \nu = 0, 1, 2$.

2. The Two-Particle-Irreducible Effective Action and Time Evolution Equations

We begin with the following Lagrangian density to describe quantum electrodynamics (QED) with electric dipoles in $2 + 1$ dimensions in the background field method [52–55],

$$
\begin{aligned}
\mathcal{L}[\Psi^*(x,\theta), \Psi(x,\theta), A(x), a(x)] \;=\; & -\frac{1}{4} F^{\mu\nu}[A + a] F_{\mu\nu}[A + a] - \frac{(\partial^\mu a_\mu)^2}{2\alpha_1} \\
& + \int_0^{2\pi} d\theta \left[\Psi^* i \frac{\partial}{\partial x^0} \Psi + \frac{1}{2m} \Psi^* \nabla_i^2 \Psi \right. \\
& \left. + \frac{1}{2I} \Psi^* \frac{\partial^2}{\partial \theta^2} \Psi - 2ed_e \Psi^* u^i \Psi F^{0i}[A + a] \right],
\end{aligned}
\tag{1}
$$

where A is the background coherent photon fields, a is the quantum fluctuations of photon fields, $F^{\mu\nu}[A] = \partial^\mu A^\nu - \partial^\nu A^\mu$ is the field strength, the α_1 is a gauge fixing parameter, the m is the mass of a dipole, the I is the moment of inertia, $u^i = (\cos\theta, \sin\theta)$ is the direction of dipoles and $2ed_e$ is the absolute value of dipole vector. The variable θ represents the degrees of freedom of rotation of dipoles in $2 + 1$ dimensions. The dipole–photon interaction term $-2ed_e \Psi^* u^i \Psi F^{0i}[A + a]$ has the similar form to that in [27]. We shall expand the electric dipole fields Ψ and Ψ^* by the angular momentum and consider only the ground state and the first excited states in energy-levels. Then we can write them as,

$$
\begin{aligned}
\Psi(x,\theta) \;=\; & \frac{1}{\sqrt{2\pi}} \left(\psi_0(x) + \psi_1(x) e^{i\theta} + \psi_{-1}(x) e^{-i\theta} \right), \\
\Psi^*(x,\theta) \;=\; & \frac{1}{\sqrt{2\pi}} \left(\psi_0^*(x) + \psi_1^*(x) e^{-i\theta} + \psi_{-1}^*(x) e^{i\theta} \right),
\end{aligned}
\tag{2}
$$

in $2 + 1$ dimensions. (In $3 + 1$ dimensions, we might expand Ψ and Ψ^* by spherical harmonics.) We can rewrite the terms in the above Lagrangian as,

$$
\int d\theta \Psi^*(x,\theta) i \frac{\partial}{\partial x^0} \Psi(x,\theta) \;=\; \psi_0^* i \frac{\partial}{\partial x^0} \psi_0 + \psi_1^* i \frac{\partial}{\partial x^0} \psi_1 + \psi_{-1}^* i \frac{\partial}{\partial x^0} \psi_{-1},
\tag{3}
$$

$$
\int d\theta \frac{1}{2m} \Psi^* \nabla_i^2 \Psi \;=\; \frac{1}{2m} \left[\psi_0^* \nabla_i^2 \psi_0 + \psi_1^* \nabla_i^2 \psi_1 + \psi_{-1}^* \nabla_i^2 \psi_{-1} \right],
\tag{4}
$$

$$
\int d\theta \frac{1}{2I} \Psi^* \frac{\partial^2}{\partial \theta^2} \Psi \;=\; \frac{-1}{2I} \left[\psi_1^* \psi_1 + \psi_{-1}^* \psi_{-1} \right].
\tag{5}
$$

We also write the dipole–photon interaction term with electric fields $F^{0i} = -E_i$ by,

$$
\begin{aligned}
\int d\theta 2ed_e \Psi^* u^i \Psi E_i \;=\; & ed_e \int d\theta \left[(E_1 - iE_2) \Psi^* e^{i\theta} \Psi + (E_1 + iE_2) \Psi^* e^{-i\theta} \Psi \right] \\
\;=\; & ed_e \left[(E_1 - iE_2)(\psi_0^* \psi_{-1} + \psi_1^* \psi_0) + (E_1 + iE_2)(\psi_0^* \psi_1 + \psi_{-1}^* \psi_0) \right],
\end{aligned}
\tag{6}
$$

with the direction of dipoles $u^i = (\cos\theta, \sin\theta)$.

Next, we show two-particle-irreducible (2PI) effective action [47–49] for electric dipole fields and photon fields. Starting with the above Lagrangian density, we write the generating functional with the gauge fixing condition for quantum fluctuation,

$$
\text{gauge fixing} : a^0 = 0,
\tag{7}
$$

and perform the Legendre transformations. Then we arrive at,

$$
\begin{aligned}
\Gamma_{2\mathrm{PI}}[A, \bar{a}^i \bar{\psi}, \bar{\psi}^*] \;=\;& \int_{\mathcal{C}} d^{d+1}x \left[-\frac{1}{4} F^{\mu\nu}[A + \bar{a}] F_{\mu\nu}[A + \bar{a}] + i\bar{\psi}_0^* \frac{\partial}{\partial x_0} \bar{\psi}_0 + \sum_{\alpha=-1,1} i\bar{\psi}_\alpha^* \frac{\partial}{\partial x_0} \bar{\psi}_\alpha \right. \\
& + \frac{1}{2m} \left(\bar{\psi}_0^* \nabla_i^2 \bar{\psi}_0 + \sum_{\alpha=-1,1} \bar{\psi}_\alpha^* \nabla_i^2 \bar{\psi}_\alpha \right) - \frac{1}{2I} \sum_{\alpha=-1,1} \bar{\psi}_\alpha^* \bar{\psi}_\alpha \\
& \left. + ed_e \sum_{\alpha=-1,1} [(E_1 + i\alpha E_2)(\bar{\psi}_0^* \bar{\psi}_\alpha + \bar{\psi}_{-\alpha}^* \bar{\psi}_0)] \right] \\
& + i\mathrm{Tr}\ln \Delta^{-1} + i\mathrm{Tr}\Delta_0^{-1}\Delta + \frac{i}{2}\mathrm{Tr}\ln D^{-1} + \frac{i}{2}\mathrm{Tr}D_0^{-1}D + \frac{\Gamma_2[\Delta, D]}{2},
\end{aligned}
\tag{8}
$$

where the $\mathcal{C}$ represents the Keldysh contour [50,51] shown in Figure 1, the spatial dimension $d = 2$, the bar represents the expectation value $\langle \cdot \rangle$ with the density matrix. The 3×3 matrix $i\Delta_0^{-1}(x, y)$ is defined as follows,

$$
\begin{aligned}
i\Delta_0^{-1}(x, y) \;\equiv\;& \left. \frac{\delta^2 \int_x \mathcal{L}}{\delta\psi^*(y)\delta\psi(x)} \right|_{a=0} \\
=\;& \begin{bmatrix} i\frac{\partial}{\partial x^0} + \frac{\nabla_i^2}{2m} - \frac{1}{2I} & ed_e(E_1 + iE_2) & 0 \\ ed_e(E_1 - iE_2) & i\frac{\partial}{\partial x^0} + \frac{\nabla_i^2}{2m} & ed_e(E_1 + iE_2) \\ 0 & ed_e(E_1 - iE_2) & i\frac{\partial}{\partial x^0} + \frac{\nabla_i^2}{2m} - \frac{1}{2I} \end{bmatrix} \delta_{\mathcal{C}}^{d+1}(x - y),
\end{aligned}
\tag{9}
$$

for $-1, 0$ and 1, and the $iD_{0,ij}^{-1}(x, y)$ is written by,

$$
\begin{aligned}
iD_{0,ij}^{-1}(x, y) \;\equiv\;& \frac{\delta^2 \int_x \mathcal{L}}{\delta a^i(x)\delta a^j(y)} \\
=\;& -\delta_{ij}\partial_x^2 \delta_{\mathcal{C}}^{d+1}(x - y),
\end{aligned}
\tag{10}
$$

where i and j run over spatial components $1, \cdots, d = 2$ in $2+1$ dimensions. The 3×3 matrix $\Delta(x, y)$ is,

$$
\Delta(x, y) \;=\; \begin{bmatrix} \Delta_{-1-1}(x, y) & \Delta_{-10}(x, y) & \Delta_{-11}(x, y) \\ \Delta_{0-1}(x, y) & \Delta_{00}(x, y) & \Delta_{01}(x, y) \\ \Delta_{1-1}(x, y) & \Delta_{10}(x, y) & \Delta_{11}(x, y) \end{bmatrix},
\tag{11}
$$

where $\Delta_{-10}(x, y) = \langle T_{\mathcal{C}} \delta\psi_{-1}(x)\delta\psi_0^*(y)\rangle$ with time-ordered product $T_{\mathcal{C}}$ in the closed-time path contour. The Green's function of dipole fields $\Delta_{-10}(x, y)$ is also written by the 2×2 matrix $\Delta_{-10}^{ab}(x, y)$ with $a, b = 1, 2$ in the contour. The Green's function for photon fields $D_{ij}(x, y)$ represents,

$$
D_{ij}(x, y) = \langle T_{\mathcal{C}} a_i(x) a_j(y)\rangle.
\tag{12}
$$

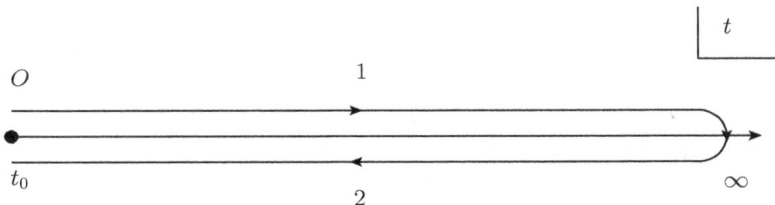

Figure 1. Closed-time path contour $\mathcal{C}$. The label "1" represents the path from t_0 to ∞ and the label "2" represents the path from ∞ to t_0.

Finally we write time evolution equations for coherent fields and quantum fluctuations. The 2PI effective action satisfies the following equations,

$$\left.\frac{\delta\Gamma_{2PI}}{\delta\Delta}\right|_{\bar{a}=0} = 0, \tag{13}$$

$$\left.\frac{\delta\Gamma_{2PI}}{\delta D}\right|_{\bar{a}=0} = 0, \tag{14}$$

$$\left.\frac{\delta\Gamma_{2PI}}{\delta\bar{a}^i}\right|_{\bar{a}=0} = \left.\frac{\delta\Gamma_{2PI}}{\delta A^i}\right|_{\bar{a}=0} = 0, \tag{15}$$

$$\left.\frac{\delta\Gamma_{2PI}}{\delta\bar{\psi}_{-1,0,1}^{(*)}}\right|_{\bar{a}=0} = 0, \tag{16}$$

due to the Legendre transformation of the generating functional. Equation (13) is written by,

$$i\Delta_0^{-1} - i\Delta^{-1} - i\Sigma = 0, \tag{17}$$

with $i\Sigma \equiv -\frac{1}{2}\frac{\delta\Gamma_2}{\delta\Delta}$. The matrix of self-energy Σ can be written by diagonal elements,

$$\Sigma = \text{diag}(\Sigma_{-1-1}, \Sigma_{00}, \Sigma_{11}), \tag{18}$$

since we can neglect the off-diagonal elements which are higher order of the coupling expansion. Equation (17) represents the Kadanoff–Baym equations for electric dipole fields in the two-energy-level approximation in $2+1$ dimensions. Similarly, the Kadanoff–Baym equation for photon fields in Equation (14) is written by,

$$iD_0^{-1} - iD^{-1} - i\Pi = 0, \tag{19}$$

with $i\Pi \equiv -\frac{\delta\Gamma_2}{\delta D}$. Equation (15) is given by,

$$\partial^\nu F_{\nu i} = J_i, \tag{20}$$

with,

$$J_1(x) = -ed_e\frac{\partial}{\partial x^0}\sum_{\alpha=-1,1}\left(\Delta_{0\alpha}(x,x) + \Delta_{\alpha0}(x,x) + \bar{\psi}_0(x)\bar{\psi}_\alpha^*(x) + \bar{\psi}_\alpha(x)\bar{\psi}_0^*(x)\right), \tag{21}$$

$$J_2(x) = -ed_e\frac{\partial}{\partial x^0}\sum_{\alpha=-1,1}\left(-i\alpha(\Delta_{0\alpha}(x,x) - \Delta_{\alpha0}(x,x) + \bar{\psi}_0(x)\bar{\psi}_\alpha^*(x) - \bar{\psi}_\alpha(x)\bar{\psi}_0^*(x))\right). \tag{22}$$

Equation (20) represents the Klein–Gordon equations for spatial dimensions $i = 1$ and 2. Equation (16) is written by,

$$\left(i\frac{\partial}{\partial x^0} + \frac{\nabla_i^2}{2m}\right)\bar{\psi}_0 + \sum_{\alpha=-1,1}ed_e(E_1 + i\alpha E_2)\bar{\psi}_\alpha = 0, \tag{23}$$

$$\left(i\frac{\partial}{\partial x^0} + \frac{\nabla_i^2}{2m} - \frac{1}{2I}\right)\bar{\psi}_\alpha + ed_e(E_1 - i\alpha E_2)\bar{\psi}_0 = 0, \tag{24}$$

and their complex conjugates. They are Schrödinger-like equations for coherent dipole fields. Equations (23) and (24) and their complex conjugates give the following probability conservation,

$$\frac{\partial}{\partial x^0}\left(\bar{\psi}_0^*\bar{\psi}_0 + \sum_{\alpha=-1,1}\bar{\psi}_\alpha^*\bar{\psi}_\alpha\right) + \frac{1}{2mi}\nabla_i\left(\bar{\psi}_0^*\nabla_i\bar{\psi}_0 - \bar{\psi}_0\nabla_i\bar{\psi}_0^* + \sum_{\alpha=-1,1}(\bar{\psi}_\alpha^*\nabla_i\bar{\psi}_\alpha - \bar{\psi}_\alpha\nabla_i\bar{\psi}_\alpha^*)\right) = 0. \quad (25)$$

We shall define $J_0(x)$ as,

$$\begin{aligned}
J_0(x) &= -ed_e\frac{\partial}{\partial x^1}\sum_{\alpha=-1,1}\left(\Delta_{0\alpha}(x,x) + \Delta_{\alpha0}(x,x) + \bar{\psi}_0(x)\bar{\psi}_\alpha^*(x) + \bar{\psi}_\alpha(x)\bar{\psi}_0^*(x)\right)\\
&\quad -ed_e\frac{\partial}{\partial x^2}\left(-i\alpha(\Delta_{0\alpha}(x,x) - \Delta_{\alpha0}(x,x) + \bar{\psi}_0(x)\bar{\psi}_\alpha^*(x) - \bar{\psi}_\alpha(x)\bar{\psi}_0^*(x))\right).
\end{aligned} \quad (26)$$

Then since we can use $\partial_0 J_0 - \nabla_i J_i = 0$ with $i = 1, 2$,

$$\partial_0 J_0 = \nabla_i J_i = -\partial^i\partial^\nu F_{\nu i} = \partial^\mu\partial^\nu F_{\nu\mu} - \partial^i\partial^\nu F_{\nu i} = \partial^0\partial^\nu F_{\nu 0},$$
$$\text{or, } \partial^\nu F_{\nu 0} = J_0, \quad (27)$$

where the time dependent term in the time integral might be interpreted as an initial charge, but it is set to be zero. This equation represents the Poisson equation for scalar potential A^0 given by $\nabla^2 A^0 = \nabla \cdot \mu$ with the vector of dipole moments $-\mu$ on the right-hand side in Equation (26). (Since the Fourier transformed $\tilde{A}^0(\mathbf{q})$ is written by $\tilde{A}^0(\mathbf{q}) \propto (q^i\tilde{\mu}_i)/\mathbf{q}^2$ with $\mu_i = \tilde{\mu}_i\delta(\mathbf{r})$, the electric field $E_j = -\nabla_j A^0(\mathbf{r})$ is proportional to $\int_{\mathbf{q}} e^{i\mathbf{q}\cdot\mathbf{r}}\frac{q^jq^i\tilde{\mu}_i}{\mathbf{q}^2}$. If we can also apply the analysis in this section to the case in $3+1$ dimensions, we find $E_j \propto \partial_j\partial_i\frac{\tilde{\mu}_i}{r}$. Then we obtain dipole–dipole interaction potential $-\tilde{\mu}_j E_j \sim \left[\frac{\tilde{\mu}_j\tilde{\mu}_j}{r^3} - \frac{3(r_i\tilde{\mu}_i)(r_j\tilde{\mu}_j)}{r^5}\right]$ in $3+1$ dimensions.)

3. Kinetic Entropy Current in the Kadanoff–Baym Equations and the H-Theorem

In this section, we derive a kinetic entropy current from the Kadanoff–Baym equations with first order approximation of the gradient expansion and show the H-theorem for the leading-order approximations in the coupling expansion based on [56–58]. The analysis in this section is similar to that in open systems (the central region connected to the left and the right region) [59]. Since $(-1, 1)$ and $(1, -1)$ components in $i\Delta_0^{-1}(x, y)$ in Equation (9) are zero, the same procedures to rewrite the Kadanoff–Baym equations as those in open systems [59–63] can be adopted. We set $t_0 \to -\infty$.

First, we shall write the Kadanoff–Baym equations in Equation (17) for each components. By multiplying the matrix Δ from the right in Equation (17) and taking the $(0, 0)$ component, we can write it as,

$$i\left(\Delta_{0,00}^{-1} - \Sigma_{00}\right)\Delta_{00} + \sum_{\alpha=-1,1}ed_e(E_1 + i\alpha E_2)\Delta_{\alpha0} = i\delta_C(x - y), \quad (28)$$

where the $(0, 0)$ component of the matrix Δ_0^{-1} represents $i\Delta_{0,00}^{-1}(x, y) = \left(i\frac{\partial}{\partial x^0} + \frac{\nabla_i^2}{2m}\right)\delta_C(x - y)$. By taking $(\alpha, 0)$ component, we can write it as,

$$i(\Delta_{0,\alpha\alpha}^{-1} - \Sigma_{\alpha\alpha})\Delta_{\alpha0} + ed_e(E_1 - i\alpha E_2)\Delta_{00} = 0. \quad (29)$$

It is convenient to introduce the Green's functions $\Delta_{g,\alpha\alpha}$ as,

$$i\Delta_{g,\alpha\alpha}^{-1} = i\Delta_{0,\alpha\alpha}^{-1} - i\Sigma_{\alpha\alpha}. \quad (30)$$

Then by using Equations (29) and (30), we can write $\Delta_{\alpha 0}$ as,

$$\Delta_{\alpha 0}(x,y) = -\frac{ed_e}{i} \int_C dw \Delta_{g,\alpha\alpha}(x,w)(E_1(w) - i\alpha E_2(w))\Delta_{00}(w,y). \tag{31}$$

Equation (31) means the propagation from y to x with zero angular momentum, change of angular momentum at w and the propagation from w to x with angular momentum $\alpha = \pm 1$. By using Equation (31), we can rewrite Equation (28) as,

$$i\int_C dw(\Delta_{0,00}^{-1}(x,w) - \Sigma_{00}(x,w))\Delta_{00}(w,y)$$

$$+i\sum_{\alpha=-1,1}(ed_e)^2 \int_C dw(E_1(x) + i\alpha E_2(x))\Delta_{g,\alpha\alpha}(x,w)(E_1(w) - i\alpha E_2(w))\Delta_{00}(w,y) = i\delta_C(x-y). \tag{32}$$

The second term on the left-hand side in Equation (32) represents the propagation from y to w with zero angular momentum, the change of the angular momentum to $\alpha = \pm 1$ at w due to the coherent electric fields, the propagation from w to x and the change of the angular momentum from $\alpha = \pm 1$ to zero due to the coherent electric fields. In a similar way to ϕ^4 theory in open systems [59], we can derive,

$$i\int_C dw\Delta_{00}(x,w)(\Delta_{0,00}^{-1}(w,y) - \Sigma_{00}(w,y))$$

$$+i\sum_{\alpha=-1,1}(ed_e)^2 \int_C dw\Delta_{00}(x,w)(E_1(w) + i\alpha E_2(w))\Delta_{g,\alpha\alpha}(w,y)(E_1(y) - i\alpha E_2(y)) = i\delta_C(x-y), \tag{33}$$

where we have used,

$$\Delta_{0\alpha}(x,y) = -\frac{1}{i}\int_C dw\Delta_{00}(x,w)(ed_e)(E_1(w) + i\alpha E_2(w))\Delta_{g,\alpha\alpha}(w,y). \tag{34}$$

The (α,α) components of the Kadanoff–Baym equations are written by,

$$i\int_C dw\left(\Delta_{0,\alpha\alpha}^{-1}(x,w) - \Sigma_{\alpha\alpha}(x,w)\right)\Delta_{\alpha\alpha}(w,y)$$

$$+i(ed_e)^2 \int_C dw(E_1(x) - i\alpha E_2(x))\Delta_{00}(x,w)(E_1(w) + i\alpha E_2(w))\Delta_{g,\alpha\alpha}(w,y) = i\delta_C(x-y), \tag{35}$$

and,

$$i\int_C dw\Delta_{\alpha\alpha}(x,w)\left(\Delta_{0,\alpha\alpha}^{-1}(w,y) - \Sigma_{\alpha\alpha}(w,y)\right)$$

$$+i(ed_e)^2 \int_C dw\Delta_{g,\alpha\alpha}(x,w)(E_1(w) - i\alpha E_2(w))\Delta_{00}(w,x)(E_1(x) + i\alpha E_2(x)) = i\delta_C(x-y), \tag{36}$$

where we have used Equations (31) and (34).

Next, we shall perform the Fourier transformation ($\int d(x-y)e^{ip\cdot(x-y)}$) with the relative coordinate $x-y$ of the $(0,0)$ and (α,α) components of the Kadanoff–Baym equations. We use the 2×2 matrix notation in the closed-time path with $a,b,c,d = 1,2$. Equations (32) and (33) are transformed as,

$$i\left(\Delta_{0,00}^{-1}(p) - \Sigma_{00}(X,p)\sigma_z + \sum_\alpha U_{\alpha\alpha}(X,p)\sigma_z\right)^{ac} \circ \Delta_{00}^{cb}(X,p) = i\sigma_z^{ab}, \tag{37}$$

$$i\Delta_{00}^{ac}(X,p) \circ \left(\Delta_{0,00}^{-1}(p) - \sigma_z\Sigma_{00}(X,p) + \sigma_z\sum_\alpha U_{\alpha\alpha}(X,p)\right)^{cb} = i\sigma_z^{ab}, \tag{38}$$

where $X = \frac{x+y}{2}$, $\sigma_z = \text{diag}(1, -1)$,

$$i\Delta_{0,00}^{-1}(p) = p^0 - \frac{\mathbf{p}^2}{2m},\tag{39}$$

and the $U_{\alpha\alpha}(X, p)$ is the Fourier transformation,

$$\begin{aligned}U_{\alpha\alpha}(X, p) &= (ed_e)^2 \int d(x-y) e^{ip\cdot(x-y)} (E_1(x) + i\alpha E_2(x))\Delta_{g,\alpha\alpha}(x, y)(E_1(y) - i\alpha E_2(y))\\ &= (ed_e)^2 \mathbf{E}(X)^2 \Delta_{g,\alpha\alpha}(X, p + \alpha\partial\zeta) + \left(\frac{\partial^2}{\partial X^2}\right),\end{aligned}\tag{40}$$

with the definition of ζ and $|\mathbf{E}|$,

$$E_1(x) + i\alpha E_2(x) = |\mathbf{E}(x)| e^{i\alpha\zeta(x)},\tag{41}$$

and,

$$(U_{\alpha\alpha}(X, p)\sigma_z)^{ac} = U_{\alpha\alpha}^{ad}(X, p)\sigma_z^{dc},\tag{42}$$

The $\circ$ is expanded by the derivative of X [64–67] as,

$$H(X, p)\circ I(X, p) = H(X, p)I(X, p) + \frac{i}{2}\{H, I\} + \left(\frac{\partial^2}{\partial X^2}\right),\tag{43}$$

with the definition of the Poisson bracket,

$$\{H, I\} \equiv \frac{\partial H}{\partial p^\mu}\frac{\partial I}{\partial X_\mu} - \frac{\partial H}{\partial X^\mu}\frac{\partial I}{\partial p_\mu}.\tag{44}$$

We find that the $U_{\alpha\alpha}$ represents the change of momenta of dipoles as shown in Figure 2a.

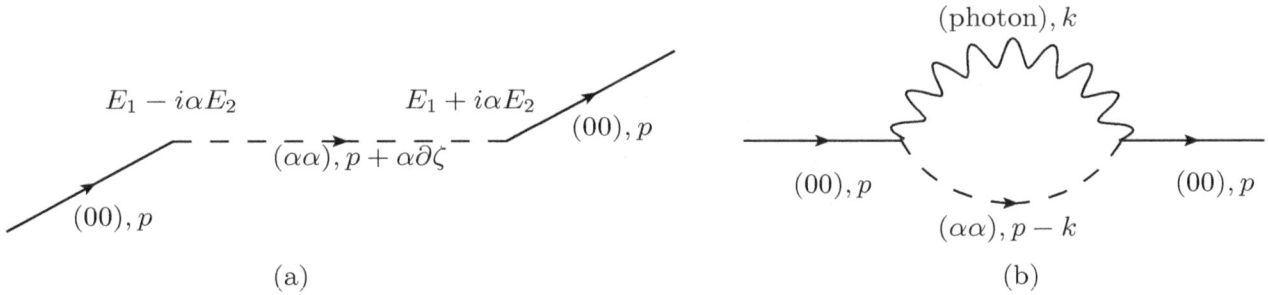

Figure 2. Diagrams of **(a)** $U_{\alpha\alpha}(X, p)$ and **(b)** self-energy $\Sigma_{00}(X, p)$.

In a similar way to [59], in the 0th and the first order in the gradient expansion in Equations (37) and (38), we can derive the following retarded Green's function,

$$\Delta_{00,R}(X, p) = \frac{-1}{p^0 - \frac{\mathbf{p}^2}{2m} - \Sigma_{00,R} + \sum_{\alpha=-1,1} U_{\alpha\alpha,R}},\tag{45}$$

with the retarded parts (the subscript 'R') $\Delta_{00,R} = i(\Delta_{00}^{11} - \Delta_{00}^{12})$, $\Sigma_{00,R} = i(\Sigma_{00}^{11} - \Sigma_{00}^{12})$ and $U_{\alpha\alpha,R} = i(U_{\alpha\alpha}^{11} - U_{\alpha\alpha}^{12})$. By taking the imaginary part of the retarded Green's function $\Delta_{00,R}(X, p)$, we can derive the spectral function $\rho_{00} = i(\Delta_{00}^{21} - \Delta_{00}^{12}) = 2i\text{Im}\Delta_{00,R}(X, p)$ which represents the information of dispersion relations. Similarly, the (α, α) components of the Kadanoff–Baym equations are written as,

$$i\left(\Delta_{0,\alpha\alpha}^{-1}(p) - \Sigma_{\alpha\alpha}(X, p)\sigma_z\right)\circ \Delta_{\alpha\alpha}(X, p) + iV_{\alpha\alpha}(X, p)\sigma_z \circ \Delta_{g,\alpha\alpha}(X, p) = i\sigma_z,\tag{46}$$

and,

$$i\Delta_{\alpha\alpha}(X,p) \circ \left(\Delta_{0,\alpha\alpha}^{-1}(p) - \sigma_z\Sigma_{\alpha\alpha}(X,p)\right) + i\Delta_{g,\alpha\alpha}(X,p) \circ \sigma_z V_{\alpha\alpha}(X,p) = i\sigma_z, \tag{47}$$

where,

$$i\Delta_{0,\alpha\alpha}^{-1}(p) = p^0 - \frac{\mathbf{p}^2}{2m} - \frac{1}{2I}, \tag{48}$$

and,

$$
\begin{aligned}
V_{\alpha\alpha}(X,p) &= (ed_e)^2 \int d(x-y)e^{ip\cdot(x-y)}(E_1(x) - i\alpha E_2(x))\Delta_{00}(x,y)(E_1(y) + i\alpha E_2(y)) \\
&= (ed_e)^2\mathbf{E}(X)^2\Delta_{00}(X,p - \alpha\partial\zeta) + \left(\frac{\partial^2}{\partial X^2}\right).
\end{aligned}
\tag{49}
$$

We can also write for $\Delta_{g,\alpha\alpha}^{cb}(X,p)$ as,

$$i\left(\Delta_{0,\alpha\alpha}^{-1}(p) - \Sigma_{\alpha\alpha}(X,p)\sigma_z\right)^{ac} \circ \Delta_{g,\alpha\alpha}^{cb}(X,p) = i\sigma_z^{ab}, \tag{50}$$

$$\Delta_{g,\alpha\alpha}^{ac}(X,p) \circ i\left(\Delta_{0,\alpha\alpha}^{-1}(p) - \sigma_z\Sigma_{\alpha\alpha}(X,p)\right)^{cb} = i\sigma_z^{ab}. \tag{51}$$

In the 0th and the first order in the gradient expansion in Equations (46) and (47), we can derive,

$$\Delta_{\alpha\alpha,R} = \Delta_{g,\alpha\alpha,R} + \Delta_{g,\alpha\alpha,R}V_{\alpha\alpha,R}\Delta_{g,\alpha\alpha,R} \tag{52}$$

with $\Delta_{\alpha\alpha,R} = i(\Delta_{\alpha\alpha}^{11} - \Delta_{\alpha\alpha}^{12})$ and $V_{\alpha\alpha,R} = i(V_{\alpha\alpha}^{11} - V_{\alpha\alpha}^{12})$. Here we have used the solution in the 0th and the first order in the gradient expansion in Equations (50) and (51) given by,

$$\Delta_{g,\alpha\alpha,R} = \frac{-1}{p^0 - \frac{\mathbf{p}^2}{2m} - \frac{1}{2I} - \Sigma_{\alpha\alpha,R}}, \tag{53}$$

with $\Sigma_{\alpha\alpha,R} = i(\Sigma_{\alpha\alpha}^{11} - \Sigma_{\alpha\alpha}^{12})$. The derivation is the same as [59]. The imaginary part of the retarded Green's function $\Delta_{\alpha\alpha,R}(X,p)$ multiplied by $2i$ represents the spectral function $\rho_{\alpha\alpha} = i(\Delta_{\alpha\alpha}^{21} - \Delta_{\alpha\alpha}^{12}) = 2i\mathrm{Im}\Delta_{\alpha\alpha,R}(X,p)$ which represents the information of dispersion relations. In addition, the Kadanoff–Baym equations for photons in Equation (19) are written by,

$$i\left(D_{0,ij}^{-1}(k) - \Pi_{ij}(X,k)\sigma_z\right)^{ac} \circ D_{jl}^{cb}(X,k) = i\delta_{il}\sigma_z^{ab}, \tag{54}$$

$$iD_{ij}^{ac}(X,k) \circ \left(D_{0,jl}^{-1}(k) - \sigma_z\Pi_{jl}(X,k)\right)^{cb} = i\delta_{il}\sigma_z^{ab}, \tag{55}$$

with,

$$iD_{0,ij}^{-1}(k) = k^2\delta_{ij}. \tag{56}$$

Next we shall derive the self-energy in the leading-order (LO) of the coupling expansion in Equation (6). The $(a,b) = (1,2)$ and $(2,1)$ component of $i\frac{\Gamma_2}{2}$ are given by,

$$
\begin{aligned}
i\frac{\Gamma_{2,\mathrm{LO}}}{2} = &-\tfrac{1}{2}(ed_e)^2 \int du\,dw \sum_{\alpha=-1,1} \Big(\Delta_{\alpha\alpha}^{21}(w,u)\Delta_{00}^{12}(u,w)(1,-\alpha i)_j\partial_u^0\partial_w^0\left(D_{jl}^{12}(u,w) + D_{lj}^{21}(w,u)\right)(1,\alpha i)_l^t \\
&+\Delta_{\alpha\alpha}^{12}(w,u)\Delta_{00}^{21}(u,w)(1,-\alpha i)_j\partial_u^0\partial_w^0\left(D_{jl}^{21}(u,w) + D_{lj}^{12}(w,u)\right)(1,\alpha i)_l^t\Big),
\end{aligned}
\tag{57}
$$

where t represents the transposition. It is convenient to rewrite,

$$D_{ij}^{ab}(k) = \left(\delta_{ij} - \frac{k_i k_j}{\mathbf{k}^2}\right) D_T^{ab}(k) + \frac{k_i k_j}{\mathbf{k}^2} D_L^{ab}(k), \tag{58}$$

$$\Pi_{ij}^{ab}(k) = \left(\delta_{ij} - \frac{k_i k_j}{\mathbf{k}^2}\right) \Pi_T^{ab}(k) + \frac{k_i k_j}{\mathbf{k}^2} \Pi_L^{ab}(k), \tag{59}$$

where T and L represent the transverse and the longitudinal part, respectively. The LO self-energy $i\Pi_{ji}^{21}(y,x) = -\frac{\delta\Gamma_{2,\mathrm{LO}}}{\delta D_{ij}^{12}(x,y)}$ is,

$$\begin{aligned}
i\Pi_{ji}^{21}(y,x) &= -i(ed_e)^2 \sum_{\alpha=-1,1} \left(\partial_x^0 \partial_y^0 \left(\Delta_{\alpha\alpha}^{21}(y,x) \Delta_{00}^{12}(x,y) \right) (1,-\alpha i)_l (1,\alpha i)_j^t \right. \\
&\qquad \left. + \partial_x^0 \partial_y^0 \left(\Delta_{00}^{21}(y,x) \Delta_{\alpha\alpha}^{12}(x,y) \right) (1,-\alpha i)_j (1,\alpha i)_l^t \right).
\end{aligned} \tag{60}$$

By Fourier-transforming with the relative coordinate $x - y$ and multiplying $\delta_{ij} - \frac{k_i k_j}{\mathbf{k}^2}$ or $\frac{k_i k_j}{\mathbf{k}^2}$, we arrive at,

$$\begin{aligned}
\Pi_T^{21}(X,k) &= -(ed_e)^2 \left(k^0\right)^2 \int_p \sum_{\alpha=-1,1} \left(\Delta_{\alpha\alpha}^{21}(X,k+p)\Delta_{00}^{12}(X,p) + \Delta_{00}^{21}(X,k+p)\Delta_{\alpha\alpha}^{12}(X,p) \right) \\
&\quad + \left(\frac{\partial^2}{\partial X^2} \right),
\end{aligned} \tag{61}$$

$$\Pi_L^{21}(X,k) = \Pi_T^{21}(X,k), \tag{62}$$

with $\int_p = \int \frac{d^{d+1}p}{(2\pi)^{d+1}}$. The second equation is due to the spatial dimension $d = 2$. Similarly, we arrive at,

$$\begin{aligned}
\Pi_T^{12}(X,k) &= -(ed_e)^2 \left(k^0\right)^2 \int_p \sum_{\alpha=1,1} \left(\Delta_{\alpha\alpha}^{12}(X,k+p)\Delta_{00}^{21}(X,p) + \Delta_{00}^{12}(X,k+p)\Delta_{\alpha\alpha}^{21}(X,p) \right) \\
&\quad + \left(\frac{\partial^2}{\partial X^2} \right),
\end{aligned} \tag{63}$$

$$\Pi_L^{12}(X,k) = \Pi_T^{12}(X,k). \tag{64}$$

The Fourier transformation of the LO self-energy $i\Sigma_{00}^{12}(x,y) = -\frac{1}{2}\frac{\delta\Gamma_{2,\mathrm{LO}}}{\delta\Delta_{00}^{21}(y,x)}$ is,

$$\Sigma_{00}^{12}(X,p) = -(ed_e)^2 \int_k \sum_{\alpha=-1,1} \left(k^0\right)^2 \Delta_{\alpha\alpha}^{12}(X,p-k) \left[D_T^{12}(X,k) + D_L^{12}(X,k) \right] + \left(\frac{\partial^2}{\partial X^2} \right). \tag{65}$$

Similarly,

$$\Sigma_{00}^{21}(X,p) = -(ed_e)^2 \int_k \sum_{\alpha=-1,1} \left(k^0\right)^2 \Delta_{\alpha\alpha}^{21}(X,p-k) \left[D_T^{21}(X,k) + D_L^{21}(X,k) \right] + \left(\frac{\partial^2}{\partial X^2} \right). \tag{66}$$

This self-energy is shown in Figure 2b. Similarly we can derive,

$$\Sigma_{\alpha\alpha}^{12}(X,p) = -(ed_e)^2 \int_k \left(k^0\right)^2 \Delta_{00}^{12}(X,p-k) \left[D_T^{12}(X,k) + D_L^{12}(X,k) \right] + \left(\frac{\partial^2}{\partial X^2} \right), \tag{67}$$

and,

$$\Sigma_{\alpha\alpha}^{21}(X, p) = -(ed_e)^2 \int_k \left(k^0\right)^2 \Delta_{00}^{21}(X, p-k) \left[D_T^{21}(X, k) + D_L^{21}(X, k)\right] + \left(\frac{\partial^2}{\partial X^2}\right). \tag{68}$$

Finally we derive a kinetic entropy current in the first order approximation in the gradient expansion and show the H-theorem in the LO approximation in the coupling expansion. By taking a difference of Equations (32) and (33), we arrive at,

$$i\left\{p^0 - \frac{\mathbf{p}^2}{2m}, \Delta_{00}^{ab}\right\} = i\left[\left(\Sigma_{00} - \sum_\alpha U_{\alpha\alpha}\right)\sigma_z \circ \Delta_{00}\right]^{ab} - i\left[\Delta_{00} \circ \sigma_z\left(\Sigma_{00} - \sum_\alpha U_{\alpha\alpha}\right)\right]^{ab}. \tag{69}$$

We use the Kadanoff–Baym Ansatz $\Delta_{00}^{12} = \frac{\rho_{00}}{i}f_{00}$, $\Delta_{00}^{21} = \frac{\rho_{00}}{i}(f_{00}+1)$, $\Sigma_{00}^{12} = \frac{\Sigma_{00,\rho}}{i}\gamma_{00}$, $\Sigma_{00}^{21} = \frac{\Sigma_{00,\rho}}{i}(\gamma_{00}+1)$, $U_{\alpha\alpha}^{12} = \frac{U_{\alpha\alpha,\rho}}{i}\gamma_{U,\alpha\alpha}$ and $U_{\alpha\alpha}^{21} = \frac{U_{\alpha\alpha,\rho}}{i}(\gamma_{U,\alpha\alpha}+1)$ with $\rho_{00} = i(\Delta_{00}^{21} - \Delta_{00}^{12}) = 2i\mathrm{Im}\Delta_{00,R}$, $\Sigma_{00,\rho} = i(\Sigma_{00}^{21} - \Sigma_{00}^{12}) = 2i\mathrm{Im}\Sigma_{00,R}$ and $U_{\alpha\alpha,\rho} = i(U_{\alpha\alpha}^{21} - U_{\alpha\alpha}^{12}) = 2i\mathrm{Im}U_{\alpha\alpha,R}$ where we just rewrite the $(1,2)$ and the $(2,1)$ components with the spectral parts ρ_{00}, $\Sigma_{00,\rho}$ and $U_{\alpha\alpha,\rho}$ and distribution functions f_{00}, γ_{00} and $\gamma_{U,\alpha\alpha}$. The distribution functions f_{00}, γ_{00} and $\gamma_{U,\alpha\alpha}$ approach the Bose–Einstein distributions near equilibrium states. In the first order approximation in the gradient expansion in Equation (69) for $(a, b) = (1, 2)$ and $(2, 1)$, we can derive,

$$f_{00} = \gamma_{00} + O\left(\frac{\partial}{\partial X}\right), \quad \text{and} \quad f_{00} = \gamma_{U,\alpha\alpha} + O\left(\frac{\partial}{\partial X}\right). \tag{70}$$

(Rewrite $(a, b) = (1, 2)$ and $(2, 1)$ components in Equation (69), then we can show the collision terms $\Delta_{00}^{21}\Sigma_{00}^{12} - \Delta_{00}^{12}\Sigma_{00}^{21} \propto f_{00} - \gamma_{00} = O\left(\frac{\partial}{\partial X}\right)$ and $f_{00} - \gamma_{U,\alpha\alpha} = O\left(\frac{\partial}{\partial X}\right)$.) We shall multiply $\ln\frac{i\Delta_{00}^{12}}{\rho_{00}}$ in $(a, b) = (1, 2)$ component in Equation (69) and $\ln\frac{i\Delta_{00}^{21}}{\rho_{00}}$ in $(2, 1)$ component in Equation (69), take the difference of them and integrate with $\int_p$. By the use of Equation (70), we arrive at,

$$\begin{aligned}
\partial_\mu s_{\text{matter},00}^\mu &= -\int_p \left(\Sigma_{00}^{21}(X, p)\Delta_{00}^{12}(X, p) - \Sigma_{00}^{12}(X, p)\Delta_{00}^{21}(X, p)\right)\ln\frac{\Delta_{00}^{12}(X, p)}{\Delta_{00}^{21}(X, p)} \\
&\quad + \sum_\alpha \int_p \left(U_{\alpha\alpha}^{21}(X, p)\Delta_{00}^{12}(X, p) - U_{\alpha\alpha}^{12}(X, p)\Delta_{00}^{21}(X, p)\right)\ln\frac{\Delta_{00}^{12}(X, p)}{\Delta_{00}^{21}(X, p)},
\end{aligned} \tag{71}$$

with the definition of entropy current $s_{\text{matter},00}^\mu$ for $(0, 0)$ component,

$$\begin{aligned}
s_{\text{matter},00}^\mu &\equiv \int_p \left[\left(\delta_0^\mu + \frac{\delta_i^\mu \mathbf{p}^i}{m} - \frac{\partial \mathrm{Re}(\Sigma_{00,R} - \sum_\alpha U_{\alpha\alpha,R})}{\partial p_\mu}\right)\frac{\rho_{00}}{i}\right. \\
&\quad \left. + \frac{\partial \mathrm{Re}\Delta_{00,R}}{\partial p_\mu}\frac{\Sigma_{00,\rho} - \sum_\alpha U_{\alpha\alpha,\rho}}{i}\right]\sigma[f_{00}],
\end{aligned} \tag{72}$$

$$\sigma[f_{00}] \equiv (1 + f_{00})\ln(1 + f_{00}) - f_{00}\ln f_{00}. \tag{73}$$

We can derive the Boltzmann entropy $\int_{\mathbf{p}}[(1+n)\ln(1+n) - n\ln n]$ with the number density $n(X, \mathbf{p})$ in the quasi-particle limit $\mathrm{Im}U_{\alpha\alpha,R} = \mathrm{Im}\Sigma_{00,R} \to 0$ in the same way as in [58]. Similarly, we can derive a kinetic entropy current for $(\alpha\alpha)$ components. >From Equations (46) and (47), we can derive

$$\begin{aligned}
i\left\{p^0 - \frac{\mathbf{p}^2}{2m} - \frac{1}{2I}, \Delta_{\alpha\alpha}^{ab}\right\} &= i\left[\Sigma_{\alpha\alpha}\sigma_z \circ \Delta_{\alpha\alpha} - \Delta_{\alpha\alpha} \circ \sigma_z\Sigma_{\alpha\alpha}\right]^{ab} \\
&\quad - i\left[V_{\alpha\alpha}\sigma_z \circ \Delta_{g,\alpha\alpha} - \Delta_{g,\alpha\alpha} \circ \sigma_z V_{\alpha\alpha}\right]^{ab}. \tag{74}
\end{aligned}$$

We use the Kadanoff–Baym Ansatz $\Delta_{\alpha\alpha}^{12} = \frac{\rho_{\alpha\alpha}}{i} f_{\alpha\alpha}$, $\Delta_{\alpha\alpha}^{21} = \frac{\rho_{\alpha\alpha}}{i}(f_{\alpha\alpha} + 1)$, $\Delta_{g,\alpha\alpha}^{12} = \frac{\Delta_{g,\alpha\alpha,\rho}}{i} \gamma_{g,\alpha\alpha}$, $\Delta_{g,\alpha\alpha}^{21} = \frac{\Delta_{g,\alpha\alpha,\rho}}{i}(\gamma_{g,\alpha\alpha} + 1)$, $\Sigma_{\alpha\alpha}^{12} = \frac{\Sigma_{\alpha\alpha,\rho}}{i} \gamma_{\alpha\alpha}$, $\Sigma_{\alpha\alpha}^{21} = \frac{\Sigma_{\alpha\alpha,\rho}}{i}(\gamma_{\alpha\alpha} + 1)$, $V_{\alpha\alpha}^{12} = \frac{V_{\alpha\alpha,\rho}}{i} \gamma_{V,\alpha\alpha}$ and $V_{\alpha\alpha}^{21} = \frac{V_{\alpha\alpha,\rho}}{i}(\gamma_{V,\alpha\alpha} + 1)$ with $\rho_{\alpha\alpha} = i(\Delta_{\alpha\alpha}^{21} - \Delta_{\alpha\alpha}^{12}) = 2i\mathrm{Im}\Delta_{\alpha\alpha,R}$, $\Sigma_{\alpha\alpha,\rho} = i(\Sigma_{\alpha\alpha}^{21} - \Sigma_{\alpha\alpha}^{12}) = 2i\mathrm{Im}\Sigma_{\alpha\alpha,R}$ and $V_{\alpha\alpha,\rho} = i(V_{\alpha\alpha}^{21} - V_{\alpha\alpha}^{12}) = 2i\mathrm{Im}V_{\alpha\alpha,R}$. In Equation (74), we can show,

$$f_{\alpha\alpha} \sim \gamma_{\alpha\alpha}, \quad \gamma_{g,\alpha\alpha} \sim \gamma_{V,\alpha\alpha}, \tag{75}$$

for distribution functions $f_{\alpha\alpha}$, $\gamma_{\alpha\alpha}$ and $\gamma_{V,\alpha\alpha}$ by writing the $(a,b) = (1,2)$ and $(2,1)$ components in the Kadanoff–Baym equations (74). We can also show,

$$\gamma_{\alpha\alpha} \sim \gamma_{g,\alpha\alpha}, \tag{76}$$

from Equations (50) and (51). We shall multiply $\ln \frac{i\Delta_{\alpha\alpha}^{12}}{\rho_{\alpha\alpha}}$ in $(a,b) = (1,2)$ component in Equation (74) and $\ln \frac{i\Delta_{\alpha\alpha}^{21}}{\rho_{\alpha\alpha}}$ in $(2,1)$ component in Equation (74), take the difference of them and integrate with $\int_p$. By using Equations (75) and (76), we arrive at,

$$
\begin{aligned}
\partial_\mu s_{\text{matter},\alpha\alpha}^\mu &= -\int_p \left(\Sigma_{\alpha\alpha}^{21}(X,p)\Delta_{\alpha\alpha}^{12}(X,p) - \Sigma_{\alpha\alpha}^{12}(X,p)\Delta_{\alpha\alpha}^{21}(X,p) \right) \ln \frac{\Delta_{\alpha\alpha}^{12}(X,p)}{\Delta_{\alpha\alpha}^{21}(X,p)} \\
&\quad + \int_p \left(V_{\alpha\alpha}^{21}(X,p)\Delta_{g,\alpha\alpha}^{12}(X,p) - V_{\alpha\alpha}^{12}(X,p)\Delta_{g,\alpha\alpha}^{21}(X,p) \right) \ln \frac{\Delta_{\alpha\alpha}^{12}(X,p)}{\Delta_{\alpha\alpha}^{21}(X,p)},
\end{aligned} \tag{77}
$$

with the definitions of entropy current $s_{\text{matter},\alpha\alpha}^\mu$ for $(\alpha\alpha)$ components,

$$
\begin{aligned}
s_{\text{matter},\alpha\alpha}^\mu &\equiv \int_p \left[\left(\delta_0^\mu + \frac{\delta_i^\mu \mathbf{p}^i}{m} - \frac{\partial \mathrm{Re}\Sigma_{\alpha\alpha,R}}{\partial p_\mu} \right) \frac{\rho_{\alpha\alpha}}{i} + \frac{\partial \mathrm{Re}\Delta_{\alpha\alpha,R}}{\partial p_\mu} \frac{\Sigma_{\alpha\alpha,\rho}}{i} \right. \\
&\quad \left. + \frac{\partial \mathrm{Re}V_{\alpha\alpha,R}}{\partial p_\mu} \frac{\Delta_{g,\alpha\alpha,\rho}}{i} - \frac{\partial \mathrm{Re}\Delta_{g,\alpha\alpha,R}}{\partial p_\mu} \frac{V_{\alpha\alpha,\rho}}{i} \right] \sigma[f_{\alpha\alpha}].
\end{aligned} \tag{78}
$$

In this derivation, we have used the same way as that in open systems in [59]. We can also derive the following equations for the Kadanoff–Baym equations for photons with the Kadanoff–Baym Ansatz $D_T^{21} = \frac{\rho_T}{i}(1 + f_T)$, $D_T^{12} = \frac{\rho_T}{i} f_T$, $D_L^{21} = \frac{\rho_L}{i}(1 + f_L)$ and $D_L^{12} = \frac{\rho_L}{i} f_L$ with distribution functions f_T and f_L and spectral functions ρ_T and ρ_L,

$$
\begin{aligned}
\partial_\mu s_{\text{photon}}^\mu &= -\frac{1}{2}\int_k \left[\Pi_T^{21}(X,k)D_T^{12}(X,k) - \Pi_T^{12}(X,k)D_T^{21}(X,k) \right] \ln \frac{D_T^{12}(X,k)}{D_T^{21}(X,k)} \\
&\quad - \frac{1}{2}\int_k \left[\Pi_L^{21}(X,k)D_L^{12}(X,k) - \Pi_L^{12}(X,k)D_L^{21}(X,k) \right] \ln \frac{D_L^{12}(X,k)}{D_L^{21}(X,k)},
\end{aligned} \tag{79}
$$

with the entropy current for photons,

$$
\begin{aligned}
s_{\text{photon}}^\mu &\equiv \int_k \left[\left(k^\mu - \frac{1}{2}\frac{\partial \mathrm{Re}\Pi_{T,R}}{\partial k_\mu} \right) \frac{D_{T,\rho}}{i} + \frac{1}{2}\frac{\partial \mathrm{Re}D_{T,R}}{\partial k_\mu} \frac{\Pi_{T,\rho}}{i} \right] \sigma[f_T] \\
&\quad + \int_k \left[\left(k^\mu - \frac{1}{2}\frac{\partial \mathrm{Re}\Pi_{L,R}}{\partial k_\mu} \right) \frac{D_{L,\rho}}{i} + \frac{1}{2}\frac{\partial \mathrm{Re}D_{L,R}}{\partial k_\mu} \frac{\Pi_{L,\rho}}{i} \right] \sigma[f_L].
\end{aligned} \tag{80}
$$

As a result, the total entropy current $s^\mu = s^\mu_{matter,00} + \sum_\alpha s^\mu_{matter,\alpha\alpha} + s^\mu_{photon}$ satisfies,

$$
\begin{aligned}
\partial_\mu s^\mu =&\ (ed_e)^2 \int_{p,k} \left(k^0\right)^2 \sum_\alpha \left[\Delta^{21}_{\alpha\alpha}(p-k)\Delta^{12}_{00}(p)D^{21}_T(k) - \Delta^{12}_{\alpha\alpha}(p-k)\Delta^{21}_{00}(p)D^{12}_T(k)\right] \\
&\times \ln \frac{\Delta^{21}_{\alpha\alpha}(p-k)\Delta^{12}_{00}(p)D^{21}_T(k)}{\Delta^{12}_{\alpha\alpha}(p-k)\Delta^{21}_{00}(p)D^{12}_T(k)} \\
&+ (ed_e)^2 \int_{p,k} \left(k^0\right)^2 \sum_\alpha \left[\Delta^{21}_{\alpha\alpha}(p-k)\Delta^{12}_{00}(p)D^{21}_L(k) - \Delta^{12}_{\alpha\alpha}(p-k)\Delta^{21}_{00}(p)D^{12}_L(k)\right] \\
&\times \ln \frac{\Delta^{21}_{\alpha\alpha}(p-k)\Delta^{12}_{00}(p)D^{21}_L(k)}{\Delta^{12}_{\alpha\alpha}(p-k)\Delta^{21}_{00}(p)D^{12}_L(k)} \\
&+ (ed_e)^2(\mathbf{E}(X))^2 \sum_\alpha \int_p \left(\Delta^{21}_{g,\alpha\alpha}(p+\alpha\partial\zeta)\Delta^{12}_{00}(p) - \Delta^{12}_{g,\alpha\alpha}(p+\alpha\partial\zeta)\Delta^{21}_{00}\right) \\
&\times \ln \frac{\Delta^{21}_{g,\alpha\alpha}(p+\alpha\partial\zeta)\Delta^{12}_{00}(p)}{\Delta^{12}_{g,\alpha\alpha}(p+\alpha\partial\zeta)\Delta^{21}_{00}(p)} \geq 0,
\end{aligned}
\tag{81}
$$

where we have used the inequality $(x-y)\ln\frac{x}{y} \geq 0$ for real variables x and y with $x > 0$ and $y > 0$. The equality is satisfied in $f_{00} = f_{\alpha\alpha} = f_T = f_L = \frac{1}{e^{p^0/T}-1}$. Here we have used $\frac{\Delta^{21}_{\alpha\alpha}}{\Delta^{12}_{\alpha\alpha}} \sim \frac{\Delta^{21}_{g,\alpha\alpha}}{\Delta^{12}_{g,\alpha\alpha}}$ with $\gamma_{g,\alpha\alpha} \sim f_{\alpha\alpha}$ in first order in the gradient expansion. We have shown the H-theorem in the LO approximation in the coupling expansion and in the first order approximation in the gradient expansion. There is no violation in the second law in thermodynamics in the dynamics.

4. Time Evolution Equations in Spatially Homogeneous Systems and Conserved Energy

In this section, we write time evolution equations in spatially homogeneous systems and show a concrete form of the conserved energy density.

It is convenient to introduce the statistical functions $F_{00} = \frac{\Delta^{21}_{00}+\Delta^{12}_{00}}{2}$, $F_{\alpha\alpha} = \frac{\Delta^{21}_{\alpha\alpha}+\Delta^{12}_{\alpha\alpha}}{2}$, $F_T = \frac{D^{21}_T+D^{12}_T}{2}$, $F_L = \frac{D^{21}_L+D^{12}_L}{2}$, which represent the information of how many particles are occupied in $(p^0, \mathbf{p})$ (particle distributions) and statistical parts, $U_{\alpha\alpha,F} = \frac{U^{21}_{\alpha\alpha}+U^{12}_{\alpha\alpha}}{2}$, $V_{\alpha\alpha,F} = \frac{V^{21}_{\alpha\alpha}+V^{12}_{\alpha\alpha}}{2}$, $\Delta_{g,\alpha\alpha,F} = \frac{\Delta^{21}_{g,\alpha\alpha}+\Delta^{12}_{g,\alpha\alpha}}{2}$, $\Sigma_{00,F} = \frac{\Sigma^{21}_{00}+\Sigma^{12}_{00}}{2}$, $\Sigma_{\alpha\alpha,F} = \frac{\Sigma^{21}_{\alpha\alpha}+\Sigma^{12}_{\alpha\alpha}}{2}$, $\Pi_{T,F} = \frac{\Pi^{21}_T+\Pi^{12}_T}{2}$ and $\Pi_{L,F} = \frac{\Pi^{21}_L+\Pi^{12}_L}{2}$. The variables of these functions are $(X^0, p^0, \mathbf{p})$ with the center-of-mass coordinate $X^0 = \frac{x^0+y^0}{2}$ and p given by the Fourier transformation with the relative coordinate $x - y$ in variables (x, y) in Green's functions and self-energy in Section 2. The statistical functions and parts are real at any time when we start with real statistical functions at initial time. The spectral functions are given by taking the difference of $(2, 1)$ and $(1, 2)$ components multiplied by i, namely $\rho_{00} = i(\Delta^{21}_{00} - \Delta^{12}_{00})$. They represent the information of which states can be occupied by particles in $(p^0, \mathbf{p})$ (dispersion relations). The spectral parts in self-energy are given by taking the difference of $(2, 1)$ and $(1, 2)$ components multiplied by i (and written by the subscript ρ), namely $\Delta_{g,\alpha\alpha,\rho} = i(\Delta^{21}_{g,\alpha\alpha} - \Delta^{12}_{g,\alpha\alpha})$, $\Sigma_{00,\rho} = i(\Sigma^{21}_{00} - \Sigma^{12}_{00})$ and so on. The spectral functions and parts are pure imaginary at any time when we start with pure imaginary spectral functions at initial time. We can use the real statistical parts labeled by the subscripts F and the pure imaginary spectral parts labeled by the subscript ρ in self-energy in the time evolution. We use the subscript 'R', 'F' and 'ρ' to represent the retarded, statistical and spectral parts in self-energy, respectively.

The Kadanoff–Baym equation for the statistical and spectral functions are given by,

$$
\begin{aligned}
&\left\{p^0 - \frac{\mathbf{p}^2}{2m} - \mathrm{Re}\Sigma_{00,R} + \sum_{\alpha=-1,1} \mathrm{Re}U_{\alpha\alpha,R}, F_{00}\right\} + \left\{\mathrm{Re}\Delta_{00,R}, \Sigma_{00,F} - \sum_\alpha U_{\alpha\alpha,F}\right\} \\
&\qquad\qquad = \frac{1}{i}\left(F_{00}\Sigma_{00,\rho} - \rho_{00}\Sigma_{00,F}\right) - \frac{1}{i}\sum_\alpha \left(F_{00}U_{\alpha\alpha,\rho} - \rho_{00}U_{\alpha\alpha,F}\right),
\end{aligned}
\tag{82}
$$

$$\left\{ p^0 - \frac{\mathbf{p}^2}{2m} - \mathrm{Re}\Sigma_{00,R} + \sum_{\alpha=-1,1} \mathrm{Re}U_{\alpha\alpha,R}, \rho_{00} \right\} + \left\{ \mathrm{Re}\Delta_{00,R}, \Sigma_{00,\rho} - \sum_{\alpha} U_{\alpha\alpha,\rho} \right\} = 0, \tag{83}$$

$$\left\{ p^0 - \frac{\mathbf{p}^2}{2m} - \frac{1}{2I} - \mathrm{Re}\Sigma_{\alpha\alpha,R}, F_{\alpha\alpha} \right\} + \left\{ \mathrm{Re}\Delta_{\alpha\alpha,R}, \Sigma_{\alpha\alpha,F} \right\} + \left\{ \mathrm{Re}V_{\alpha\alpha,R}, \Delta_{g,\alpha\alpha,F} \right\} - \left\{ \mathrm{Re}\Delta_{g,\alpha\alpha,R}, V_{\alpha\alpha,F} \right\}$$
$$= \frac{1}{i} \left(F_{\alpha\alpha}\Sigma_{\alpha\alpha,\rho} - \rho_{\alpha\alpha}\Sigma_{\alpha\alpha,F} \right) - \frac{1}{i} \left(\Delta_{g,\alpha\alpha,F} V_{\alpha\alpha,\rho} - \Delta_{g,\alpha\alpha,\rho} V_{\alpha\alpha,F} \right), \tag{84}$$

$$\left\{ p^0 - \frac{\mathbf{p}^2}{2m} - \frac{1}{2I} - \mathrm{Re}\Sigma_{\alpha\alpha,R}, \rho_{\alpha\alpha} \right\} + \left\{ \mathrm{Re}\Delta_{\alpha\alpha,R}, \Sigma_{\alpha\alpha,\rho} \right\}$$
$$+ \left\{ \mathrm{Re}V_{\alpha\alpha,R}, \Delta_{g,\alpha\alpha,\rho} \right\} - \left\{ \mathrm{Re}\Delta_{g,\alpha\alpha,R}, V_{\alpha\alpha,\rho} \right\} = 0, \tag{85}$$

$$\left\{ p^0 - \frac{\mathbf{p}^2}{2m} - \frac{1}{2I} - \mathrm{Re}\Sigma_{\alpha\alpha,R}, \Delta_{g,\alpha\alpha,F} \right\} + \left\{ \mathrm{Re}\Delta_{g,\alpha\alpha,R}, \Sigma_{\alpha\alpha,F} \right\}$$
$$= \frac{1}{i} \left(\Delta_{g,\alpha\alpha,F}\Sigma_{\alpha\alpha,\rho} - \Delta_{g,\alpha\alpha,\rho}\Sigma_{\alpha\alpha,F} \right), \tag{86}$$

$$\left\{ p^0 - \frac{\mathbf{p}^2}{2m} - \frac{1}{2I} - \mathrm{Re}\Sigma_{\alpha\alpha,R}, \Delta_{g,\alpha\alpha,\rho} \right\} + \left\{ \mathrm{Re}\Delta_{g,\alpha\alpha,R}, \Sigma_{\alpha\alpha,\rho} \right\} = 0, \tag{87}$$

$$\left\{ p^2 - \mathrm{Re}\Pi_{R,T}, F_T \right\} + \left\{ \mathrm{Re}D_{R,T}, \Pi_{F,T} \right\} = \frac{1}{i} \left(F_T\Pi_{\rho,T} - \rho_T\Pi_{F,T} \right), \tag{88}$$

$$\left\{ p^2 - \mathrm{Re}\Pi_{R,T}, \rho_T \right\} + \left\{ \mathrm{Re}D_{R,T}, \Pi_{\rho,T} \right\} = 0, \tag{89}$$

and longitudinal parts given by changing the label T to L in Equations (88) and (89). We can use Equation (69) in the previous section to derive Equations (82) and (83), for example.

We can write,

$$U_{\alpha\alpha,F}(X,p) = (ed_e)^2 \mathbf{E}(X)^2 \Delta_{g,\alpha\alpha,F}(p+\alpha\partial\zeta), \quad U_{\alpha\alpha,\rho}(X,p) = (ed_e)^2 \mathbf{E}(X)^2 \Delta_{g,\alpha\alpha,\rho}(p+\alpha\partial\zeta), \tag{90}$$
$$V_{\alpha\alpha,F}(X,p) = (ed_e)^2 \mathbf{E}(X)^2 F_{00}(p-\alpha\partial\zeta), \quad V_{\alpha\alpha,\rho}(X,p) = (ed_e)^2 \mathbf{E}(X)^2 \rho_{00}(p-\alpha\partial\zeta). \tag{91}$$

In case we start with initial condition $E_2(X^0 = 0) = 0$, $\partial_0 E_2(X^0 = 0) = 0$ and symmetric Green's functions for $\alpha \to -\alpha$ in spatially homogeneous systems, we can use $\partial\zeta = 0$ in the above equations at any time. We can write the self-energy as,

$$\Sigma_{00,F}(p) = -(ed_e)^2 \sum_{\alpha=-1,1} \int_k \left(k^0\right)^2 \left[F_{\alpha\alpha}(p-k)(F_T(k)+F_L(k)) + \frac{1}{4}\frac{\rho_{\alpha\alpha}(p-k)}{i}\frac{\rho_T(k)+\rho_L(k)}{i} \right], \tag{92}$$

$$\Sigma_{00,\rho}(p) = -(ed_e)^2 \sum_{\alpha=-1,1} \int_k \left(k^0\right)^2 \left[F_{\alpha\alpha}(p-k)(\rho_T(k)+\rho_L(k)) + \rho_{\alpha\alpha}(p-k)(F_T(k)+F_L(k)) \right], \tag{93}$$

$$\Sigma_{\alpha\alpha,F}(p) = -(ed_e)^2 \int_k \left(k^0\right)^2 \left[F_{00}(p-k)(F_T(k)+F_L(k)) + \frac{1}{4}\frac{\rho_{00}(p-k)}{i}\frac{\rho_T(k)+\rho_L(k)}{i} \right], \tag{94}$$

$$\Sigma_{\alpha\alpha,\rho}(p) = -(ed_e)^2 \int_k \left(k^0\right)^2 \left[F_{00}(p-k)(\rho_T(k)+\rho_L(k)) + \rho_{00}(p-k)(F_T(k)+F_L(k)) \right], \tag{95}$$

$$\Pi_{T,F}(k) = \Pi_{L,F}(k) = -(ed_e)^2 \left(k^0\right)^2 \sum_{\alpha=-1,1} \int_p \left[F_{\alpha\alpha}(k+p) F_{00}(p) - \frac{1}{4} \frac{\rho_{\alpha\alpha}(k+p)}{i} \frac{\rho_{00}(p)}{i} \right.$$

$$\left. + F_{00}(k+p) F_{\alpha\alpha}(p) - \frac{1}{4} \frac{\rho_{00}(k+p)}{i} \frac{\rho_{\alpha\alpha}(p)}{i} \right], \tag{96}$$

$$\Pi_{T,\rho}(k) = \Pi_{L,\rho}(k) = -(ed_e)^2 \left(k^0\right)^2 \sum_{\alpha=-1,1} \int_p \left[\rho_{\alpha\alpha}(k+p) F_{00}(p) - F_{\alpha\alpha}(k+p)\rho_{00}(p) \right.$$

$$\left. + \rho_{00}(k+p) F_{\alpha\alpha}(p) - F_{00}(k+p)\rho_{\alpha\alpha}(p) \right], \tag{97}$$

where we have omitted the label of the center-of-mass cordinate X in Green's functions and self-energy. We find that the $\Pi_{T,F}(k) = \Pi_{L,F}(k)$ are symmetric ($\Pi_{T,F}(-k) = \Pi_{T,F}(k)$) under $k \to -k$ and that $\Pi_{T,\rho} = \Pi_{L,\rho}$ are anti-symmetric ($\Pi_{T,\rho}(-k) = -\Pi_{T,\rho}(k)$) under $k \to -k$, for any Green's functions for dipole fields. When we prepare initial conditions with symmetric $F_{T,L}$ and anti-symmetric $\rho_{T,L}$ for photons, we can derive symmetric $F_{T,L}$ and anti-symmetric $\rho_{T,L}$ at any time. In addition, since $\Pi(k)$'s are proportional to $(k^0)^2$, there is no mass gap for incoherent photons for the leading-order self-energy in the coupling expansion. The velocity of gapless modes of incoherent photons will decrease when we increase the density of dipoles in this theory.

Finally, we show the energy density E_{tot}. In the spatially homogeneous system in the $2+1$ dimensions, we can derive $\frac{\partial E_{\text{tot}}}{\partial X^0} = 0$ with the energy density given by,

$$E_{\text{tot}} \equiv \frac{1}{2I} \sum_{\alpha=-1,1} \bar{\psi}_\alpha^* \bar{\psi}_\alpha + \frac{1}{2} (\partial_0 A_i)^2 + \int_p p^0 \left(F_{00} + \sum_{\alpha=-1,1} F_{\alpha\alpha} \right) + \frac{1}{2} \int_p \left(p^0 \right)^2 (F_T + F_L)$$

$$+ 2(ed_e)^2 \mathbf{E}^2 \sum_{\alpha=-1,1} \int_p \left(F_{00}(p)\mathrm{Re}\Delta_{g,\alpha\alpha,R}(p+\alpha\partial\zeta) + \mathrm{Re}\Delta_{00,R}(p)\Delta_{g,\alpha\alpha,F}(p+\alpha\partial\zeta) \right)$$

$$- \int_p \left(\mathrm{Re}\Sigma_{00,R}F_{00} + \mathrm{Re}\Delta_{00,R}\Sigma_{00,F} \right) - \sum_{\alpha=-1,1} \int_p \left(\mathrm{Re}\Sigma_{\alpha\alpha,R}F_{\alpha\alpha} + \mathrm{Re}\Delta_{\alpha\alpha,R}\Sigma_{\alpha\alpha,F} \right)$$

$$- \frac{1}{2} \int_p \left(\mathrm{Re}\Pi_{R,T}F_T + \mathrm{Re}D_{R,T}\Pi_{F,T} + \mathrm{Re}\Pi_{R,L}F_L + \mathrm{Re}D_{R,L}\Pi_{F,L} \right), \tag{98}$$

where we have used the KB equations in this section, the Klein–Gordon Equation (20) and the Schödinger-like Equations (23) and (24) in Section 2. The first term represents the contribution of nonzero angular momenta for coherent dipole fields. The second term represents the contribution by electric fields $E_i = \partial_0 A_i$. The third and the fourth terms represent the contribution by quantum fluctuations for dipoles and photons, respectively. When the temperature is nonzero $T \neq 0$ at equilibrium states and the spectral width in the spectral functions is small enough, statistical functions which are proportional to the Bose–Einstein distributions $\frac{1}{e^{p^0/T}-1}$ give temperature-dependent terms mT^2 for dipole fields and $\propto T^3$ for photon fields in $2+1$ dimensions. The fifth term represents the potential energy in processes in Figure 2a. The sixth, seventh and eighth terms represent the potential energy in processes in Figure 2b. The coefficients in the sixth and seventh terms are not $\frac{1}{3}$ but 1. While the factor 1 might look like a contradiction with the preceding research in [68,69] which suggest that the factor $\frac{1}{3}$ appears in the interaction with 3-point-vertex, the factor 1 appears due to time derivative $(\partial^0)^2$ in self-energy for dipole fields and photon fields.

5. Dynamics of Coherent Fields

In this section, we show that our Lagrangian describes the super-radiance phenomena in time evolution equations of coherent fields. We shall assume that all the coherent fields are independent of x^1 (dependent on x^0 and x^2). We also assume the symmetry for $\alpha = -1$ and $\alpha = 1$, namely $\bar{\psi}_1^{(*)} = \bar{\psi}_{-1}^{(*)}$, $\Delta_{01} = \Delta_{0-1}$, and $\Delta_{10} = \Delta_{-10}$. We set initial conditions $E_2 = 0$ and $\partial_0 E_2 = 0$ at $x^0 = 0$.

We define $Z \equiv 2|\bar{\psi}_1|^2 - |\bar{\psi}_0|^2$. It is possible to derive the following equations from time evolution Equations (20), (23) and (24) with their complex conjugates for background coherent fields in Section 2.

$$\partial_0 Z = i4 e d_e E_1 \left(\bar{\psi}_1^* \bar{\psi}_0 - \bar{\psi}_0^* \bar{\psi}_1 \right), \tag{99}$$

$$\partial_0 \left(\bar{\psi}_1^* \bar{\psi}_0 \right) = \frac{i}{2I} \bar{\psi}_1^* \bar{\psi}_0 + i e d_e E_1 Z \tag{100}$$

$$\left[(\partial_0)^2 - (\partial_2)^2 \right] E_1 = -2 e d_e (\partial_0)^2 \left[\bar{\psi}_1^* \bar{\psi}_0 + \bar{\psi}_0^* \bar{\psi}_1 + \Delta_{01}(x,x) + \Delta_{10}(x,x) \right]. \tag{101}$$

We have used moderately varying spatial dependence $|\nabla_i^2 \bar{\psi}_{-1,0,1}/m| \ll |\partial_0 \bar{\psi}_{-1,0,1}|$. We derive aspects of the super-radiance and the Higgs mechanism in the above three equations.

5.1. Super-Radiance

In this section, we show the super-radiance in time evolution equations for coherent fields with the rotating wave approximations neglecting non-resonant terms and quantum fluctuations. We have used the derivations in [70,71] for background coherent fields.

We shall consider only $k^0 = \frac{1}{2I}$ in this section and we expand the electric field E_1 and the transition rate $\bar{\psi}_0 \bar{\psi}_1^*$ as,

$$E_1(x^0, x^2) = \frac{1}{2} \epsilon(x^0, x^2) e^{-i(k^0 x^0 - k^0 x^2)} + \frac{1}{2} \epsilon^*(x^0, x^2) e^{i(k^0 x^0 - k^0 x^2)}, \tag{102}$$

$$\bar{\psi}_1 \bar{\psi}_0^* = \frac{1}{2} R(x^0, x^2) e^{-i(k^0 x^0 - k^0 x^2)}, \tag{103}$$

We consider the following case,

$$|\partial_0 \epsilon| \ll |k^0 \epsilon|, \quad |\partial_0 R| \ll |k^0 R|,$$
$$|\partial_2 \epsilon| \ll |k^0 \epsilon|. \tag{104}$$

Neglect non-resonant terms like $e^{\pm 2ik^0 x^0}$ and quantum fluctuations (Green's functions Δ_{01} and Δ_{10}) (the rotating wave approximation). Then from Equations (99)–(101), we arrive at the Maxwell–Bloch equations,

$$\frac{\partial \epsilon}{\partial x^0} + \frac{\partial \epsilon}{\partial x^2} = i e d_e k^0 R, \tag{105}$$

$$\frac{\partial Z}{\partial x^0} = i e d_e (\epsilon R^* - \epsilon^* R), \tag{106}$$

$$\frac{\partial R}{\partial x^0} = -i e d_e \epsilon Z. \tag{107}$$

We assume that ϵ, Z and R are independent of the spatial coordinate of the x^2 direction. We shall change $\epsilon \to i\epsilon$ in the above equations and assume real functions $R = R^*$ and $\epsilon = \epsilon^*$. Then we can write,

$$\frac{\partial \epsilon}{\partial x^0} = e d_e k^0 R, \tag{108}$$

$$\frac{\partial Z}{\partial x^0} = -2 e d_e \epsilon R, \tag{109}$$

$$\frac{\partial R}{\partial x^0} = e d_e \epsilon Z. \tag{110}$$

We find the conservation law with the definition $B^2 \equiv 2R^2 + Z^2$,

$$\frac{\partial}{\partial x^0} B^2 = \frac{\partial}{\partial x^0} \left(2R^2 + Z^2 \right) = 0. \tag{111}$$

The relation $\frac{\partial B}{\partial x^0} = 0$ represents the probability conservation since we can rewrite $B^2 = (2|\bar{\psi}_1|^2 + |\bar{\psi}_0|^2)^2$ by Equation (103) and $Z \equiv 2|\bar{\psi}_1|^2 - |\bar{\psi}_0|^2$. We also find the following conservation law,

$$\frac{\partial}{\partial x^0}\left[\frac{1}{2}\epsilon^2 + \frac{1}{2}k^0 Z\right] = 0, \tag{112}$$

which represents the energy conservation. By this relation, we might be able to estimate the maximum energy density of electric fields,

$$\left(\frac{1}{2}\epsilon^2\right)_{\text{max}} = -\frac{1}{2}k^0 Z_{\text{min}} = \frac{1}{2}k^0 B, \tag{113}$$

in case there is no external energy supply. We derive the following solutions in Equations (108)–(110),

$$R(x^0) = \frac{1}{\sqrt{2}} B\sin\theta(x^0), \quad Z(x^0) = B\cos\theta(x^0), \tag{114}$$

$$\theta(x^0) = \theta_0 + \sqrt{2}ed_e \int_0^{x^0} dx'^0 \epsilon(x'^0), \tag{115}$$

with $\frac{\partial\theta}{\partial x^0} = \sqrt{2}ed_e\epsilon$ and the constant B in a similar way to [71]. The $\theta(x^0)$ swings around the position $\theta = \pi$ with the frequency $\Omega = ed_e\sqrt{k^0 B}$ in case we start with initial conditions at around $\theta_0 \sim \pi$ $(|\bar{\psi}_1|^2 = 0)$, since we can rewrite Equation (108) as

$$\frac{\partial^2\theta(x^0)}{\partial(x^0)^2} = (ed_e)^2 k^0 B\sin\theta(x^0). \tag{116}$$

The B is the order of the number density of dipoles.

We introduce the damping term $\frac{1}{L}\epsilon$ for the release of radiation and the propagation length L in Equation (108). We can write,

$$\frac{\partial\epsilon}{\partial x^0} + \frac{1}{L}\epsilon = \frac{ed_e k^0}{\sqrt{2}} B\sin\theta(x^0). \tag{117}$$

In $\kappa = \frac{1}{L} \gg$ time derivative, we can neglect the first term in the above equations, then

$$\frac{\partial\theta}{\partial x^0} = \frac{(ed_e)^2 k^0 B}{\kappa}\sin\theta(x^0). \tag{118}$$

The solution is,

$$\theta(x^0) = 2\tan^{-1}\left[\exp\left(\frac{(ed_e)^2 k^0 B x^0}{\kappa}\right)\tan\frac{\theta_0}{2}\right], \tag{119}$$

and,

$$\epsilon = \frac{1}{\sqrt{2}ed_e\tau_R} \times \left[\cosh\left(\frac{x^0 - \tau_0}{\tau_R}\right)\right]^{-1} \tag{120}$$

with $\tau_R = \frac{\kappa}{(ed_e)^2 k^0 B}$ and $\tau_0 = -\tau_R\ln(\tan\frac{\theta_0}{2})$. The $\tau_R \propto 1/B \sim 1/N$ with the number of dipoles N represents the relaxation time of electric fields in the super-radiance. When N dipoles decay within time scales $1/N$, the intensity of electric fields becomes the order N^2 (super-radiant decay with correlation among dipoles), not N (spontaneous decay without correlation among dipoles).

5.2. Higgs Mechanism and Tachyonic Instability

In this section, we rewrite time evolution equations for coherent fields with only real functions. We assume the spatially homogeneous case. We do not adopt the rotating wave approximation in this section. We show how coherent electric fields E_1 are affected by $Z = 2|\bar{\psi}_1|^2 - |\bar{\psi}_0|^2$.

In Equation (101), the second derivatives of coherent fields on the right-hand side are written by,

$$\frac{ed_e}{2I^2}\left(\bar{\psi}_1^*\bar{\psi} + \bar{\psi}_0^*\bar{\psi}_1\right) + \frac{2(ed_e)^2 Z}{I}E_1,$$

where we have used Equation (100). As a result, we arrive at,

$$\left[(\partial_0)^2 - (\partial_2)^2 - \frac{2(ed_e)^2 Z}{I}\right]E_1 = \frac{\mu_1}{4I^2} + \frac{2(ed_e)^2 E_1}{I}\int_p \left(2F_{11}(X,p) - F_{00}(X,p) - \Delta_{g,11,F}(X,p)\right)$$

$$+ \frac{(ed_e)^2}{I^2}E_1 \int_p \left(\text{Re}\Delta_{g,11,R}(X,p)F_{00}(X,p) + \Delta_{g,11,F}(X,p)\text{Re}\Delta_{00,R}(X,p)\right)$$

$$+ \frac{(ed_e)^2}{2I^2}\frac{\partial E_1}{\partial X^0}\int_p\left(\frac{\partial F_{00}}{\partial p^0}\frac{\Delta_{g,11,\rho}}{i} + \frac{\rho_{00}}{i}\frac{\partial\Delta_{g,11,F}}{\partial p^0}\right) + \frac{(ed_e)^2}{4I^2}E_1\frac{\partial}{\partial X^0}\int_p\left(\frac{\partial F_{00}}{\partial p^0}\frac{\Delta_{g,11,\rho}}{i} + \frac{\rho_{00}}{i}\frac{\partial\Delta_{g,11,F}}{\partial p^0}\right), \quad (121)$$

with the x^1 direction of the dipole moment (density) given by $\mu_1 = 2ed_e\left(\bar{\psi}_1^*\bar{\psi}_0 + \bar{\psi}_0^*\bar{\psi}_1\right)$, $F_{11}(X,p) = \frac{\Delta_{11}^{21}(X,p)+\Delta_{11}^{12}(X,p)}{2}$, $F_{00}(X,p) = \frac{\Delta_{00}^{21}(X,p)+\Delta_{00}^{12}(X,p)}{2}$ and $\Delta_{g,11,F}(X,p) = \frac{\Delta_{g,11}^{21}(X,p)+\Delta_{g,11}^{12}(X,p)}{2}$. In the Appendix A we have shown the detailed derivation for the second, third, fourth and fifth terms in the above equations. We have assumed the self-energy $\Sigma_{00} = \Sigma_{11} = 0$ in deriving the time derivatives of Δ_{10} and Δ_{01} in Equation (101). Even if we include contributions of self-energy in Equation (121), they are higher order $O\left((ed_e)^4\right)$ in the coupling expansion. We have neglected higher order terms in the gradient expansion for quantum fluctuations. In Equation (121), we leave the $-(\partial_2)^2 E_1$ term on the left-hand side in the above equation to compare with the sign of $-\frac{2(ed_e)^2 Z}{I}E_1$ term. We find the Higgs mechanism with the mass squared $-\frac{2(ed_e)^2 Z}{I}$ in the case of the normal population $Z = 2|\bar{\psi}_1|^2 - |\bar{\psi}_0|^2 < 0$. On the other hand, the tachyonic instability appears in the inverted population $Z > 0$ in the above equation. Then the electric field E_1 will increase exponentially until Z becomes negative. In Equation (121), the second term on the right-hand side is proportional to $2F_{11} - F_{00} - \Delta_{g,11,F}$. Near equilibrium states, we might find $F_{00} > 2F_{11} - \Delta_{g,11,F}$, where statistical functions F_{11}, F_{00} and $\Delta_{g,11,F}$ are proportional to the Bose–Einstein distribution $\frac{1}{e^{p^0/T}-1}$ plus $\frac{1}{2}$ (with the Kadanoff–Baym ansatz) with different dispersion relations $p^0 \sim \frac{\mathbf{p}^2}{2m}$ for F_{00} and $p^0 \sim \frac{\mathbf{p}^2}{2m} + \frac{1}{2I}$ for F_{11} and $\Delta_{g,11,F}$, due to the energy difference $\frac{1}{2I} - \frac{0}{2I}$ between the ground state and first excited states. So the $2F_{11} - F_{00} - \Delta_{g,11,F}$ in the second term is negative near the equilibrium states, which might mean no tachyonic unstable terms appear from quantum fluctuations near equilibrium states. The contributions of quantum fluctuations on the right-hand side written by statistical functions (second, third, fourth and fifth terms) vanish at zero temperature $T = 0$. Quantum fluctuations represent finite temperature effects at equilibrium states, although we need not restrict ourselves to only the equilibrium case. We have shown general contributions of quantum fluctuations in both equilibrium and non-equilibrium case in this paper.

Finally we shall consider remaining equations for coherent dipole fields. By using Equations (99) and (100) and the definitions of real functions $\mu_1 = 2ed_e(\bar{\psi}_1^*\bar{\psi}_0 + \bar{\psi}_0^*\bar{\psi}_1)$, $P = ied_e(\bar{\psi}_1^*\bar{\psi}_0 - \bar{\psi}_0^*\bar{\psi}_1)$ and $Z = 2|\bar{\psi}_1|^2 - |\bar{\psi}_0|^2$, we can also derive,

$$\partial_0 Z = 4E_1 P, \quad (122)$$

$$\partial_0 \mu_1 = \frac{P}{I}, \quad (123)$$

$$\partial_0 P = -\frac{\mu_1}{4I} - 2(ed_e)^2 E_1 Z. \quad (124)$$

We can show $\partial_0(2|\bar{\psi}_1|^2 + |\bar{\psi}_0|^2) = 0$ by using these three equations. In these equations with initial conditions $E_1 > 0$, $Z > 0$ (inverted population), $P = 0$ and $\mu_1 = 0$, the P and the μ_1 decrease at around the initial time and Z starts to decrease due to $E_1 P < 0$. In initial conditions $E_1 > 0$, $Z < 0$ (normal population), $P = 0$ and $\mu_1 = 0$, the P and the μ_1 increase at around the initial time and Z starts to increase due to $E_1 P > 0$. The absolute values of Z decrease at around the initial time. We find that there is no term of quantum fluctuations in Equations (122)–(124).

We can solve Equations (121)–(124) with real functions in this section and the Kadanoff–Baym equations with real statistical functions and pure imaginary spectral functions in Section 4, simultaneously.

6. Discussion

In this paper, we have derived time evolution equations, namely the Klein–Gordon equations for coherent photon fields, the Schrödinger-like equations for coherent electric dipole fields and the Kadanoff–Baym equations for quantum fluctuations, starting with the Lagrangian in quantum electrodynamics with electric dipoles in $2 + 1$ dimensions. We have adopted the two-particle-irreducible effective action technique with the leading-order self-energy of the coupling expansion. We find that electric dipoles change their angular momenta due to coherent electric fields $E_1 \pm i\alpha E_2$ with $\alpha = \pm 1$. They also change momenta and angular momenta by scattering with incoherent photons. The proof of H-theorem is possible for these processes as shown in Section 3. Our analysis provides the dynamics of both the order parameters with coherent fields and quantum fluctuations for incoherent particles.

In Section 2, we adopt two-energy level approximation for the angular momenta of dipoles. Then, we find that the $i\Delta_0^{-1}$ is written by 3×3 matrix with zero $(-1, 1)$ and $(1, -1)$ components. The form of the matrix is similar to 3×3 matrix in the analysis in open systems, the central region, left and right reservoirs as in [59,61–63]. Hence we can simplify the Kadanoff–Baym equations for dipole fields in an isolated system with the same procedures as those in open systems. The difference between QED with dipoles and ϕ^4 theory in open systems is that the coherent electric field changes the momenta of dipoles when the phase $\alpha\zeta$ in $E_1 \pm i\alpha E_2$ with $\alpha = \pm 1$ is dependent on space–time. The space dependence of coherent electric fields might disappear in the time evolution due to the instability by the lower entropy of the system, then electric fields will change angular momenta of dipoles but not change momenta p due to $\partial\zeta = 0$. We can also trace the dynamics with $\partial\zeta = 0$. By setting the initial conditions with the symmetry $\alpha \to -\alpha$, namely $\bar{\psi}_\alpha^{(*)} = \bar{\psi}_{-\alpha}^{(*)}$, $\Delta_{\alpha 0} = \Delta_{-\alpha 0}$ and $\Delta_{0\alpha} = \Delta_{0-\alpha}$, with initial conditions $E_2 = 0$ and $\partial_0 E_2 = 0$ in spatially homogeneous systems in $\partial^\nu F_{\nu 2} = J_2$ in Equation (20), we can show $E_2 = 0$ at any time. Then we can use $\partial\zeta = 0$. This condition simplifies numerical simulations in the Kadanoff–Baym equations since we need not estimate the momentum shift $p \to p \pm \alpha\partial\zeta$ in the finite-size lattice for the momentum space. As a result, the simulations for Kadanoff–Baym equations for dipoles and photons will be similar to those in QED with charged bosons in [72].

In Section 3, we have introduced a kinetic entropy current and shown the H-theorem in the leading-order of the coupling expansion with ed_e. This entropy approaches the Boltzmann entropy in the limit of zero spectral width as in [58]. The mode-coupling processes between dipoles and photons produce entropy. When there are deviations between (00) and $(\alpha\alpha)$ components of distribution functions, entropy production occurs. Entropy production stops when the Bose–Einstein distribution is realized in the dynamics of Kadanoff–Baym equations.

We can also derive the energy shifts in dispersion relations due to nonzero electric fields by using the retarded Green's functions in Section 3. The 0th order equations for retarded Green's functions are given by,

$$\left(p^0 - \frac{\mathbf{p}^2}{2m} + 2(ed_e)^2 E_1^2 \Delta_{g,11,R} \right) \Delta_{00,R} = -1, \tag{125}$$

$$\left(p^0 - \frac{\mathbf{p}^2}{2m} - \frac{1}{2I} \right) \Delta_{11,R} + (ed_e)^2 E_1^2 \Delta_{00,R} \Delta_{g,11,R} = -1, \tag{126}$$

with $\Delta_{g,11,R} = \frac{-1}{p^0 - \frac{\mathbf{p}^2}{2m} - \frac{1}{2I}}$. Multiply $p^0 - \frac{\mathbf{p}^2}{2m} - \frac{1}{2I}$, take the imaginary parts in the above equations and remember the imaginary parts of retarded Green's functions are the spectral functions, then we find,

$$W \begin{bmatrix} \rho_{00} \\ \rho_{11} \end{bmatrix} = 0, \text{ with,}$$

$$W = \begin{bmatrix} \left(p^0 - \frac{\mathbf{p}^2}{2m} - \frac{1}{2I} \right)\left(p^0 - \frac{\mathbf{p}^2}{2m} \right) - 2(ed_e)^2 E_1^2 & 0 \\ -(ed_e)^2 E_1^2 & \left(p^0 - \frac{\mathbf{p}^2}{2m} - \frac{1}{2I} \right)^2 \end{bmatrix} \tag{127}$$

By setting determinant $|W|$ to be zero, we find the following solutions for dispersion relations,

$$p^0 = \frac{\mathbf{p}^2}{2m} + \frac{1}{4I} \pm \frac{1}{2}\sqrt{\frac{1}{4I^2} + 8(ed_e)^2 E_1{}^2}. \tag{128}$$

Here we assumed the symmetry for $\alpha = \pm 1$ for Green's functions and zero self-energy $\Sigma_{00} = \Sigma_{11} = 0$. We find how electric fields shift two energy levels 0 and $\frac{1}{2I}$. The above energy shift is similar to the energy shift given in [27] in $3+1$ dimensions due to nonzero electric fields.

In Section 5.1, we have derived the super-radiance from time evolution equations for coherent fields. We find that it is possible to derive the Maxwell–Bloch equations from our Lagrangian with the probability conservation law and the energy conservation law. Super-radiant decay with intensity of the order $\propto N^2$ (N: The number of dipoles) appears in a similar way to [70,71]. It is possible to derive the maximum energy of electric fields by use of Equation (113). We know that the moment of inertia of water molecule is $I = 2m_H R^2$ with $m_H = 940\,\text{MeV}$ with $R = 0.96 \times 10^{-10}$ m. Hence the $k^0 = \frac{1}{2I} = 1.1 \times 10^{-3}$ eV. Since $B = \frac{N}{V} = 3.3 \times 10^{28}$ /m^3 for liquid water, we find

$$\frac{1}{2}\epsilon_{\text{max}}^2 = \frac{1}{2}k^0 B = 1.8 \times 10^{25} \text{ eV/m}^3. \tag{129}$$

When we multiply the volume of all microtubules (MTs) in a brain,

$$V_{\text{MT}} = \pi \times 15\text{nm}^2 \times 1000\text{nm} \times 2000 \text{ MTs/neuron} \times 10^{11} \text{ neurons/brain} = 1.4 \times 10^{-7} \text{ m}^3, \tag{130}$$

we can arrive at,

$$\frac{1}{2}\epsilon_{\text{max}}^2 V_{\text{MT}} = 0.41 \text{ J} = 0.1 \text{ cal}. \tag{131}$$

If we maintain our brain 100 s without energy supply, we need at least 0.1×10^{-2} cal/s or 86 cal/day to maintain the ordered states of memory. We can compare 86 cal/day with 4000 cal/day $=$ 2000 kcal/day $\times$ 0.2 (energy consumption rate of brain) $\times$ 0.01 (energy rate to maintain the ordered system). The 86 cal/day is within the 4000 cal/day, which is consistent with our experiences. In this derivation, we have used coefficients in $2+1$ dimensions and the number density of water molecules in $3+1$ dimensions.

In Section 5.2, we have derived time evolution equations for electric field E_1. The Higgs mechanism appears in this equation in normal population $Z < 0$. As a result, the dynamical mass generation occurs with the maximum mass $\Omega_{\text{Higgs}} = 2ed_e\sqrt{k^0 B} = 30k^0$ where the number density of dipoles is $B = 2|\bar{\psi}_1|^2 + |\bar{\psi}_0|^2 = \frac{N}{V}$. The period is $2\pi/\Omega_{\text{Higgs}} = 1.3 \times 10^{-13}$ s. In normal population $Z < 0$, the Meissner effect appears with the penetrating length $1/\Omega_{\text{Higgs}} = 6.3$ μm. On the other hand, the tachyonic instability occurs in inverted population $Z > 0$. The electric field E_1 increases exponentially with $\exp(\Omega X^0)$ (with $\Omega \leq \Omega_{\text{max}}$) where the time scale is $1/\Omega_{\text{max}} = 2.1 \times 10^{-14}$ s with $\Omega_{\text{max}} = \Omega_{\text{Higgs}}$. Due to energy conservation, since Z decreases as the absolute value of the electric field increases, tachyonic instability stops in $Z < 0$.

We have prepared for numerical simulations with time evolution equations, namely the Schödinger-like equations for coherent electric dipole fields, the Klein–Gordon equations for coherent electric fields and the Kadanoff–Baym equations for quantum fluctuations. Our simulations might describe the dynamics towards equilibrium states for quantum fluctuations and the dynamics of super-radiant states for coherent fields. Our analysis is also extended to simulations in open systems by preparing the left and the right reservoirs like those in [59] or networks [73].

7. Conclusions

We have derived the Schrödinger equations for coherent electric dipole fields, the Klein–Gordon equations for coherent electric fields and the Kadanoff–Baym equations for quantum fluctuations in QED with electric dipoles in $2+1$ dimensions. It is possible to derive equilibration for quantum fluctuations and super-radiance for background coherent fields simultaneously. Total energy consumption to maintain super-radiance in microtubules is consistent with energy consumption in our experiences. We can describe dynamical information transfer with super-radiance via microtubules without violation of the second law in thermodynamics. We have also derived the Higgs mechanism in normal population and the tachyonic instability in inverted population. These dynamical properties might be significant to form and maintain coherent domains composed of dipoles and photons. We are ready to describe memory formation processes towards equilibrium states in $2+1$ dimensions with equations in this paper. Furthermore, our approach might pave the way to understand the dynamical thinking processes with memory recalling in QBD by investigating the case in open systems with the Kadanoff–Baym equations. This work will be extended to the $3+1$ dimensional analysis to describe memory formation processes in numerical simulations. We should derive the Schödinger-like equations, the Klein–Gordon equations and the Kadanoff–Baym equations by starting with the single Lagrangian in QED with electric dipoles in $3+1$ dimensions in the future study. These equations in $3+1$ dimensions will describe more realistic and practical dynamics in QBD.

Author Contributions: Conceptualization, A.N, S.T. and J.A.T.; methodology, A.N.; software, A.N.; validation, A.N., S.T. and J.A.T.; formal analysis, A.N.; investigation, A.N.; resources, S.T. and J.A.T.; data curation, A.N.; writing-original draft preparation, A.N.; writing-review and editing, A.N., S.T. and J.A.T.; visualization, A.N.; supervision, S.T. and J.A.T.; project administration, S.T. and J.A.T.; funding acquisition, S.T. and J.A.T.

Acknowledgments: J.A.T. is grateful for research support received from NSERC (Canada).

Appendix A. Quantum Fluctuations in the Klein–Gordon Equations

In this section, we shall derive the second, third, fourth and fifth terms involving quantum fluctuations on the right-hand side in Equation (121) in spatially homogeneous systems. They correspond to the following term,

$$-2ed_e(\partial_0)^2 \left[\Delta_{10}(x,x) + \Delta_{01}(x,x) \right],$$

in Equation (101) with the symmetry $\Delta_{10} = \Delta_{-10}$ and $\Delta_{01} = \Delta_{0-1}$. It appears in taking the time derivative of J_1 (given by Equation (21)) in Equation (20). Here $\Delta_{10}(x,x)$ and $\Delta_{01}(x,x)$ can be rewritten by,

$$\Delta_{10}(x,x) = -\frac{ed_e}{i} \int_w \Delta_{g,11}(x,w) E_1(w) \Delta_{00}(w,x), \tag{A1}$$

$$\Delta_{01}(x,x) = -\frac{ed_e}{i} \int_w \Delta_{00}(x,w) E_1(w) \Delta_{g,11}(w,x), \tag{A2}$$

where we have used Equations (31) and (34) by setting $E_2 = 0$. We rewrite second time derivatives of $\Delta_{10}(x,x)$ and $\Delta_{01}(x,x)$.

We shall rewrite Equation (30) without self-energy $\Sigma_{\alpha\alpha}$ as,

$$\left[i\frac{\partial}{\partial x^0} + \frac{\nabla_i^2}{2m} - \frac{1}{2I} \right] \Delta_{g,11}(x,w) = i\delta_C(x-w), \tag{A3}$$

$$\left[-i\frac{\partial}{\partial x^0} + \frac{\nabla_i^2}{2m} - \frac{1}{2I} \right] \Delta_{g,11}(w,x) = i\delta_C(w-x), \tag{A4}$$

where we have multiplied $\Delta_{g,11}$ from the right and left of Equation (30). By using the above equations and Equations (32) and (33) with Equations (A1) and (A2) and $\Delta_{0,00}^{-1}(x,y) = \left(i\frac{\partial}{\partial x^0} + \frac{\nabla_i^2}{2m} \right) \delta_C(x-y)$, we can show

$$\frac{\partial}{\partial x^0} \Delta_{10}(x,x) = ed_e \left[\left[\left(-\frac{\nabla_i^2}{2m} + \frac{1}{2I} \right) \Delta_{g,11} + i\delta_C \right] E_1 \Delta_{00} \right.$$
$$\left. + \Delta_{g,11} E_1 \frac{\nabla_i^2}{2m} \Delta_{00} + 2\Delta_{g,11} ed_e E_1 \Delta_{01} E_1 - \Delta_{g,11} E_1 i\delta_C \right]$$
$$= ed_e \left[\left(\frac{1}{2I} \Delta_{g,11} + i\delta_C \right) E_1 \Delta_{00} + 2\Delta_{g,11} ed_e E_1 \Delta_{01} E_1 - \Delta_{g,11} E_1 i\delta_C \right], \tag{A5}$$

$$\frac{\partial}{\partial x^0} \Delta_{01}(x,x) = ed_e \left[(-2ed_e E_1 \Delta_{10} + i\delta_C) E_1 \Delta_{g,11} + \Delta_{00} E_1 \left(-\frac{1}{2I} \Delta_{g,11} i\delta_C \right) \right], \tag{A6}$$

where δ_C represents the delta function in the closed-time path. Here the terms proportional to ∇_i^2 are cancelled in spatially homogeneous systems. By use of the above two equations, we can show

$$\frac{\partial}{\partial x^0} (\Delta_{10} + \Delta_{01}) = \frac{1}{2iI} (\Delta_{10} - \Delta_{01}), \tag{A7}$$

and,

$$\frac{\partial^2}{\partial(x^0)^2} (\Delta_{10} + \Delta_{01}) = \frac{ed_e}{2iI} \left[\left(\frac{\Delta_{g,11}}{2I} + i\delta_C \right) E_1 \Delta_{00} + 2\Delta_{g,11} ed_e E_1 \Delta_{01} E_1 - \Delta_{g,11} E_1 i\delta_C \right.$$
$$\left. - (-2ed_e E_1 \Delta_{10} + i\delta_C) E_1 \Delta_{g,11} - \Delta_{00} E_1 \left(-\frac{1}{2I} \Delta_{g,11} - i\delta_C \right) \right]$$
$$= \frac{ed_e}{2iI} \left[2iE_1 (\Delta_{00} - \Delta_{g,11}) + \frac{1}{2I} (\Delta_{g,11} E_1 \Delta_{00} + \Delta_{00} E_1 \Delta_{g,11}) \right.$$
$$\left. + 2ed_e (\Delta_{g,11} E_1 \Delta_{01} E_1 + E_1 \Delta_{10} E_1 \Delta_{g,11}) \right]. \tag{A8}$$

Since we can rewrite Equations (35) or (36) by multiplying $i\Delta_{g,11}$ as,

$$
\begin{aligned}
i\Delta_{g,11} - i\Delta_{11} &= ed_e\Delta_{g,11}E_1\Delta_{01} \\
&= ed_e\Delta_{10}E_1\Delta_{g,11}, \tag{A9}
\end{aligned}
$$

we arrive at,

$$
\frac{\partial^2}{\partial(x^0)^2}\left(\Delta_{10} + \Delta_{01}\right) = -\frac{1}{4I^2}\left(\Delta_{10} + \Delta_{01}\right) + \frac{ed_eE_1}{I}\left(\Delta_{00} - 2\Delta_{11} + \Delta_{g,11}\right), \tag{A10}
$$

where we have used Equations (A1) and (A2).

Finally by rewriting the statistical parts (subscript 'F') of $\Delta_{10} + \Delta_{01}$ with Equations (A1) and (A2), and using $E_1(w) = E_1(x) + (w^0 - x^0)\partial_0 E_1(x)$ in,

$$
\left[\int dw\left[\Delta_{g,11}(x,w)E_1(w)\Delta_{00}(w,x) + \Delta_{00}(x,w)E_1(w)\Delta_{g,11}(w,x)\right]\right]_F,
$$

and the relation in the first order in the gradient expansion,

$$
\begin{aligned}
\left[\int dw\Delta_{g,11}(x,w)\Delta_{00}(w,x)\right]_F &= \int_p\left(\frac{\Delta_{g,11,R}(x,p)}{i}F_{00}(x,p) + \Delta_{g,11,F}\frac{\Delta_{00,A}}{i}\right. \\
&\left. +\frac{i}{2}\left\{\frac{\Delta_{g,11,R}(x,p)}{i}, F_{00}(x,p)\right\} + \frac{i}{2}\left\{\Delta_{g,11,F}, \frac{\Delta_{00,A}}{i}\right\}\right), \tag{A11}
\end{aligned}
$$

with the advanced (subscript 'A') $\Delta_{00,A} = i(\Delta_{00}^{11} - \Delta_{00}^{21}) = \mathrm{Re}\Delta_{00,R} - \frac{\rho_{00}}{2}$ and the retarded $\Delta_{00,R} = i(\Delta_{00}^{11} - \Delta_{00}^{12}) = \mathrm{Re}\Delta_{00,R} + \frac{\rho_{00}}{2}$, we can derive the third, fourth and fifth terms on the right-hand side in Equation (121).

References

1. Day, J.J.; Sweatt, J.D. DNA methylation and memory formation. *Nat. Neurosci.* **2010**, *13*, 1319. [CrossRef] [PubMed]
2. Adolphs, R. The unsolved problems of neuroscience. *Trends Cogn. Sci.* **2015**, *19*, 173–175. [CrossRef] [PubMed]
3. Kukushkin, N.V.; Carew, T.J. Memory takes time. *Neuron* **2017**, *95*, 259–279. [CrossRef] [PubMed]
4. Jibu, M.; Yasue, K. *Quantum Brain Dynamics and Consciousness*; John Benjamins: Amsterdam, The Netherlands, 1995.
5. Vitiello, G. *My Double Unveiled: The Dissipative Quantum Model of Brain*; John Benjamins: Amsterdam, The Netherlands, 2001; Volume 32.
6. Sabbadini, S.A.; Vitiello, G. Entanglement and Phase-Mediated Correlations in Quantum Field Theory. Application to Brain-Mind States. *Appl. Sci.* **2019**, *9*, 3203. [CrossRef]
7. Lashley, K.S. *Brain Mechanisms and Intelligence: A Quantitative Study of Injuries to the Brain*; PsycBOOKS: Chicago, IL, USA, 1929.
8. Pribram, K.H. Languages of the brain: Experimental paradoxes and principles in neuropsychology. *Nerv. Ment. Dis.* **1973**, *157*, 69–70.
9. Pribram, K. *Brain and Perception: Holonomy and Structure in Figural Processing*; Lawrence Erlbaum Associates: Hillsdale, NJ, USA, 1991.
10. Umezawa, H. *Advanced Field Theory: Micro, Macro, and Thermal Physics*; AIP: College Park, MD, USA, 1995.
11. Nambu, Y.; Lasinio, G.J. Dynamical Model of Elementary Particles Based on an Analogy with Superconductivity. I. *Phys. Rev.* **1961**, *112*, 345. [CrossRef]
12. Goldstone, J. Field theories with «Superconductor» solutions. *Il Nuovo Cim. (1955–1965)* **1961**, *19*, 154–164. [CrossRef]

13. Goldstone, J.; Salam, A.; Weinberg, S. Broken symmetries. *Phys. Rev.* **1962**, *127*, 965. [CrossRef]

14. Ricciardi, L.M.; Umezawa, H. Brain and physics of many-body problems. *Kybernetik* **1967**, *4*, 44–48. [CrossRef]

15. Stuart, C.; Takahashi, Y.; Umezawa, H. On the stability and non-local properties of memory. *J. Theor. Biol.* **1978**, *71*, 605–618. [CrossRef]

16. Stuart, C.; Takahashi, Y.; Umezawa, H. Mixed-system brain dynamics: Neural memory as a macroscopic ordered state. *Found. Phys.* **1979**, *9*, 301–327. [CrossRef]

17. Fröhlich, H. Bose condensation of strongly excited longitudinal electric modes. *Phys. Lett. A* **1968**, *26*, 402–403. [CrossRef]

18. Fröhlich, H. Long-range coherence and energy storage in biological systems. *Int. J. Quantum Chem.* **1968**, *2*, 641–649. [CrossRef]

19. Fröhlich, H. Long range coherence and the action of enzymes. *Nature* **1970**, *228*, 1093–1093. [CrossRef]

20. Fröhlich, H. Selective long range dispersion forces between large systems. *Phys. Lett. A* **1972**, *39*, 153–154. [CrossRef]

21. Fröhlich, H. Evidence for Bose condensation-like excitation of coherent modes in biological systems. *Phys. Lett. A* **1975**, *51*, 21–22. [CrossRef]

22. Fröhlich, H. Long-range coherence in biological systems. *La Riv. Del Nuovo Cim. (1971–1977)* **1977**, *7*, 399–418. [CrossRef]

23. Davydov, A.; Kislukha, N. Solitons in One-Dimensional Molecular Chains. *Phys. Status Solidi (B)* **1976**, *75*, 735–742. [CrossRef]

24. Tuszyński, J.; Paul, R.; Chatterjee, R.; Sreenivasan, S. Relationship between Fröhlich and Davydov models of biological order. *Phys. Rev. A* **1984**, *30*, 2666. [CrossRef]

25. Del Giudice, E.; Doglia, S.; Milani, M.; Vitiello, G. Spontaneous symmetry breakdown and boson condensation in biology. *Phys. Lett. A* **1983**, *95*, 508–510. [CrossRef]

26. Del Giudice, E.; Doglia, S.; Milani, M.; Vitiello, G. A quantum field theoretical approach to the collective behaviour of biological systems. *Nucl. Phys. B* **1985**, *251*, 375–400. [CrossRef]

27. Del Giudice, E.; Preparata, G.; Vitiello, G. Water as a free electric dipole laser. *Phys. Rev. Lett.* **1988**, *61*, 1085. [CrossRef] [PubMed]

28. Del Giudice, E.; Smith, C.; Vitiello, G. Magnetic Flux Quantization and Josephson Systems. *Phys. Scr.* **1989**, *40*, 786–791. [CrossRef]

29. Jibu, M.; Yasue, K. A physical picture of Umezawa's quantum brain dynamics. *Cybern. Syst. Res.* **1992**, *92*, 797–804.

30. Jibu, M.; Yasue, K. Intracellular quantum signal transfer in Umezawa's quantum brain dynamics. *Cybern. Syst.* **1993**, *24*, 1–7. [CrossRef]

31. Jibu, M.; Hagan, S.; Hameroff, S.R.; Pribram, K.H.; Yasue, K. Quantum optical coherence in cytoskeletal microtubules: Implications for brain function. *Biosystems* **1994**, *32*, 195–209. [CrossRef]

32. Jibu, M.; Yasue, K. What is mind?- Quantum field theory of evanescent photons in brain as quantum theory of consciousness. *INF* **1997**, *21*, 471–490.

33. Dicke, R.H. Coherence in spontaneous radiation processes. *Phys. Rev.* **1954**, *93*, 99. [CrossRef]

34. Gross, M.; Haroche, S. Superradiance: An essay on the theory of collective spontaneous emission. *Phys. Rep.* **1982**, *93*, 301–396. [CrossRef]

35. Preparata, G. Quantum field theory of superradiance. *Probl. Fundam. Mod. Phys.* **1990**, 303. [CrossRef]

36. Preparata, G. *QED Coherence in Matter*; World Scientific: Singapore, 1995.

37. Enz, C.P. On Preparata's theory of a superradiant phase transition. *Helv. Phys. Acta* **1997**, *70*, 141–153.

38. Vitiello, G. Dissipation and memory capacity in the quantum brain model. *Int. J. Mod. Phys.* **1995**, *9*, 973–989. [CrossRef]

39. Vitiello, G. Classical chaotic trajectories in quantum field theory. *Int. J. Mod. Phys. B* **2004**, *18*, 785–792. [CrossRef]

40. Zheng, J.M.; Pollack, G.H. Long-range forces extending from polymer-gel surfaces. *Phys. Rev. E* **2003**, *68*, 031408. [CrossRef] [PubMed]

41. Del Giudice, E.; Voeikov, V.; Tedeschi, A.; Vitiello, G. The origin and the special role of coherent water in living systems. *F. Cell* **2014**, 95–111. [CrossRef]

42. Tegmark, M. Importance of quantum decoherence in brain processes. *Phys. Rev. E* **2000**, *61*, 4194. [CrossRef]

43. Craddock, T.J.; Tuszynski, J.A.; Hameroff, S. Cytoskeletal signaling: Is memory encoded in microtubule lattices by CaMKII phosphorylation? *PLoS Comput. Biol.* **2012**, *8*, e1002421. [CrossRef]

44. Baym, G.; Kadanoff, L.P. Conservation laws and correlation functions. *Phys. Rev.* **1961**, *124*, 287. [CrossRef]

45. Kadanoff, L.P.; Baym, G. *Quantum Statistical Mechanics: Green's Function Methods in Equilibrium Problems*; WA Benjamin: Los Angeles, CA, USA, 1962.

46. Baym, G.; Kadanoff, L.P. Self-Consistent Approximations in Many-Body Systems. *Phys. Rev.* **1962**, *127*, 1391. [CrossRef]

47. Cornwall, J.M.; Jackiw, R.; Tomboulis, E. Effective action for composite operators. *Phys. Rev. D* **1974**, *10*, 2428. [CrossRef]

48. Niemi, A.J.; Semenoff, G.W. Finite-temperature quantum field theory in Minkowski space. *Ann. Phys.* **1984**, *152*, 105–129. [CrossRef]

49. Calzetta, E.; Hu, B.L. Nonequilibrium quantum fields: Closed-time-path effective action, Wigner function, and Boltzmann equation. *Phys. Rev. D* **1988**, *37*, 2878. [CrossRef] [PubMed]

50. Schwinger, J. Brownian motion of a quantum oscillator. *J. Math. Phys.* **1961**, *2*, 407–432. [CrossRef]

51. Keldysh, L.V. Diagram technique for nonequilibrium processes. *Sov. Phys. Jetp.* **1965**, *20*, 1018–1026.

52. Kluberg-Stern, H.; Zuber, J. Renormalization of non-Abelian gauge theories in a background-field gauge. I. Green's functions. *Phys. Rev. D* **1975**, *12*, 482. [CrossRef]

53. Abbott, L.F. The background field method beyond one loop. *Nucl. Phys. B.* **1981**, *185*, 189–203. [CrossRef]

54. Abbott, L.F. Introduction to the background field method. *Acta Phys. Pol. B.* **1981**, *13*, 33–50.

55. Wang, Q.; Redlich, K.; Stöcker, H.; Greiner, W. From the Dyson–Schwinger to the transport equation in the background field gauge of QCD. *Nucl. Phys. A* **2003**, *714*, 293–334. [CrossRef]

56. Ivanov, Y.B.; Knoll, J.; Voskresensky, D. Resonance transport and kinetic entropy. *Nucl. Phys. A* **2000**, *672*, 313–356. [CrossRef]

57. Kita, T. Entropy in nonequilibrium statistical mechanics. *J. Phys. Soc. Jpn.* **2006**, *75*, 114005–114005. [CrossRef]

58. Nishiyama, A. Entropy production in 2D $\lambda\phi4$ theory in the Kadanoff–Baym approach. *Nucl. Phys. A* **2010**, *832*, 289–313. [CrossRef]

59. Nishiyama, A.; Tuszynski, J.A. Non-Equilibrium $\phi4$ in open systems as a toy model of quantum field theory of the brain. *Ann. Phys.* **2018**, *398*, 214. [CrossRef]

60. Myöhänen, P.; Stan, A.; Stefanucci, G.; van Leeuwen, R. A many-body approach to quantum transport dynamics: Initial correlations and memory effects. *EPL (Europhys. Lett.)* **2008**, *84*, 67001. [CrossRef]

61. Myöhänen, P.; Stan, A.; Stefanucci, G.; Van Leeuwen, R. Kadanoff-Baym approach to quantum transport through interacting nanoscale systems: From the transient to the steady-state regime. *Phys. Rev. B* **2009**, *80*, 115107. [CrossRef]

62. Wang, J.S.; Agarwalla, B.K.; Li, H.; Thingna, J. Nonequilibrium Green's function method for quantum thermal transport. *Front. Phys.* **2014**, *9*, 673–697. [CrossRef]

63. Dražić, M.S.; Cerovski, V.; Zikic, R. Theory of time-dependent nonequilibrium transport through a single molecule in a nonorthogonal basis set. *Int. J. Quantum Chem.* **2017**, *117*, 57–73. [CrossRef]

64. Stratonovich, R.L. Gauge Invariant Generalization of Wigner Distribution. *Dok. Akad. Nauk SSSR* **1956**, *109*, 72–75.

65. Fujita, S. *Introduction to Non-Equilibrium Quantum Statistical Mechanics*; Krieger Pub Co: Malabar, FL, USA, 1966.

66. Groenewold, H.J. *On the Principles of Elementary Quantum Mechanics*; Springer: The Netherlands, 1946; Volume 12, pp. 1–56.

67. Moyal, J.E. Quantum mechanics as a statistical theory. *Math. Proc. Camb. Philos. Soc.* **1949**, *45*, 99–124. [CrossRef]

68. Knoll, J.; Ivanov, Y.B.; Voskresensky, D.N. Exact conservation laws of the gradient expanded Kadanoff–Baym equations. *Ann. Phys.* **2001**, *293*, 126–146. [CrossRef]

69. Ivanov, Y.B.; Knoll, J.; Voskresensky, D. Self-consistent approach to off-shell transport. *Phys. At. Nucl.* **2003**, *66*, 1902–1920. [CrossRef]

70. Bonifacio, R.; Preparata, G. Coherent spontaneous emission. *Phys. Rev. A* **1970**, *2*, 336. [CrossRef]

71. Benedict, M.G. *Super-Radiance: Multiatomic Coherent Emission*; Routledge: London, UK, 2018.

72. Nishiyama, A.; Tuszynski, J.A. Nonequilibrium quantum electrodynamics: Entropy production during equilibration. *Int. J. Mod. Phys. B* **2018**, *32*, 1850265. [CrossRef]

73. Nishiyama, A.; Tuszynski, J.A. Non-Equilibrium $\phi4$ theory for networks: Towards memory formations with quantum brain dynamics. *J. Phys. Commun.* **2019**, *3*, 055020. [CrossRef]

10

Analysis of Entropy Production in Structured Chemical Reactors: Optimization for Catalytic Combustion of Air Pollutants

Mateusz Korpyś [1,*], **Anna Gancarczyk** [1], **Marzena Iwaniszyn** [1], **Katarzyna Sindera** [1], **Przemysław J. Jodłowski** [2] and **Andrzej Kołodziej** [1,3]

[1] Institute of Chemical Engineering, Polish Academy of Sciences, Bałtycka 5, 44-100 Gliwice, Poland; anna.g@iich.gliwice.pl (A.G.); miwaniszyn@iich.gliwice.pl (M.I.); katarzyna.sindera@iich.gliwice.pl (K.S.); ask@iich.gliwice.pl (A.K.)

[2] Faculty of Chemical Engineering and Technology, Cracow University of Technology, Warszawska 24, 31-155 Kraków, Poland; pjodlowski@pk.edu.pl

[3] Faculty of Civil Engineering and Architecture, Opole University of Technology, Katowicka 48, 45-061 Opole, Poland

* Correspondence: matkor@iich.gliwice.pl

Abstract: Optimization of structured reactors is not without some difficulties due to highly random economic issues. In this study, an entropic approach to optimization is proposed. The model of entropy production in a structured catalytic reactor is introduced and discussed. Entropy production due to flow friction, heat and mass transfer and chemical reaction is derived and referred to the process yield. The entropic optimization criterion is applied for the case of catalytic combustion of methane. Several variants of catalytic supports are considered including wire gauzes, classic (long-channel) and short-channel monoliths, packed bed and solid foam. The proposed entropic criterion may indicate technically rational solutions of a reactor process that is as close as possible to the equilibrium, taking into account all the process phenomena such as heat and mass transfer, flow friction and chemical reaction.

Keywords: entropy production; optimization; reactor modelling; irreversible thermodynamics

1. Introduction

At the industrial level, optimization of chemical processes, including those based on structured catalytic reactors, is an inherent issue of the design procedure. Process optimization considers the prices of raw materials, energy, products and installations (apparatus); the prices may change rapidly and unpredictably due to market fluctuations, even at the negotiation stage. Therefore, process optimization is usually regarded as being within the engineering domain, it is in fact more connected with business and economic issues. These issues usually exceed the knowledge of an engineer or a scientist and require input from other individuals.

Structured reactors are very important in chemistry and catalysis [1–3]. The process design, i.e., the apparatus and the process conditions, has to secure some economic profitability in spite of potential changes of costs. Regardless of possible economic fluctuations (excluding any collapses), the process has to be profitable during the following years.

A review of the literature provides hints about recommended flow velocities, process temperatures and catalyst carriers. The data originate from the long-standing technical and economic experience of engineers and entrepreneurs. Recently, a new generation of structured catalytic reactors has been introduced into industry, and there is a paucity of knowledge and experience about their optimization. Moreover, the inner-structure design of the reactors is complicated because many geometrical parameters need to be optimized.

In the literature, different criteria can be found, which help identify optimal operating conditions of chemical reactors. "The technical" or "engineering" optimization, with which this work deals, focuses on reactor optimization in terms of fluid velocities, process (reaction) temperature, structured catalyst carrier shape and dimensions. This kind of optimization has begun in energetics due to the introduction of compact heat exchangers that usually exploit a combination of fins, turbulence mixers and other features. In the current literature, even more sophisticated criteria are proposed for multiparameter optimization of different equipment such as heat exchangers. So far, similar criteria for catalytic reactors have been derived. The comprehensive performance evaluation criteria (PEC) use three components: transport coefficients, reaction kinetics and pressure drop [4,5]. Another approach is the comparison of reactor length (or catalyst mass) with the resulting flow resistance as shown in [4,6]. For heat exchanger optimization, there are also evaluation criteria based on entropy production during the process, as presented, e.g., by London [7] and Bejan [8], who also predicted the extension of entropic criteria to chemical reactors. Entropy in economic analysis is treated as trade-off factor and can be a substitute of currency [9]. The application of entropic criterion can also be found in [10–12].

The aim of the study is to propose a highly simplified approach, based on irreversible thermodynamics, suitable for engineering optimization of chemical reactors. The entropic criterion is proposed to optimize structured catalytic reactors. The assumed model process is the catalytic combustion of methane.

2. Theoretical Background

To derive the equations governing entropy production, the reactor model must be specified. For the purposes of this paper, the one-dimensional plug-flow model (neglecting axial dispersion) in the steady-state was assumed. Due to the very thin catalyst layer deposited on the structured carrier, the internal diffusional resistance can be neglected.

Mass balance of reactant A, in the flowing fluid, per unit surface area of the reactor cross-section, is as follows:

$$w_0 \frac{dC_A}{dx} + k_C S_v (C_A - C_{AS}) = 0 \tag{1}$$

The initial conditions are: (i) $x = 0$; $C_A = C_{A0}$ and (ii) the reactant A, mass transferred from the gas bulk to the catalyst surface is balanced by the first-order catalytic reaction:

$$k_C (C_A - C_{AS}) = k_r C_{AS}. \tag{2}$$

Deriving concentration of A, at the catalyst surface from Equation (2), Equation (1) becomes:

$$-w_0 \frac{dC_A}{dx} = S_v \frac{k_C k_r}{k_C + k_r} C_A, \tag{3}$$

and, after integration, local concentration C_{Ax} and the reactor length L, required for the outlet concentration C_{AL} are:

$$C_{Ax} = C_{A0} \exp\left(-\frac{x}{w_0} \frac{S_v k_C k_r}{k_C + k_r}\right), \tag{4}$$

$$L = \frac{w_0}{S_v} \frac{k_C + k_r}{k_C k_r} \ln\left(\frac{C_{A0}}{C_{AL}}\right). \tag{5}$$

The energy balance may be presented (assuming no heat losses to the environment) as:

$$w_0 \varrho c_p \frac{dT}{dx} + \alpha S_v (T - T_S) = 0, \tag{6}$$

the initial conditions: at $x = 0$, $T = T_0$.

The mass and heat transfer in a heterogeneous catalytic reactor are strictly bound up (released reaction heat depends on the reactants mass transferred to the catalyst), thus

$$q = \alpha(T_S - T) = -\Delta H_R J_A = -\Delta H_R k_C (C_A - C_{AS}). \tag{7}$$

The above equations assume an isothermal process. In reality, the process is adiabatic. However, the concentration of organic air pollutants is usually low. For the volatile organic compounds (VOCs), a concentration of very few ppm is typical; for methane, it depends on the kind of source and may be within 1–1000 ppm. The level of concentrations of 100 ppm and higher can be treated by homogeneous combustion in, e.g., reverse-flow reactors due to important reaction heat. Thus, we assumed the concentration of methane at 200 ppm as rational for our analysis. In such a case, the adiabatic temperature rise is about 6 K, so the temperature increase along the reactor can be securely neglected.

Entropy production is an increase of system entropy due only to the irreversible phenomena [13]. This means that there is no entropy production at equilibrium or during a quasi-static process that runs infinitely close to the equilibrium. Any industrial process runs far from the equilibrium, and it produces entropy at irreversible conditions. In irreversible thermodynamics, entropy production is derived as the product of flux J_i and the driving force $\Delta \pi$ (causing the stream) divided by absolute temperature T [13,14]:

$$S_i = \frac{J_i \Delta \pi}{T}. \tag{8}$$

Assuming that the stream J_i is proportional to the driving force:

$$J_i = k_i \Delta \pi, \tag{9}$$

entropy production is proportional to the square of the driving force, thus it increases rapidly with the distance from the equilibrium:

$$S_i = \frac{k_i (\Delta \pi)^2}{T}. \tag{10}$$

In this paper, entropy production is considered due to the following irreversible phenomena:

- heat transfer between the gas phase and the catalyst surface (further denoted as H);
- diffusional mass transfer between the gas phase and the catalyst surface (denoted as D);
- irreversible catalytic reaction (denoted as R);
- flow friction, i.e., work performed against the flow resistance (denoted as F).

Total entropy production (per 1 mole of reactant A consumed in the reactor) is the sum of all the components:

$$S_P = S_H + S_D + S_R + S_F. \tag{11}$$

The above-mentioned components of entropy production are gathered in Table 1.

In the first column, basic equations of local entropy production are presented. In the second and third columns, the equations for the stream and the driving force are presented, respectively, derived using the reactor model. The last column presents reactor-integrated entropy production per 1 mole of substrate A consumed (e.g., burned) in the reactor. Detailed derivations, simple in fact, are not presented for reason of conciseness. The last position in Table 1, *flow friction* needs further comment. The entropy source considered is the volume fluid flow. The stream (flux) is the flow velocity and the driving force is the pressure gradient. The entropy produced is tantamount to viscous dissipation of pumping energy. This approach seems more friendly for engineers than viscous momentum flux often presented by irreversible thermodynamics; the flux is the pressure tensor and the driving force is the velocity gradient [11].

The impact of the reaction rate constant, k_r, and the heat and mass transfer coefficients, α and k_C, respectively, on the entropy produced by the heat (S_H) and mass (S_D) transfer is illustrated in Figure 1 for the combustion process and exemplary k_r and k_C values. The heat and mass transfer coefficients are bound by the Chilton–Colburn analogy [15], Equation (12), which allows the influence of mass transport on S_H to be determined.

$$j = \frac{Nu}{RePr^{1/3}} = \frac{Sh}{ReSc^{1/3}}. \tag{12}$$

$$S_D = Rln\left(1+\frac{k_r}{k_C}\right), \tag{13}$$

$$S_H = \frac{k_r}{k_C+k_r}\left[\frac{(-\Delta H_R)^2(C_{A0}-C_{AL})D_A Sc^{1/3}}{2\lambda T^2 \, Pr^{1/3}}\right]. \tag{14}$$

Table 1. Local and reactor-averaged components of entropy produced.

Entropy, σ_i	Flux, J_i	Driving Force, $\Delta\pi$	Entropy, Reactor Average Value, S_i (per mol of Substrate A)
Heat transfer (H) $\sigma_H = -\frac{q}{T^2}\nabla T$	Heat flux $q = -\Delta H_R J_A = $ $= \alpha(T_s-T)$	Temperature gradient $(T_s - T) = $ $= \frac{k_C(-\Delta H_R)(C_A-C_{AS})}{\alpha}$	$S_H = \frac{k_C k_r}{k_C+k_r}$ $\frac{(-\Delta H_R)^2(C_{A0}-C_{AL})}{2\alpha T^2}$
Mass transfer (D) $\sigma_D = -\sum_i \frac{J_i}{T}\nabla\mu_i$	Diffusive mass flux $J_A = k_C(C_A - C_{AS}) = $ $= k_{Cr}C_A$	Chemical potential gradient $\nabla\mu_A = RT\frac{\mu_A-\mu_{AS}}{s_{ef}}$	$S_D = Rln\left(\frac{k_C+k_r}{k_C}\right)$
Reaction (R) $\sigma_R = -\frac{A r_A S_v}{T}$	Reaction rate $r_A = k_r C_{AS} = k_{Cr}C_A$	Chemical affinity $A = -\sum_i \nu_i\mu_i = $ $= -\Delta G_R^{o,T} - RT\sum_i \nu_i ln y_i$	$S_R = \frac{A}{T}$
Flow friction (F) $\sigma_F = \frac{W}{TF_cL} = -\frac{w\nabla P}{T}$	Fluid stream w	Pressure gradient $-\nabla P$	$S_F = \frac{f}{2T}\frac{w_0^3\varrho}{\varepsilon^3 k_{Cr}}\frac{ln\left(\frac{C_{A0}}{C_{AL}}\right)}{(C_{A0}-C_{AL})}$

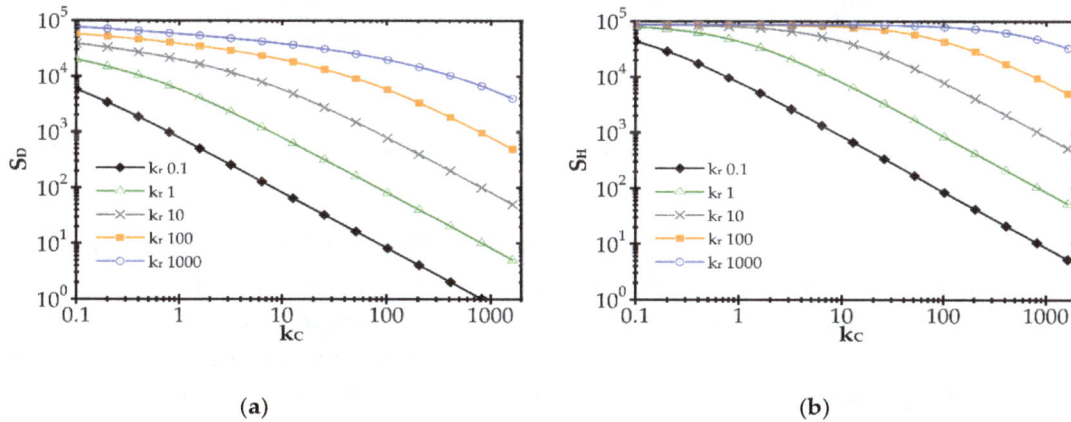

Figure 1. Impact of the mass transfer coefficient and reaction rate on entropy production due to: (a) mass transfer and (b) heat transfer.

In Figure 1, a distinct increase of entropy produced with the reaction rate constant, k_r, is observed. Conversely, entropy decreases with the mass transfer coefficient, k_C (due to heat, S_H, and mass, S_D, transfer). A rapid chemical reaction (i.e., high k_r) generates intense mass transport of substrates to the catalyst surface and adequate heat transfer in the opposite direction. The faster the reaction, the further the process runs from the equilibrium. When the transfer coefficients are small compared to the reaction rate, the concentration and temperature gradients are large, and even the substrates concentration on the catalyst goes to zero. Entropy production is large, being proportional to the square of the driving force (concentration or temperature gradient, cf. Equation (10)).

The impact of the mass transfer coefficient is opposite. The higher the transfer coefficient for a given reaction rate, the lower the temperature and concentration gradients are and the closer to the equilibrium the process runs. Smaller driving forces lead to lower entropy according to Equation (10). However, when analysing the plots in Figure 1, the impact of mass transfer intensification is distinct only if k_C is close to the k_r value. If k_C is much smaller than k_r, slight transfer enhancement will give nothing as the concentration and temperature gradients are still large (zero concentration at the catalyst surface). The gradients start to decrease as the reaction and transfer become comparable.

Obviously, the values of k_r and especially of k_C in Figure 1, may not be found in reality as the plots presented are theoretical, to illustrate the common impact of transfer and reaction rates on entropy production.

3. Catalyst Supports Considered

The aim of this study is to show the optimal adjustment of the catalyst carrier geometry, as well as its transfer and friction characteristics to the catalytic reaction kinetics. The catalyst performance (reaction kinetics) is treated as a model parameter only. Therefore, analysed catalyst supports were selected on the basis of similar value of specific surface area. This means that, in all considered cases, approximately, the same area was available for active layer catalyst deposition. For comparison, monolith and packed bed are also examined.

Correlations for the heat transfer and Fanning friction factor were derived experimentally and presented in detail in our earlier papers [4,16]. A photo of catalyst supports considered in the study is presented in Figure 2, and a summary of equations for Fanning friction factor, Nusselt number and Sherwood number of investigated supports are presented in Table 2 and compared in Figure 3.

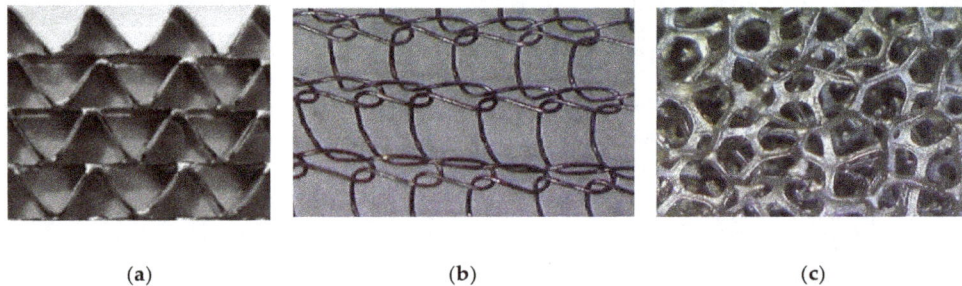

(a)   (b)   (c)

Figure 2. Catalyst supports: (**a**) triangular short-channel structure, (**b**) wire gauze, and (**c**) nickel chromium foam.

The kinetic tests were performed experimentally. Two different catalyst deposition methods were applied: (1) for Pd/ZrO_2, the incipient wetness (IW) method [20] and (2) for Pd/Al_2O_3, sonochemical (SC) method [4]. The kinetic studies were conducted in the temperature range of 373–823 K [20]. Kinetic data are presented in Table 3. As was found in [21], the sonochemical method allows higher catalyst activity to be obtained in comparison to the incipient wetness method.

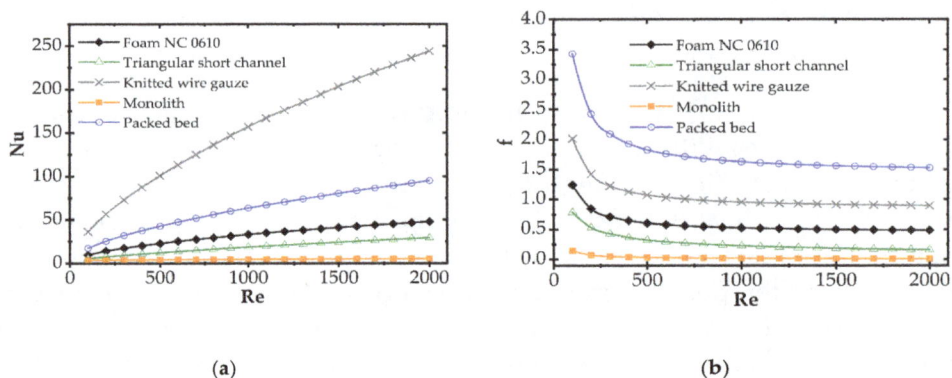

(a)   (b)

Figure 3. (**a**) Average Nusselt number and (**b**) Fanning friction factor for considered catalyst carriers.

Table 2. Correlations used to calculate flow resistance, heat and mass transfer for analysed catalyst supports.

Structure Description	Correlations
Wire gauze [4]	$f = 118.09/Re + 0.836$ $Nu = 2.19Re^{0.636} Pr^{1/3}$ $Sh = 2.19Re^{0.636} Sc^{1/3}$ $S_v = 1355$ $\varepsilon = 0.97$
Triangular short channel [16]	$(fRe) = 13.33 + 11.59(L^+)^{-0.514}$ $Nu = \left(3.11 + 0.45(L^*)^{-0.61}\right)\left(0.55(PrL^*)^{-0.15}\right)$ $Sh = \left(3.11 + 0.45(L^{*M})^{-0.61}\right)\left(0.55(PrL^{*M})^{-0.15}\right)$ $S_v = 1314$ $\varepsilon = 0.95$
Nickel chromium foam (NC 0610), Recemat® (Dodewaard, The Netherlands); [4]	$f = 79.9/Re + 0.445$ $Nu = 0.96Re^{0.53} Pr^{1/3}$ $Sh = 0.96Re^{0.53} Sc^{1/3}$ $S_v = 1298$ $\varepsilon = 0.89$
Monolith [17]	$(fRe) = 14.23\left(1 + 0.045/L^+\right)^{0.5}$ $Nu = 3.608(1 + 0.095/L^*)^{0.45}$ $Sh = 3.608\left(1 + 0.095/L^{*M}\right)^{0.45}$ $S_v = 1339$ $\varepsilon = 0.72$
Packed bed [18,19]	$f = \dfrac{(\varepsilon-1)[600\eta(\varepsilon-1)-7D_h\varrho w]}{8D_h\varepsilon\varrho w}$ $Nu = 2 + 1.1Re^{0.6} Pr^{1/3}$ $Sh = 2 + 1.1Re^{0.6} Sc^{1/3}$ $S_v = 1240$ $\varepsilon = 0.38$

Table 3. Kinetic data of tested catalysts.

Catalyst	Pre-Exponential Coefficient in Arrhenius Equation, k_∞, m s^{-1}	Activation Energy, Ea, kJ mol^{-1}
Slow kinetic, incipient wetness (IW) Pd/ZrO$_2$	252.49	62.79
Fast kinetic, sonochemical (SC) Pd/Al$_2$O$_3$	$1.07 \cdot 10^{10}$	110.4

4. Results and Discussion

Plots referring to analysis of entropy production were constructed assuming reactor length required for 90% conversion and show the entropy produced per 1 kmole of methane combusted in the reactor under given process conditions. Entropy production is presented as a function of process temperature and the Reynolds number. Entropy is produced due to the four components denoted as R—reaction, H—heat transfer, D—diffusional mass transfer and F—flow friction. The subscript $HDFR$ means total entropy produced due to the H, D, F and R components.

The components of entropy production (according to Table 1) for the knitted wire gauze are compared for the methane catalytic combustion process vs. process temperature (Figure 4) and the Reynolds number (Figure 5) for the fast (Pd/Al$_2$O$_3$) and slow (Pd/ZrO$_2$) kinetics assuming initial methane concentration of 200 ppm in both cases.

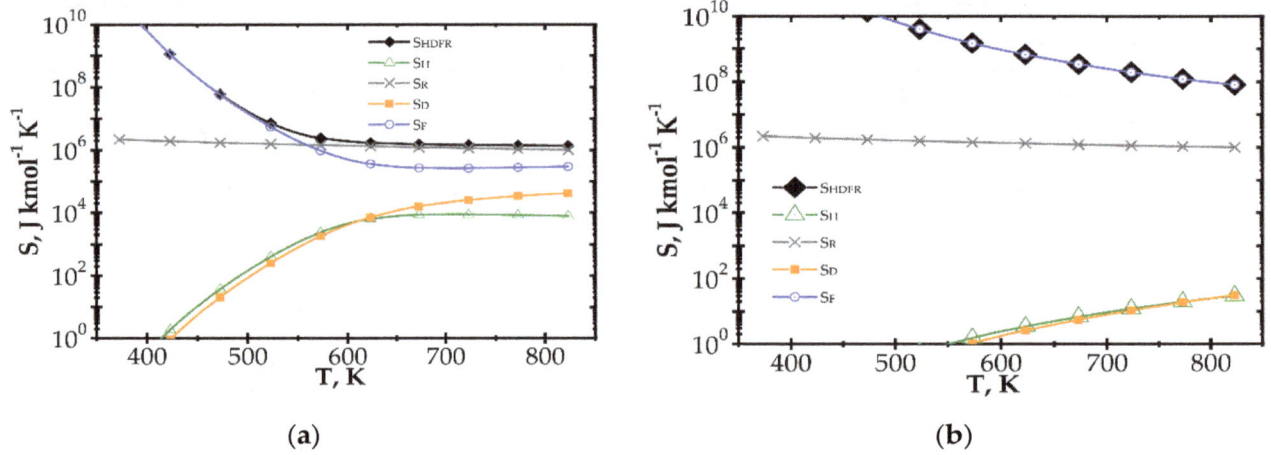

Figure 4. Comparison of entropy production components vs. process temperature for knitted wire gauze, Re = 1000, CH_4 inlet concentration: 200 ppm: (**a**) fast kinetics, Pd/Al_2O_3 and (**b**) slow kinetics, Pd/ZrO_2.

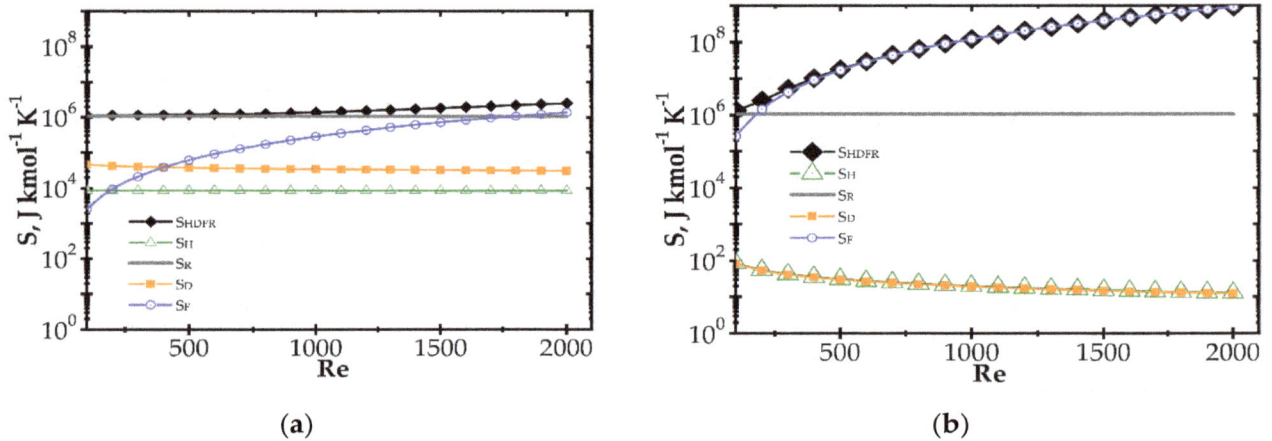

Figure 5. Comparison of entropy production components vs. Reynolds number for knitted wire gauze, T = 773 K, CH_4 inlet concentration: 200 ppm: (**a**) fast kinetics, Pd/Al_2O_3 and (**b**) slow kinetics, Pd/ZrO_2.

When analysing the Pd/Al_2O_3 catalyst (Figure 4a) within the lower temperature range, entropy due to flow friction, S_F, is the major component, and it is close to the total entropy production S_{HDFR}. The heat and mass transport components, S_H and S_D, play less important roles. However, for higher temperatures, the kinetics become much faster, causing significant shortening of reactor length necessary to attain 90% conversion. The share of flow friction entropy decreases; simultaneously, the entropy components due to heat and mass transport play more important roles. For the highest temperature range analysed, total entropy S_{HDFR} is close to the reaction component S_R, while the remaining components are comparable. Increased entropy production due to heat and mass transport at higher temperatures is a result of faster reaction rate. This leads to lower methane concentration on the catalyst surface, and thus to higher temperature and concentration gradients, in consequence of more intense entropy production (cf. Table 1, Equation (10) and Figure 1).

For the Pd/ZrO_2 catalyst (Figure 4b), total entropy production is close to the flow friction component in the whole temperature range analysed. The transport component S_D, S_H are minor due to low gradients (a result of slow kinetics), and even the reaction component S_R is much lower than the flow friction one, S_F.

Figure 5 illustrates entropy production as a function of the Reynolds number for knitted wire gauze assuming a rather high temperature of 773 K. The transport components S_D and S_H are almost constant within the whole Re range analysed. The flow friction component S_F increases with Re, reaching an even higher value than S_R, especially in the case of the Pd/ZrO$_2$ catalyst. Moreover, in Figure 5b, the total entropy produced is close to the flow friction component, with a minor role played by the remaining components.

Large entropy production is due to the irreversible reaction of methane catalytic combustion. Moreover, this entropy component is almost the same per mole of reactant, regardless of process conditions (T, Re and catalyst); analysis of the equation for S_R (Table 1) should render this as no surprise. Chemical affinity is close to the standard Gibbs energy of reaction (at the process temperature) $\Delta G_R^{o,T}$, because the sum of the concentration logarithms is minor. For optimization purposes, the place of the minimum total entropy production reflects the process optimum, making the precise value less important. Analysis of Figures 4 and 5 shows that the S_R component is nearly constant within the ranges studied. Note that reaction component, S_R, is the lowest possible entropy that can be produced in the chemical reactor. For engineering purposes, such as process optimization, the remaining components are more interesting because they make entropy production higher than that due to chemical reaction (S_R) and they are dependent on the physical properties of carriers. For slow reaction, there is no difference between the analysed approaches, because, in this case, flow resistance plays a major role (cf. Figures 4b and 5b) and the minimum is not observed within the considered temperature range. In summarising the catalytic structures displaying close specific surface area S_v (i.e., similar catalyst amount), S_R will be neglected during next analysis.

Analysis of entropy production due to the heat and mass transfer and flow friction (denoted as S_{HDF}) is presented in Figures 6 and 7 presents S_{HDF} as a function of the Reynolds number and process temperature for the five catalyst supports considered. In the following figures, minimal entropy production for each support is shown; these points give optimal process conditions for particular catalyst supports.

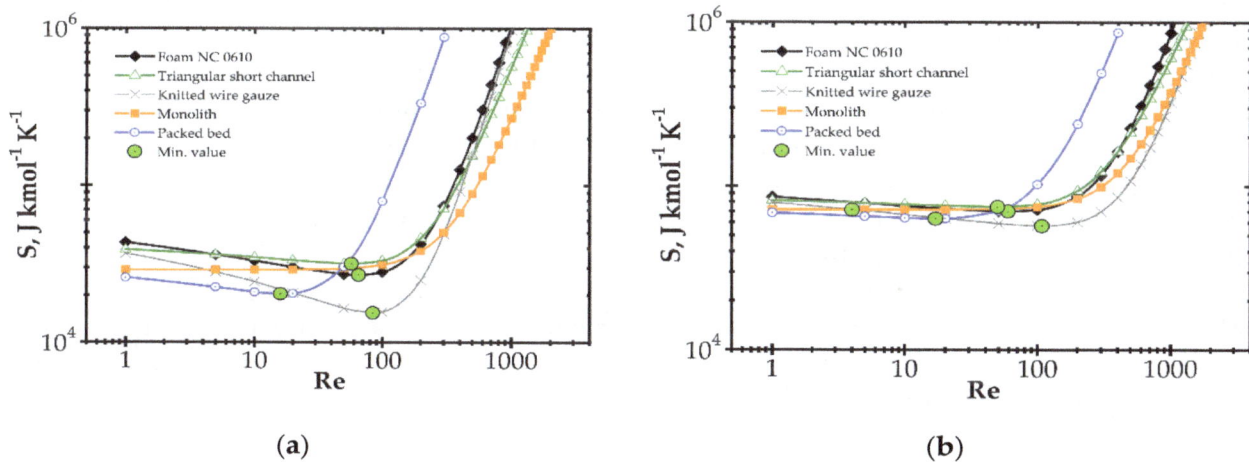

(a)   (b)

Figure 6. Entropy production vs. Reynolds number for different catalyst supports for the fast kinetics, Pd/Al$_2$O$_3$ at temperature: (a) 573 K and (b) 773 K.

In Figure 6, entropy is presented for two selected temperatures, moderate (573 K) and high (773 K). For the moderate temperature of 573 K (Figure 6a), packed bed seems the best for Re < 20. For Re < 500, knitted wire gauze is optimal (minimum value at Re = 84) in that this results in the lowest entropy production and the most profitable behaviour within this analysis. For a higher Reynolds number, monolith displays the lowest entropy production, undoubtedly due to its lowest flow resistance. For higher temperatures of 773 K (Figure 6b), the impact of transfer properties is more pronounced as a result of faster reaction rate, and knitted wire gauze appears to be the best with classic and short-channel monoliths. Packed bed produces the largest entropy in almost the entire Reynolds range, due to the highest flow resistance. For the higher temperature (773 K), the minima are generally slightly shifted to higher Reynolds numbers and entropy production is several times higher.

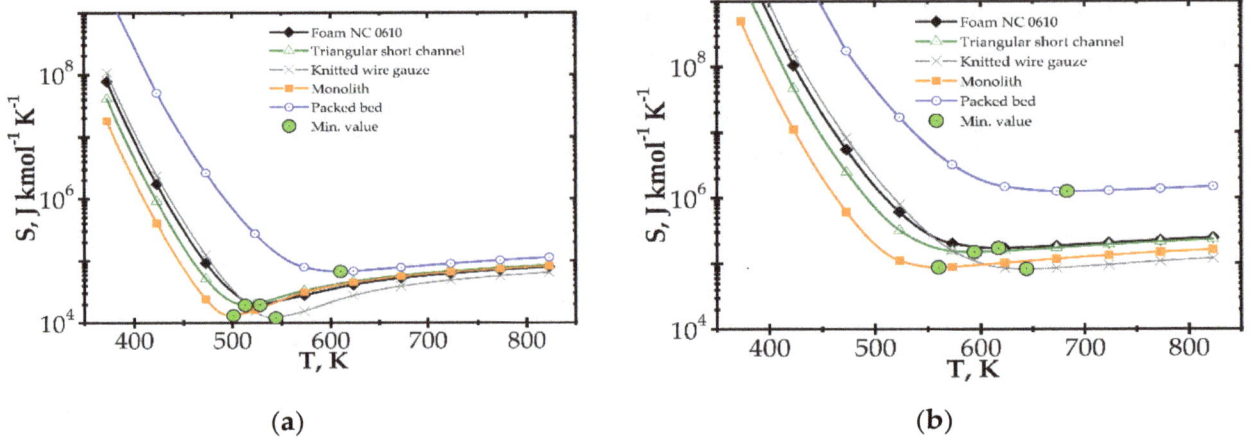

Figure 7. Entropy production vs. temperature for the fast kinetics, Pd/Al_2O_3 for different catalyst supports at Reynolds number: (**a**) Re = 100 and (**b**) Re = 500.

When considering temperature influence on entropy production (Figure 7), the same conclusions may be derived. Low process temperature is favourable for the classic monolith, while for higher temperatures, wire gauze and monolith seem to be the best choice. For Re = 100 (Figure 7a), above 650 K, all the internals display close entropy production. Interestingly, all the structures except packed bed show minima within the narrow range of 500–540 K. For Re = 500 (Figure 7b), entropy produced is higher, especially for packed bed. The minima are shifted towards higher temperatures by 60–100 K. Above 600 K, knitted gauze and monolith are the best.

Analogous plots for slow kinetics (Figures 8 and 9) show quite different behaviour. Here, the reactor is long due to the slow reaction rate. Moreover, slow reaction does not require intense heat and mass transfer. Concentration and temperature differences between the flowing fluid and catalyst surface are very small; entropy production due to transfer is small compared to that due to flow friction. Consequently, entropy produced for the slow kinetics is ordered identically to the friction factors (Figure 3b) vs. the Reynolds number and process temperature. Flow friction is the main entropy source (when neglecting chemical reaction). For slow kinetics, entropy production characteristic considered for all the internals is similar. The shift observed (towards higher or lower entropy produced) results mainly from the flow resistance. All the curves are nearly parallel, and only slight convergence is observed for low Re as a result of different transport properties. The internals displaying the lowest flow resistance (monolith and short-channel structure, cf. Figure 3) offer the lowest entropy production, while those of high flow resistance (packed bed, cf. Figure 3) produce larger entropy, so are less profitable.

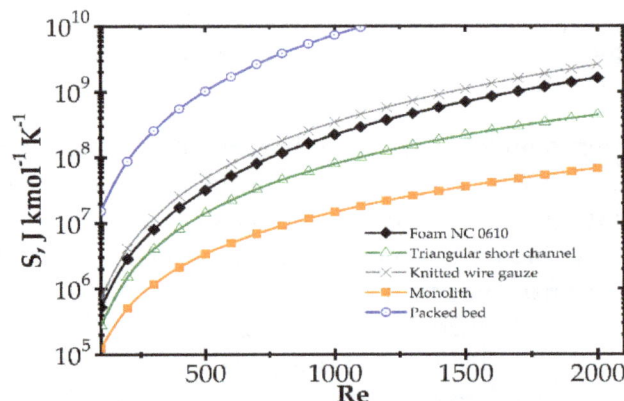

Figure 8. Entropy production vs. Reynolds number for different catalyst supports for the slow kinetics, Pd/ZrO_2 at temperature 673 K.

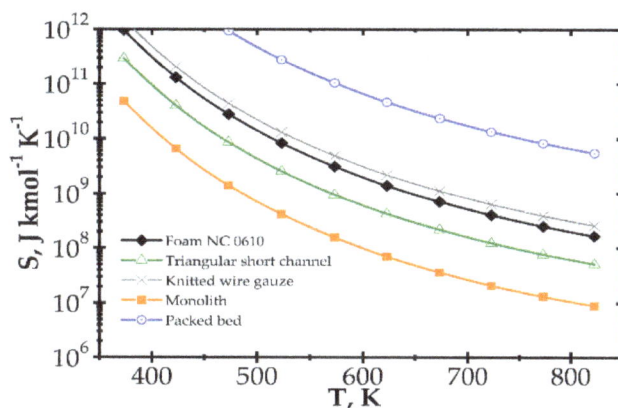

Figure 9. Entropy production vs. temperature for the slow kinetics, Pd/ZrO$_2$ for different catalyst supports at Reynolds number 1500.

5. Conclusions

The results obtained by entropy analysis indicate that wire gauze is the best choice for the Pd/Al$_2$O$_3$ catalyst and the packed bed is the worst one. In the case of the Pd/ZrO$_2$ catalyst, the best carriers are monolith and short-channel structures, while the worst solution is a packed bed. However, meeting the efficiency criteria cannot be regarded as the ultimate verdict. Any process has its own characteristics and limitations. It is rare for a process to occur separately, as it is usually part of a larger installation. For example, process temperature is limited by catalyst thermal deactivation, and the flow resistance may be limited by the gas pressure available. Therefore, each process needs to be considered individually, and any overall limiting parameters must also be taken into consideration during optimization.

The entropy-based optimization methodology is able to optimize reactor structure (indicating the best geometry, specific surface, etc.), as well as the process temperature and fluid velocity for considered reaction kinetics. The criterion, ensuring the minimum entropy production, ignores the reactor cost and is able to indicate the best structure from among the considered ones, as well as the optimal working conditions of a reactor (e.g., temperature and flow velocity).

Irreversible chemical reaction produces almost the same entropy, per mole of reactant, regardless of the process conditions. Therefore, it can be safely neglected during entropic optimization. The hypothesis is confirmed by analysis presented in Figure 4. For proper results, entropy produced by heat transfer, mass transfer and flow friction should be accounted for.

The gauze structures are assessed as being very effective due to their satisfactory transfer and friction properties. The monolith and short triangular channel display good efficiency for slow kinetics (Pd/ZrO$_2$ catalyst) due to their low flow resistance. The packed bed usually appears as an unsatisfactory solution.

For fast kinetics (Pd/Al$_2$O$_3$ catalyst), the transfer properties of the catalyst support are the most important for low entropy production. The intense transfer properties of, e.g., knitted wire gauze, make the support excellent for such processes. The impact of flow resistance is minor as, for a fast reaction not hampered by insufficient transfer rate, the reactor is very short.

For slow kinetics (Pd/ZrO$_2$ catalyst), the reactor is long. The impact of flow resistance becomes important. In contrast, heat and mass transfer contributions to entropy production are minor. Heat and mass transfer resistance is low, so temperatures (concentrations) gradients between fluid and catalyst surface are low, and the process runs near to the equilibrium.

The optimization methodology presented in this study obviously requires further development, including thorough experimental industrial and economic application. In spite of this, the entropic criterion seems able to indicate technically rational solutions of the reactor process considering the heat and mass transfer, flow resistance and reaction kinetics.

Author Contributions: Conceptualization, A.K. and M.K.; methodology, A.K.; formal analysis, A.K. and M.K.; investigation, A.G., M.I., and K.S.; writing—original draft preparation, A.K. and M.K.; writing—review and editing, M.K., M.I., and A.G.; supervision, A.K., A.G., and P.J.J. All authors have read and agreed to the published version of the manuscript.

Acknowledgments: The authors are sincerely grateful for the constructive suggestions and advices provided by Andrzej Burghardt. We regret to inform that A. Burghardt passed away on 1 March 2020.

Nomenclature

A	chemical affinity, J mol^{-1}
C_A	reagent concentration, mol m^{-3}
c_p	heat capacity, J kg^{-1} K^{-1}
D_A	diffusivity, m^2 s^{-1}
D_h	hydraulic diameter, $= 4\varepsilon S_v^{-1}$, m
F_c	reactor cross-sectional area, m^2
f	Fanning friction factor, $= \varrho w_0^2 L(2\Delta P D_h \varepsilon^2)^{-1}$
J_A	diffusional mass flux, mol s^{-1} m^{-2}
J_i	stream (flux) of irreversible process, Equation (9)
k_C	mass transfer coefficient, m s^{-1}
k_r	kinetic rate constant of the first-order reaction, referred to the catalyst surface area, m s^{-1}
k_∞	pre-exponential coefficient in Arrhenius equation, m s^{-1}
$k_{Cr} =$ $k_C k_r/(k_C+k_r)$	combined transfer-reaction coefficient, m s^{-1}
L	bed length, m
ΔP	pressure drop, Pa/m
q	heat flux, W m^{-2}
R	gas constant, J mol^{-1} K^{-1}
r_A	reaction rate, mol m^{-2} s^{-1}
S	entropy production rate, J K^{-1}mol^{-1}
s_{ef}	film thickness, m
S_v	specific surface area, m^2 m^{-3}
T	temperature, K
W	pumping power, W
w_0	superficial fluid velocity, m s^{-1}
y_i	mole fraction
ΔH_R	reaction enthalpy, J mol^{-1}
ΔG_R	reaction Gibbs energy, J mol^{-1}

Greek symbols

α	heat transfer coefficient, W m^{-2} K^{-1}
ε	porosity
η	dynamic viscosity, Pa s
λ	thermal conductivity, W m^{-1} K^{-1}
μ	chemical potential, J mol^{-1}
v	stoichiometric coefficient
$\Delta\pi$	driving force of irreversible process
ϱ	density, kg m^{-3}
σ	entropy production per m^3 of reactor volume, W m^{-3} K^{-1}

Dimensionless numbers

L^+	dimensionless length for the hydrodynamic entrance region, $= LD_h^{-1}\mathrm{Re}^{-1}$
L^*	dimensionless length for the thermal entrance region, $= LD_h^{-1}\mathrm{Re}^{-1}\mathrm{Pr}^{-1}$
L^{*M}	dimensionless length for the mass transfer entrance region, $= LD_h^{-1}\mathrm{Re}^{-1}\mathrm{Sc}^{-1}$

Pr	Prandtl number, $= \eta c_p \lambda^{-1}$
Re	Reynolds number, $= w0 D h \varrho \eta^{-1} \varepsilon^{-1}$
Sc	Schmidt number, $= \eta \varrho{-1} D_A^{-1}$
Sh	Sherwood number, $= k_C D_h D_A^{-1}$

Subscripts

A	key reactant
D	entropy production due to mass transfer
F	entropy production due to flow friction
H	entropy production due to heat transfer
P	total entropy production
R	entropy production due to chemical reaction
S	catalyst surface
x	reactor arbitrary axial coordinate
$0, L$	reactor inlet, outlet

References

1. Vilé, G.; Richard-Bildstein, S.; Lhuillery, A.; Rueedi, G. Electrophile, Substrate Functionality, and Catalyst Effects in the Synthesis of α-Mono and Di-Substituted Benzylamines via Visible-Light Photoredox Catalysis in Flow. *ChemCatChem* **2018**, *10*, 3786–3794. [CrossRef]

2. Amini-Rentsch, L.; Vanoli, E.; Richard-Bildstein, S.; Marti, R.; Vilé, G. A Novel and Efficient Continuous-Flow Route To Prepare Trifluoromethylated N -Fused Heterocycles for Drug Discovery and Pharmaceutical Manufacturing. *Ind. Eng. Chem. Res.* **2019**, *58*, 10164–10171. [CrossRef]

3. Ramirez, A.; Hueso, J.L.; Mallada, R.; Santamaria, J. Microwave-activated structured reactors to maximize propylene selectivity in the oxidative dehydrogenation of propane. *Chem. Eng. J.* **2020**, *393*, 124746. [CrossRef]

4. Gancarczyk, A.; Iwaniszyn, M.; Piątek, M.; Korpyś, M.; Sindera, K.; Jodłowski, P.J.; Łojewska, J.; Kołodziej, A. Catalytic Combustion of Low-Concentration Methane on Structured Catalyst Supports. *Ind. Eng. Chem. Res.* **2018**, *57*, 10281–10291. [CrossRef]

5. Giani, L.; Groppi, G.; Tronconi, E. Mass-Transfer Characterization of Metallic Foams as Supports for Structured Catalysts. *Ind. Eng. Chem. Res.* **2005**, *44*, 4993–5002. [CrossRef]

6. Kołodziej, A.; Łojewska, J. Prospect of compact afterburners based on metallic microstructures. Design and modelling. *Top. Catal.* **2007**, *42–43*, 475–480. [CrossRef]

7. London, A.L. Economics and the second law: An engineering view and methodology. *Int. J. Heat Mass Transf.* **1982**, *25*, 743–751. [CrossRef]

8. Bejan, A. *Advanced Engineering Thermodynamics*, 1st ed.; John Wiley and Sons: New York, NY, USA, 1988.

9. London, A.L.; Shah, R.K. Costs of Irreversibilities in Heat Exchanger Design. *Heat Transf. Eng.* **1983**, *4*, 59–73. [CrossRef]

10. Zimparov, V. Extended performance evaluation criteria for enhanced heat transfer surfaces: Heat transfer through ducts with constant wall temperature. *Int. J. Heat Mass Transf.* **2000**, *43*, 3137–3155. [CrossRef]

11. Kjelstrup, S.; Johannessen, E.; Rosjorde, A.; Nummedal, L.; Bedeaux, D. Minimizing the entropy production of the methanol producing reaction in a methanol reactor. *Int. J. Thermodyn.* **2000**, *3*, 147–153.

12. Nummedal, L.; Kjelstrup, S.; Costea, M. Minimizing the Entropy Production Rate of an Exothermic Reactor with a Constant Heat-Transfer Coefficient: The Ammonia Reaction. *Ind. Eng. Chem. Res.* **2003**, *42*, 1044–1056. [CrossRef]

13. De Groot, S.R.; Mazur, P. *Non-Equilibrium Thermodynamics*; From the Series in Physics; North-Holland Publishing Company: Amsterdam, The Netherlands, 1969.

14. Wei, J. Irreversible thermodynamics in engineering. *Ind. Eng. Chem.* **1966**, *58*, 55–60. [CrossRef]

15. Chilton, T.H.; Colburn, A.P. Mass Transfer (Absorption) Coefficients Prediction from Data on Heat Transfer and Fluid Friction. *Ind. Eng. Chem.* **1934**, *26*, 1183–1187. [CrossRef]

16. Iwaniszyn, M.; Ochońska, J.; Gancarczyk, A.; Jodłowski, P.; Knapik, A.; Łojewska, J.; Janowska-Renkas, E.; Kołodziej, A. Short-channel structured reactor as a catalytic afterburner. *Top. Catal.* **2013**, *56*, 273–278. [CrossRef]

17. Hawthorn, R.D. Afterburner catalysts effects of heat and mass transfer between gas and catalyst surface. *AIChE Symp. Ser.* **1974**, *70*, 428–438.

18. Bird, R.B.; Stewart, W.E.; Lightfoot, E.N. *Transport Phenomena*, 2nd ed.; Wiley International Edition; Wiley: New York, NY, USA, 2007; ISBN 9780470115398.

19. Wakao, N.; Kaguei, S. *Heat and Mass Transfer in Packed Beds*; Routledge: New York, NY, USA, 1982.

20. Jodłowski, P.J.; Jędrzejczyk, R.J.; Gancarczyk, A.; Łojewska, J.; Kołodziej, A. New method of determination of intrinsic kinetic and mass transport parameters from typical catalyst activity tests: Problem of mass transfer resistance and diffusional limitation of reaction rate. *Chem. Eng. Sci.* **2017**, *162*, 322–331. [CrossRef]

21. Jodłowski, P.; Jędrzejczyk, R.; Chlebda, D.; Dziedzicka, A.; Kuterasiński, Ł.; Gancarczyk, A.; Sitarz, M. Non-Noble Metal Oxide Catalysts for Methane Catalytic Combustion: Sonochemical Synthesis and Characterisation. *Nanomaterials* **2017**, *7*, 174. [CrossRef] [PubMed]

Permissions

All chapters in this book were first published in MDPI; hereby published with permission under the Creative Commons Attribution License or equivalent. Every chapter published in this book has been scrutinized by our experts. Their significance has been extensively debated. The topics covered herein carry significant findings which will fuel the growth of the discipline. They may even be implemented as practical applications or may be referred to as a beginning point for another development.

The contributors of this book come from diverse backgrounds, making this book a truly international effort. This book will bring forth new frontiers with its revolutionizing research information and detailed analysis of the nascent developments around the world.

We would like to thank all the contributing authors for lending their expertise to make the book truly unique. They have played a crucial role in the development of this book. Without their invaluable contributions this book wouldn't have been possible. They have made vital efforts to compile up to date information on the varied aspects of this subject to make this book a valuable addition to the collection of many professionals and students.

This book was conceptualized with the vision of imparting up-to-date information and advanced data in this field. To ensure the same, a matchless editorial board was set up. Every individual on the board went through rigorous rounds of assessment to prove their worth. After which they invested a large part of their time researching and compiling the most relevant data for our readers.

The editorial board has been involved in producing this book since its inception. They have spent rigorous hours researching and exploring the diverse topics which have resulted in the successful publishing of this book. They have passed on their knowledge of decades through this book. To expedite this challenging task, the publisher supported the team at every step. A small team of assistant editors was also appointed to further simplify the editing procedure and attain best results for the readers.

Apart from the editorial board, the designing team has also invested a significant amount of their time in understanding the subject and creating the most relevant covers. They scrutinized every image to scout for the most suitable representation of the subject and create an appropriate cover for the book.

The publishing team has been an ardent support to the editorial, designing and production team. Their endless efforts to recruit the best for this project, has resulted in the accomplishment of this book. They are a veteran in the field of academics and their pool of knowledge is as vast as their experience in printing. Their expertise and guidance has proved useful at every step. Their uncompromising quality standards have made this book an exceptional effort. Their encouragement from time to time has been an inspiration for everyone.

The publisher and the editorial board hope that this book will prove to be a valuable piece of knowledge for researchers, students, practitioners and scholars across the globe.

List of Contributors

Ben Akih-Kumgeh
Department of Mechanical and Aerospace Engineering, Syracuse University, 263 Link Hall, Syracuse, NY 13244, USA

Henning Struchtrup
Mechanical Engineering, University of Victoria, Victoria, BC V8W 2Y2, Canada

Prasanna Ponnusamy and Johannes de Boor
Institute of Materials Research, German Aerospace Center (DLR), D-51170 Köln, Germany

Eckhard Müller
Institute of Materials Research, German Aerospace Center (DLR), D-51170 Köln, Germany
Institute of Inorganic and Analytical Chemistry, Justus Liebig University Gießen, D-35392 Gießen, Germany

Alexander Ryzhov and Henni Ouerdane
Center for Energy Science and Technology, Skolkovo Institute of Science and Technology, 3 Nobel Street, Skolkovo, Moscow Region 121205, Russia

Ilia A. Luchnikov
Center for Energy Science and Technology, Skolkovo Institute of Science and Technology, 3 Nobel Street, Skolkovo, Moscow Region 121205, Russia
Moscow Institute of Physics and Technology, Institutskii Per. 9, Dolgoprudny, Moscow Region 141700, Russia

Pieter-Jan Stas
Department of Applied Physics, Stanford University 348 Via Pueblo Mall, Stanford, CA 94305, USA

Sergey N. Filippov
Moscow Institute of Physics and Technology, Institutskii Per. 9, Dolgoprudny, Moscow Region 141700, Russia
Valiev Institute of Physics and Technology of Russian Academy of Sciences, Nakhimovskii Pr. 34, Moscow 117218, Russia
Steklov Mathematical Institute of Russian Academy of Sciences, Gubkina St. 8, Moscow 119991, Russia

Ioulia Chikina
LIONS, NIMBE, CEA, CNRS, Universitè Paris-Saclay, CEA Saclay, 91191 Gif-sur-Yvette, France

Valeri Shikin
ISSP, RAS, Chernogolovka, 142432 Moscow, Russia

Andrey Varlamov
CNR-SPIN, c/o DICII-Universitá di Roma Tor Vergata, Via del Politecnico, 1, 00133 Roma, Italy

Nicolás Pérez and Gabi Schierning
Institute for Metallic Materials, IFW-Dresden, 01069 Dresden, Germany

Constantin Wolf and Alexander Kunzmann
Institute for Metallic Materials, IFW-Dresden, 01069 Dresden, Germany
Institute of Materials Science, TU Dresden, 01062 Dresden, Germany

Jens Freudenberger
Institute for Metallic Materials, IFW-Dresden, 01069 Dresden, Germany
Institute of Materials Science, TU Bergakademie Freiberg, 09599 Freiberg, Germany

Maria Krautz and Bruno Weise
Institute for Complex Materials, IFW-Dresden, 01069 Dresden, Germany

Kornelius Nielsch
Institute for Metallic Materials, IFW-Dresden, 01069 Dresden, Germany
Institute of Materials Science, TU Dresden, 01062 Dresden, Germany
Institute of Applied Physics, TU Dresden, 01062 Dresden, Germany

Armin Feldhoff
Institute of Physical Chemistry and Electrochemistry, Leibniz University Hannover, Callinstraße 3A, D-30167 Hannover, Germany

Christophe Goupil and Eric Herbert
Université de Paris, Laboratoire Interdisciplinaire des Energies de Demain (LIED), UMR 8236 CNRS, F-75013 Paris, France

Akihiro Nishiyama and Shigenori Tanaka
Graduate School of System Informatics, Kobe University, 1-1 Rokkodai, Nada-ku, Kobe 657-8501, Japan

Jack A. Tuszynski
Department of Oncology, University of Alberta, Cross Cancer Institute, Edmonton, AB T6G 1Z2, Canada
Department of Physics, University of Alberta, Edmonton, AB T6G 2J1, Canada
DIMEAS, Corso Duca degli Abruzzi, 24, Politecnico di Torino, 10129 Turin, TO, Italy

Mateusz Korpyś, Anna Gancarczyk, Marzena Iwaniszyn and Katarzyna Sindera
Institute of Chemical Engineering, Polish Academy of Sciences, Bałtycka 5, 44-100 Gliwice, Poland

Przemysław J. Jodłowski
Faculty of Chemical Engineering and Technology, Cracow University of Technology, Warszawska 24, 31-155 Kraków, Poland

Andrzej Kołodziej
Institute of Chemical Engineering, Polish Academy of Sciences, Bałtycka 5, 44-100 Gliwice, Poland
Faculty of Civil Engineering and Architecture, Opole University of Technology, Katowicka 48, 45-061 Opole, Poland

Index